W0230176

Bacterial Pathogenesis

METHODS IN MOLECULAR BIOLOGY™

John M. Walker, SERIES EDITOR

METHODS IN MOLECULAR BIOLOGY™

Bacterial Pathogenesis

Methods and Protocols

Edited by

Frank R. DeLeo

and

Michael Otto

Laboratory of Human Bacterial Pathogenesis,
Rocky Mountain Laboratories,
National Institute of Allergy and Infectious Diseases,
National Institutes of Health, Hamilton, MT

 Humana Press

Editors
Frank R. DeLeo
Pathogen Host-Cell Biology Section
Laboratory of Human Bacterial
  Pathogenesis
Rocky Mountain Laboratories
National Institute of Allergy and
  Infectious Diseases
National Institutes of Health
Hamilton, MT
fdeleo@niaid.nih.gov

Michael Otto
Pathogen Molecular Genetics Section
Laboratory of Human Bacterial
  Pathogenesis
Rocky Mountain Laboratories
National Institute of Allergy and
  Infectious Diseases
National Institutes of Health
Hamilton, MT
motto@niaid.nih.gov

Series Editor
John M. Walker
School of Life Sciences
University of Hertfordshire
Hatfield, Herts., AL10 9AB
UK

ISBN: 978-1-58829-740-2   e-ISBN: 978-1-60327-032-8

Library of Congress Control Number: 2007941654

Cover illustration: Background: Scanning electron micrograph of human neutrophil lysis caused by *Staphylococcus aureus*. Inset box: *Coxiella burnetti* vacuole in African green monkey kidney (Vero) fibroblasts. Scanning electron microscopy was performed by David W. Dorward, Ph.D. (background) and Elizabeth R. Fischer (inset box), Microscopy Unit, Research Technologies Section/RTB, Rocky Mountain Laboratories, National Institute of Allergy and Infectious Diseases, National Institutes of Health.

Printed on acid-free paper

9 8 7 6 5 4 3 2 1

springer.com

Preface

Bacterial infections are a leading cause of morbidity and mortality worldwide. For example, *Mycobacterium tuberculosis* alone caused 1,644,000 deaths worldwide in 2002 (World Health Organization, World Health Report, 2002). Mortality attributed to bacterial infections is not limited to developing countries, as there were greater than 93,000 deaths in the USA due to bacterial infections in the year 2000 (World Health Organization). Notably, antibiotic resistance in bacterial pathogens such as *M. tuberculosis* and *Staphylococcus aureus* is increasing at an alarming rate. There is also a broad spectrum of niches for pathogenic bacteria in the human host and routes of infection vary considerably. Thus, an enhanced understanding of pathogenesis mechanisms is needed to facilitate the development of new prophylactics and treatments for bacterial diseases. To that end, numerous in vitro methods, in vivo animal model systems, and cutting-edge genomics assays have been developed to study bacterial pathogenesis or to identify potential therapeutic targets. A comprehensive source of state-of-the-art protocols and methods for studying bacterial pathogenesis would be a welcome asset to microbiologists, immunologists, cell biologists, infectious disease clinicians, and other basic science researchers interested in studying pathogenic bacteria, and related disease processes.

The aim of *Bacterial Pathogenesis: Methods and Protocols* is to provide in-depth and detailed methods to investigate pathogenesis for a wide range of bacteria, including *Anaplasma phagocytophilum*, *Borrelia burgdorferi*, *Borrelia hermsii*, *Brucella abortus*, *Coxiella burnetti*, *Francisella tularensis*, *Helicobactor pylori*, *Salmonella enterica*, *Staphylococcus aureus*, *Staphylococcus epidermidis*, *Streptococcus pyogenes*, *Mycobacterium tuberculosis*, *Neisseria meningitidis*, *Pseudomonas aeruginosa*, and *Yersinia pestis*. The categories of pathogens for which methods are provided (e.g., vector-borne, obligate intracellular, Gram-negative/positive, and biosafety-level 3 containment), although not all-inclusive, are described so that they can be modified to study most bacteria. Part I details methods to study specific mechanisms of bacterial pathogenesis, such as growth and passage of *Borrelia* spp. through ticks, molecular genetic analysis of *Staphylococcus epidermidis* biofilms, and investigation of quorum-sensing by *Pseudomonas aeruginosa*. Part II provides protocols for studying host–pathogen interactions, such as methods for growth of obligate intracellular pathogens and provides protocols for evaluating pathogen uptake and killing by host cells. The section includes detailed methods for studying pathogen transcriptomes during interaction with host cells. Part III describes animal models of infection and includes a recently developed model of streptococcal pharyngitis in non-human primates. Part IV concludes with novel approaches for the identification of therapeutic targets designed to control infections and up-to-date molecular typing methods for *Staphylococcus aureus*.

We thank John M. Walker, Series Editor, and Humana Press for the opportunity to put together a superb collection of articles and the authors for contributing outstanding chapters. Finally, we are grateful for the support from the Intramural Research Program of the National Institutes of Health, National Institutes of Allergy and Infectious Diseases.

Michael Otto
Frank R. DeLeo

Contents

Contributors

LEFA E. ALKSNE • *Wyeth Research, Pearl River, NY*

LEE-ANN H. ALLEN • *Inflammation Program and the Departments of Medicine and Microbiology, University of Iowa and the VA Medical Center, Iowa City, IA*

MICHAEL A. APICELLA • *Department of Microbiology, University of Iowa College of Medicine, Iowa City, IA*

JAMES M. BATTISTI • *Division of Biological Sciences, University of Montana, Missoula, MT*

DORI L. BORJESSON • *Department of Pathology, Microbiology and Immunology, School of Veterinary Medicine, University of California, Davis, CA*

JULIE A. BOYLAN • *Laboratory of Zoonotic Pathogens, Rocky Mountain Laboratories, National Institute of Allergy and Infectious Diseases, National Institutes of Health, Hamilton, MT*

JEAN CELLI • *Tularemia Pathogenesis Section, Laboratory of Intracellular Parasites, Rocky Mountain Laboratories, National Institute of Allergy and Infectious Diseases, National Institutes of Health, Hamilton, MT*

MICHAEL S. CHAUSSEE • *Division of Basic Biomedical Sciences, The Stanford School of Medicine of the University of South Dakota, Vermillion, SD*

MICHELLE A. CHAUSSEE • *Division of Basic Biomedical Sciences, The Stanford School of Medicine of the University of South Dakota, Vermillion, SD*

JASON E. COMER • *Plague Section, Laboratory of Zoonotic Pathogens, Rocky Mountain Laboratories, National Institute of Allergy and Infectious Diseases, National Institutes of Health, Hamilton, MT*

FRANK R. DELEO • *Pathogen Host-Cell Biology Section, Laboratory of Human Bacterial Pathogenesis, Rocky Mountain Laboratories, National Institute of Allergy and Infectious Diseases, National Institutes of Health, Hamilton, MT*

DAVID W. DORWARD • *Research Technologies Section/RTB, Rocky Mountain Laboratories, National Institute of Allergy and Infectious Diseases, National Institutes of Health, Hamilton, MT*

PAUL M. DUNMAN • *Department of Pathology and Microbiology, University of Nebraska Medical Center, Omaha, NE*

SUSANNE ENGELMANN • *Institut für Mikrobiologie, Ernst-Moritz-Arndt-Universität, Greifswald, Germany*

J. ROSS FITZGERALD • *Centre for Infectious Diseases, Chancellor's Building, New Royal Infirmary, University of Edinburgh, Edinburgh, Scotland, UK*

KEVIN P. FRANCIS • *Xenogen Corporation, Alameda, CA*

FRANK C. GHERARDINI • *Laboratory of Zoonotic Pathogens, Rocky Mountain Laboratories, National Institute of Allergy and Infectious Diseases, National Institutes of Health, Hamilton, MT*

MICHAEL GIVSKOV • *Section of Molecular Microbiology, BioCentrum-DTU, The Technical University of Denmark, Kongens Lyngby, Denmark*

CAITRIONA M. GUINANE • *Centre for Infectious Diseases, Chancellor's Building, New Royal Infirmary, University of Edinburgh, Edinburgh, Scotland, UK*

MICHAEL HECKER • *Institut für Mikrobiologie, Ernst-Moritz-Arndt-Universität, Greifswald, Germany*

ROBERT A. HEINZEN • *Coxiella Pathogenesis Section, Laboratory of Intracellular Parasites, Rocky Mountain Laboratories, National Institute of Allergy and Infectious Diseases, National Institutes of Health, Hamilton, MT*

B. JOSEPH HINNEBUSCH • *Plague Section, Laboratory of Zoonotic Pathogens, Rocky Mountain Laboratories, National Institute of Allergy and Infectious Diseases, National Institutes of Health, Hamilton, MT*

JAGATH L. KADURUGAMUWA • *Xenogen Corporation, Alameda, CA*

STAFFAN KJELLEBERG • *Centre for Marine Bio-Innovation, and The School of Biotechnology and Biomolecular Sciences, The University of New South Wales, Sydney, Australia*

JOHN F. KOKAI-KUN • *Biosynexus Incorporated, Gaithersburg, MD*

BARRY N. KREISWIRTH • *Public Health Research Institute Tuberculosis Center, International Center for Public Health, Newark, NJ*

ELLEN A. LORANGE • *Plague Section, Laboratory of Zoonotic Pathogens, Rocky Mountain Laboratories, National Institute of Allergy and Infectious Diseases, National Institutes of Health, Hamilton, MT*

BARUN MATHEMA • *Public Health Research Institute Tuberculosis Center, International Center for Public Health, Newark, NJ*

DIANE MCDOUGALD • *Centre for Marine Bio-Innovation, and The School of Biotechnology and Biomolecular Sciences, The University of New South Wales, Sydney, Australia*

EMILY J. MCDOWELL • *Division of Basic Biomedical Sciences, The Stanford School of Medicine of the University of South Dakota, Vermillion, SD*

JOSE MEDIAVILLA • *Public Health Research Institute Tuberculosis Center, International Center for Public Health, Newark, NJ*

JAMES M. MUSSER • *Center for Molecular and Translational Human Infectious Diseases Research, The Methodist Hospital Research Institute, Houston, TX*

MICHAEL OTTO • *Pathogen Molecular Genetics Section, Laboratory of Human Bacterial Pathogenesis, Rocky Mountain Laboratories, National Institute of Allergy and Infectious Diseases, National Institutes of Health, Hamilton, MT*

SANDRA J. RAFFEL • *Laboratory of Zoonotic Pathogens, Rocky Mountain Laboratories, National Institute of Allergy and Infectious Diseases, National Institutes of Health, Hamilton, MT*

SCOTT A. RICE • *Centre for Marine Bio-Innovation, and The School of Biotechnology and Biomolecular Sciences, The University of New South Wales, Sydney, Australia*

PATRICIA A. ROSA • *Laboratory of Zoonotic Pathogens, Rocky Mountain Laboratories, National Institute of Allergy and Infectious Diseases, National Institutes of Health, Hamilton, MT*

TOM G. SCHWAN • *Laboratory of Zoonotic Pathogens, Rocky Mountain Laboratories, National Institute of Allergy and Infectious Diseases, National Institutes of Health, Hamilton, MT*

JEFFREY G. SHANNON • *Coxiella Pathogenesis Section, Laboratory of Intracellular Parasites, Rocky Mountain Laboratories, National Institute of Allergy and Infectious Diseases, National Institutes of Health, Hamilton, MT*

OLIVIA STEELE-MORTIMER • *Salmonella Host-Cell Interaction Section, Laboratory of Intracellular Parasites, Rocky Mountain Laboratories, National Institute of Allergy and Infectious Diseases, National Institutes of Health, Hamilton, MT*

PHILIP E. STEWART • *Laboratory of Zoonotic Pathogens, Rocky Mountain Laboratories, National Institute of Allergy and Infectious Diseases, National Institutes of Health, Hamilton, MT*

DAN E. STURDEVANT • *Research Technologies Section/RTB, Rocky Mountain Laboratories, National Institute of Allergy and Infectious Diseases, National Institutes of Health, Hamilton, MT*

PAUL SUMBY • *Center for Molecular and Translational Human Infectious Diseases Research, The Methodist Hospital Research Institute, Houston, TX*

ANNE H. TART • *Center for Molecular and Translational Human Infectious Diseases Research, The Methodist Hospital Research Institute, Houston, TX*

JOVANKA M. VOYICH • *Department of Veterinary Molecular Biology, Montana State University, Bozeman, MT*

CUONG VUONG • *Department of Anesthesiology, University of Massachusetts Medical School, Worcester, MA*

List of Color Plates

I

MECHANISMS OF BACTERIAL PATHOGENESIS

1

Isolation and Characterization of Lipopolysaccharides

Michael A. Apicella

Summary

Lipopolysaccharide (LPS) is the signature glycolipid isolated from almost all Gram-negative bacteria. LPSs are well known for their ability to elicit the release of cytokines from eukaryotic cells including macrophages, neutrophils, and epithelial cells. LPS can be isolated free of contaminating nucleic acids and proteins by various techniques. In this review, we outline approaches for the isolation and preparation of LPSs for structural studies as well as preparation of very highly purified material for biological studies. Methods are also provided for the analysis of the purity and the structural composition of the LPSs. Finally, three methods for the isolation of lipid A are described.

Key Words: Lipopolysaccharide; lipooligosaccharide; phenol; proteinase K; SDS–PAGE; Lipid A.

1. Introduction

Lipopolysaccharides (LPSs) are a family of bacterial glycolipids which consist of a lipid that has been designated lipid A and a carbohydrate component of variable length. The lipid A molecules are frequently substituted with phosphates, phosphoethanolamines, and sugars. The first description of LPS was in 1941, when Shear determined that the component of the *Serratia marcescens* cell wall responsible for tumor-destroying activity was composed of a polysaccharide and a lipid *(1)*. He was also the first to designate this material "lipopolysaccharide." LPS is embedded in the external layer of the bacterial outer membrane and is linked through 2-keto-3-deoxyoctulosonic acid to a carbohydrate component that extends into the surrounding environment *(2)*.

From: *Methods in Molecular Biology, vol. 431: Bacterial Pathogenesis*
Edited by: F. DeLeo and M. Otto © Humana Press, Totowa, NJ

The carbohydrate portion of LPS consists of a relatively conserved core region linked to a series of repeating four to six sugar units (the O-antigen). These units can form polysaccharides with Mr in excess of 65 kDa *(3,4)*. Lipooligosaccharides (LOSs) are analogous to LPSs and have carbohydrate components that are smaller (3200–7200 kDa) than those of LPS and tend to be more intricately branched than the LPS carbohydrates *(5)*. These structures can be isolated from *Haemophilus influenzae*, *Neisseria gonorrhoeae*, *Neisseria meningitidis*, *Haemophilus ducreyi*, *Branhamella catarrhalis*, and *Bordetella pertussis*.

The purpose of this chapter is to describe methods for the isolation and analysis of LPSs and LOSs. Various methods for the isolation of these glycolipids have evolved since the 1930s. These include extraction with trichloracetic acid *(6)*, extraction with ether *(7)*, extraction with water *(8)*, extraction with EDTA *(9)*, extraction with pyridine *(10)*, extraction with phenol *(11)*, and extraction after solubilization with sodium docecyl sulfate (SDS) *(12)*.

Two procedures can suffice for the isolation of LPS preparations of high purity. These are the modified phenol–water method *(13)* and a modification of the proteinase K method and phenol extraction, which allows extraction of a LPS preparation free of contaminating lipoproteins *(14)*.

Because it is effective in removing contaminating proteins, the latter method is useful for studying biological interactions with LPS. The phenol–water technique, which was first described by Westphal and Jann *(11)*, takes advantage of the amphopathic nature of the LPS and the solubility of the majority of bacterial proteins in phenol. Polysaccharides, mucopolysaccharides, LPSs with O-side chains, and nucleic acids are usually soluble in aqueous solutions and insoluble in phenol. Phenol is a weak acid; its dissociation constant at $18\,^{\circ}\mathrm{C}$ in water is $\sim 1.2 \times 10^{-10}$. Mixtures of phenol and water have a high dielectric constant. These facts form the basis of a method of partition of proteins and polysaccharides and/or nucleic acids between phenol and water. Minor modifications have been made to the basic protocol described by Westphal et al. *(15)* and by Johnson and Perry *(16)*. These changes result in LPS and LOS preparations that have less contamination with nucleic acids and produce somewhat greater yields. One major limitation of the phenol–water extraction method is that LPS with truncated polysaccharide components or the more hydrophobic, shorter chain LOSs frequently partition into the phenol phase, as Erwin and co-workers *(17)* have isolated the LOS of *H. influenzae aegyptius* from the phenol phase.

The combination of proteinase K digestion of bacterial proteins followed by nuclease digestion and phenol water extraction results in an LPS preparation of very high quality, free of contaminating proteins and nucleic acids.

2. Materials

2.1. LOS and LPS Acrylamide Gel Electrophoresis

1. 30% Acrylamide Stock: Dissolve 29.2 g acrylamide and 0.8 g bis-acrylamide in 40 mL of distilled water and q.s. to 100 mL. Filter into dark bottle and store at 4 °C. This stock should not be used after 2 weeks.
2. Resolving Buffer, 1.88 *M* Tris–HCl (pH 8.8): Dissolve 22.78 g Trizma base in 70 mL of distilled water, adjust pH to 8.8 with concentrated HCl and q.s. to 100 mL. Store at 4 °C. This stock should not be used after 2 weeks.
3. Spacer Buffer, 1.25 *M* Tris–HCl (pH 6.8): Dissolve 15.12 g Trizma base in 70 mL of distilled water, adjust pH to 6.8 with concentrated HCl and q.s. to 100 mL. Store at 4 °C. This stock should not be used after 2 weeks.
4. Sample Buffer: Add 0.727 g Trizma base, 0.034 g EDTA, and 2.0 g SDS to 70 mL distilled water, adjust pH to 6.8 with concentrated HCl, q.s. to 100 mL with distilled water, filter sterilize, and store at room temperature.
5. Dye Buffer: Combine 2.5 mL of sample buffer, 2.0 mL of glycerol, 400 µL of β-mercaptoethanol, and 200 µL of a saturated solution of bromophenol blue. This buffer should be made fresh for each run. A stock solution can be made without adding the β-mercaptoethanol. This should be added just prior to use.
6. Ammonium Persulfate: Dissolve 0.05 g ammonium persulfate in 1 mL of distilled water. This solution should be made fresh for each experiment.
7. Reservoir Buffer, Tris–glycine buffer (pH 8.3): Dissolve 115.2 g glycine, 24 g Trizma base, and 8.0 g SDS in 8 L of distilled water. No pH adjustment should be necessary if correct reagents and quantities are added.

3. Methods

3.1. Phenol–Water Technique (16)

1. Place 5 g of freeze- or acetone-dried bacteria in a mortar and pestle and ground until a very fine powder is formed. Suspend powder in 25 mL of 50 m*M* sodium phosphate (pH 7.0) containing 5 m*M* EDTA. Allow suspension to completely hydrate before proceeding.
2. Stir suspension in a shearing mixer such as a blender at top speed for 1 min. Add hen egg lysozyme (100 mg) to the suspension and stir overnight at 4 °C.
3. Place suspension at 37 °C for 20 min and then stir at top speed in a blender for 3 min. Increase volume of the suspension to 100 mL with 50 m*M* sodium phosphate (pH 7.0) containing 20 m*M* MgCl$_2$. Add ribonuclease A and deoxyribonuclease I to a final concentration of 1 µg/mL. Incubate the suspension for 60 min at 37 °C and then for 60 min at 60 °C (*see* **Note 1**).
4. Place the bacterial uspension in a 70 °C water bath until the temperature equilibrates. Add an equal volume of 90% (w/v) phenol that has been preheated to 70 °C and mixed thoroughly. The resulting mixture is rapidly cooled by stirring for 15 min in an ice water bath.

5. Centrifuge the phenol–bacterial suspension mixture at 18,000 × g for 15 min at 4 °C. A sharp interface occurs between the aqueous, phenol, and interface layers. There is occasionally a sediment. The aqueous phase should be carefully removed by aspiration and retained. The phenol layer can be discarded.

6. Dialyze the aqueous layer against frequent changes of distilled water until no detectable phenol odor remains. The dialyzate can then be lyophilized and stored at this point before further processing (*see* **Note 2**).

7. Sediment suspension at 1100 × g for 5 min. Discard sediment and centrifuge the supernatant fractions at 105,000 × g for 2 h. Save the gel-like pellet and discard supernatant.

8. Resuspend the pellet in the original volume of distilled water and repeat the process of low- and high-speed centrifugation until the desired purity is obtained. The final pellet can be resuspended in distilled water and lyophilized.

3.2. Proteinase K Digested, Phenol–Water Extracted LPS

The phenol–water extraction method can contain small amounts of lipoproteins that could confound experiments designed to measure only LPS biological activity. These contaminating proteins can be removed by an alternative procedure described below, which incorporates proteinase degradation prior to the phenol extraction.

1. Suspend dried bacterial cells (500 mg), which have been ground with a mortar and pestle, in 15 mL of 10 mM Tris–Cl buffer (pH 8.0), containing 2% SDS, 4% 2-mercaptoethanol, and 2 mM MgCl$_2$. Vortex the mixture and place it in a 65 °C water bath until the bacterial cells are solubilized. Occasional vortexing may be necessary during the heating process.

2. Add 1 mL of proteinase K (100 µg/mL) in solubilization solution to the cell mixture and keep the sample at 65 °C for an additional hour. Subsequently place sample in a 37 °C water bath overnight.

3. Add 2 mL of 3 M sodium acetate to the proteinase K-digested cell suspension and mix sample thoroughly (the salt will enhance the LPS precipitation by the ethanol). Add 40 mL of cold absolute ethanol to the cell suspension and allow precipitate to form overnight at –20 °C. Centrifuge the mixture at 4000 × g for 15 min and decant and discard supernatant.

4. Raise the precipitate in 9 mL of distilled water. Add 1 mL of 3 M sodium acetate and vortex the mixture. Add 20 mL of cold absolute ethanol and vortex the mixture again. The suspension is allowed to precipitate overnight at –20 °C. This precipitation is repeated one additional time. The purpose of the repeated ethanol precipitations is to remove residual SDS from the LPS preparation.

5. After the final centrifugation, suspend precipitate in 9 mL of 10 mM Tris–Cl (pH 7.4), and add 0.5 mL of DNase I (100 µg/mL) and 0.5 mL of RNase (25 µg/mL).

Place this mixture at 37 °C for 4 h to allow digestion of any residual contaminating nucleic acids.

6. The final step is a phenol extraction of the LPS nuclease-treated mixture to assure removal of all residual protein contamination. Place the LPS mixture in a 65 °C water bath for 30 min, add an equal volume of 90% phenol preheated to 65 °C, and allow to set at 65 °C for 15 min. Place the mixture in an ice bath and cool quickly to 4 °C.

7. Centrifuge the cooled mixture at 6000 × *g* for 15 min. Remove the aqueous top layer and re-extract the phenol layer with an equal volume of distilled water. Heat sample again to 65 °C for 15 min and then place in ice water. After centrifugation at 6000 × *g* for 15 min, add top aqueous layer to the first aqueous extraction, and dialyze against multiple changes of distilled water over 2 days. Discard the bottom phenol layer. After dialysis, the LPS can be lyophilized and stored indefinitely at room temperature.

3.3. Small-Scale Preparations of LPS or LOS

Occasionally, analytical studies such as western blot analysis of LPS or LOS will require preparation of same quantities of these glycolipids from multiple bacterial samples. Two methods have been described which can be used for such isolations *(18,19)*. One relies on a modified phenol–water extraction *(20)*, and the second utilizes SDS solubility of bacteria followed by enzymatic degradation of bacterial proteins *(19)*. Neither preparation gives a highly purified preparation, but both enrich for LPS or LOS and can serve as adequate substitutes in acrylamide gel and western blot studies. These two methods are presented below.

3.3.1. Rapid Isolation Micro Method for LPS (20)

1. Centrifuge a bacterial suspension (10^8 colony-forming units/mL in 2 mL PBS) in a 15-mL tube at 10,000 × *g* for 5 min. Wash the pellet once in PBS (pH 7.2) containing 0.15 m*M* $CaCl_2$ and 0.5 m*M* $MgCl_2$. Resuspend washed cells in 300 µL of water and transfer to a 1-dram vial containing a stir bar.

2. Add an equal volume of hot (65–70 °C) 90% phenol. Stir the mixture vigorously at 65–70 ° for 15 min. Chill suspension on ice, transfer to a 1.5-mL polypropylene tube, and centrifuge at 8500 × *g* for 15 min.

3. Transfer supernate to a 15-mL conical centrifuge tube. Re-extract phenol phase with 300 µL of distilled water. Pool the aqueous phases.

4. Add sodium acetate to 0.5 *M* final concentration. Add 10 volume of 95% ethanol and place sample at –20 °C overnight in order to precipitate the LPS. Centrifuge precipitate at 2000 × *g* at 4 °C for 10 min. Discard supernatant.

5. Suspend pellet in 100 µL of distilled water and transfer to a 1.5 mL polypropylene tube. Repeat ethanol precipitation. Dry the precipitate and resuspend with 50 µL of distilled water. Store at –20 °C indefinitely.

3.3.2. LPS Microextraction Using Proteinase K Digestion (19)

1. Harvest organisms grown on solid medium with a sterile swab and suspend in 10 mL of cold phosphate-buffered saline (PBS) (pH 7.2) to a turbidity of 0.4 absorbance at 650 nm. Centrifuge a portion (1.5 mL) of this suspension for 1.5 min in a microfuge at $14,000 \times g$.
2. Solubilize the pellet in 50 µL of lysing buffer containing 2% SDS, 4% 2-mercaptoethanol, 10% glycerol, 1 M Tris–Cl (pH 6.8), and bromophenol blue. Heat sample at 100 °C for 10 min.
3. To digest bacterial proteins, add 25 µg of proteinase K in 10 µL of lysing buffer to each boiled lysate and incubate at 60 °C for 60 min.
4. Use the preparation in acrylamide gel electrophoresis or for western blots in volumes ranging from 0.5 to 2 µL (*see* **Note 3**).

3.4. LOS and LPS Acrylamide Gel Electrophoresis

Acrylamide gel electrophoresis of LOS and LPS can be an exasperating procedure because of day-to-day variation in resolution. To ensure consistency of these gels, acid wash all glassware and rinse well before drying; all reagents must be of the highest quality, and the quality of water should be in the 18 ohm range. Finally, stock solutions should not be used past the time recommended.

1. To prepare two 14% resolving gels, combine 18.45 mL of 30% acrylamide, 7.9 mL of resolving buffer, and 12.57 mL of distilled water. Place solution in a vacuum flask and degas for 15 min.
2. Add 0.3 mL ammonium persulfate solution and N,N,N´,N´-tetramethylethylene-diamine (TEMED) (10 µL/50 mL gel solution). The resolving gel should be poured between the glass plates as soon as the TEMED is added and the solution gently mixed. Overlay the gel with 2 mm distilled water and allow to polymerize for at least 2 h. This will make a total of 39.56 mL resolving gel, which is sufficient for two 12 × 14 cm slab gels with a 0.75-mm spacer. Double the volumes if you are using 1.5-mm spacers. To make 16% resolving gels, mix 10.55 mL of 30% acrylamide, 3.95 mL resolving buffer, and 4.95 mL of distilled water. The quantities of the other reagents are unchanged.
3. To prepare the spacer gel, combine 2 mL of 30% acrylamide, 2 mL of spacer buffer, and 15.6 mL of distilled water. Degas the solution for 15 min and add 0.2 mL of ammonium persulfate, 0.2 mL of 10% SDS, and 10 µL of TEMED. Mix gently. The total volume of the spacer gel is 20.01 mL.
4. Remove the layer of distilled water over the resolving gel and pour the spacer gel. Insert at this time the comb chosen for the run. Allow the spacer gel to polymerize for at least 1 h but preferably overnight.

5. Solubilize the LPS or LOS in sample buffer to the desired concentration (0.1–1 mg/mL should be sufficient). Dilute 1:1 in dye buffer, boil for 5 min in a water bath, and allow to cool. Load 5–10 μL per well.
6. To electrophorese, remove the comb and rinse the sample wells with reservoir buffer. Fill the wells with reservoir buffer and add samples allowing them to sink to the bottom. Place gels into the chamber and electrophorese at constant current under the following conditions; for one 12 × 14 cm slab gel, use 10–12 mA through the spacer gel and then raise to 15 mA through the resolving gel. Total run time will be approximately 5 h. For two 12 × 14 cm slab gels, use 20 mA through spacer gel and then raise to 30 mA through resolving gel. The total run time will be 5–6 h.
7. Stain the gels using the silver stain method described by Tsai and Frasch *(21)* or prepare the gels for western blotting *(22)*.

3.5. Silver Staining Technique for LPS and LOS Gels (23)

LPS and LOS bands can be rapidly identified in acrylamide gels using the silver stain method of Tsai et al. *(21)*. This method is very sensitive and requires careful attention to cleanliness of the glassware used. Only a glass tray should be used for the procedure, and it is best to have dedicated trays for each step in the procedure. Glassware should be pre-cleaned before each study with nitric acid and washed thoroughly with distilled water. Gloves should be worn while performing the procedure. Mild rotary agitation (70 rpm) on a platform rocker is required for each step. The rinses and washes are crucial and should not be reduced in time or number.

1. As soon as the electrophoresis is complete, place the gel in a fixing solution consisting of 40% ethanol and 5% acetic acid in distilled water overnight. Add 0.9% periodic acid (1.8 g/200 mL) to fixing solution.
2. Rinse three times with distilled water. Transfer to a separate dish and wash three additional times with distilled water (500–1000 mL), agitating for 10 min each time.
3. In a separate dish, pour in freshly prepared staining reagent (this should be made during the last distilled water wash). This is made by mixing in the following order in a hood, 28 mL of 0.1N NaOH, 2.1 mL of concentrated NH_4OH, 5 mL of 20% silver nitrate (add silver nitrate drop-wise), and 115 mL distilled water.
4. Agitate for 10 min. Transfer the gel to a separate dish and rinse gel three times with distilled water.
5. Transfer gel to a separate dish and add fresh formaldehyde developer (50 mg anhydrous citric acid, 0.5 mL of 37% formaldehyde). For best results, the gel should be developed in the dark using a Kodak GBX-2 safelight.
6. The bands will develop over the next 10–15 min. Background staining will intensify in proportion to the amount of time the gel is left in the developer. Stop the reaction by rinsing the gel in water, transferring to a separate dish and adding rapid fix (10 mL in 100 mL of distilled water).

3.6. Isolation of Lipid A (24)

Isolation of lipid A takes advantage of the labile linkage between the lipid A backbone and the ketodeoxyoctanoate (KDO) in the LPS core. Mild acid hydrolysis and heat are sufficient to disrupt this linkage. The lipid A is insoluble in water and forms a precipitate that can be readily extracted by centrifugation. Three methods are given in this section. The first is the classical method, which has been most widely used. The second method allows extraction of small amounts of relatively pure lipid A directly from whole cells without a prior LPS extraction. The third method uses a relatively mild hydrolysis step to preserve acid labile phosphorylation sites and head groups on the lipid A backbone. This method is particularly useful in isolating lipid A for structural analysis.

3.6.1. Acetic Acid Hydrolysis Method

1. Lipid A can be isolated by mild acid hydrolysis of the purified LPS. Solubilize LPS or LOS in aqueous 0.02% triethylamine and add acetic acid to a final concentration of 1.5% (v/v).
2. Heat the mixture for 2 h at 100 °C and then cool. Quantitatively precipitate lipid A by adding 1 M HCl to a final pH of 1.5.
3. Centrifuge insoluble lipid A at 2000 × g, wash three times with cold distilled water, and lyophilize.

3.6.2. Microlipid A Extraction from Whole Cells (25)

1. Suspend lyophilized crude or freshly washed bacterial cells (10 mg) in 400 µL of isobutyric acid–ammonium hydroxide 1 M (5:3 v/v) and keep for 2 h at 100 °C in a screw-cap test tube under magnetic stirring.
2. Cool mixture in ice water and centrifuge at 2000 × g for 15 min. Dilute supernatant with water (1:1 v/v) and lyophilize.
3. Wash the sample twice with 400 µL of methanol and centrifuge at 2000 × g for 15 min.
4. Soluble insoluble lipid A and extract once in 100–200 µL of a mixture of chloroform–methanol–water (3:1.5:0.25 v/v). For 1 mg samples, 100 µL of the different solvent mixtures are used at each step (25).

3.6.3. Lipid A Isolation (26)

1. Dissolve 5 mg of LPS in 500 µL of 10 mM sodium acetate (pH adjusted to 4.5 with 4 M HCl) containing 1% SDS and then place in an ultrasound bath until the sample is dissolved.
2. Heat sample at 100 °C for 1 h. Dry the mixture by Speed Vac and remove SDS by washing the sample with 100 µL of distilled water and 500 µL of acidified ethanol (prepared by combining 100 µL of 4 M HCl with 20 mL of 95% ethanol) followed by centrifugation at 2000 × g for 10 min.

3. Wash the sample again with 500 µL of non-acidified 95% ethanol and centrifuge at 2000 × *g* for 10 min.
4. Repeat centrifugation and washing steps. Finally, lyophilize the sample to yield fluffy white solid lipid A *(26)*.

4. Notes

1. Occasionally, the suspension will become gelatinous at this stage. If this occurs, stir for 3 min in the blender before phenol extraction.
2. The dialyzate is still a relatively crude LPS preparation that needs further purification by centrifugation. This will remove residual nucleic acids and any capsular materials that remain in the sample. The preparation is suspended in distilled water at concentrations of 25–35 mg/mL. At times, this crude LPS can be difficult to suspend. It should be vortexed vigorously to obtain a smooth suspension. Gentle sonication may be required to obtain an even distribution of the LPS suspension.
3. Purity of the LPS and LOS preparations: After isolation of the LOS or LPS, the degree of purity should be ascertained. Nucleic acid and protein contamination are the principle concerns. Spectral analysis from 245 nm thru 290 nm using purified LPS or LOS at 1 mg/mL can be performed to determine nucleic acid and protein contamination. Agarose gel electrophoresis with ethidium bromide staining can also be useful as a highly sensitive means of determining the degree of nucleic acid contamination. Acrylamide gel electrophoresis of the LOS/LPS preparation will define the physical characteristics of the preparation *(21)*. This method utilized in a number of laboratories studying LPS and LOS is described in **Subheading 3.4**. The pore size of the gel can be adjusted according to the preparation under study and can range from 10 to 16% acrylamide.

References

1. Shear, M. J. (1941) Effect of concentrate from *B. prodigiosus* filtrate on subcutaneous primary induced mouse tumors. *Cancer Res.* **1**, 732–741.
2. Luderitz, O., Staub, A. M., and Westphal, O. (1966) Immunochemistry of O and R antigens of *Salmonella* and related *Enterobacteriaceae*. *Bacteriol. Rev.* **30**, 192–255.
3. Goldman, R. C. and Leive, L. (1980) Heterogeneity of antigenic-side-chain length in lipopolysaccharide from *Escherichia coli* 0111 and *Salmonella typhimurium* LT2. *Eur. J. Biochem.* **107**, 145–153.
4. Palva, E. T. and Makela, P. H. (1980) Lipopolysaccharide heterogeneity in Salmonella typhimurium analyzed by sodium dodecyl sulfate/polyacrylamide gel electrophoresis. *Eur. J. Biochem.* **107,** 137–143.
5. Preston, A., Mandrell, R. E., Gibson, B. W., and Apicella, M. A. (1996) The lipooligosaccharides of pathogenic gram-negative bacteria. *Crit. Rev. Microbiol.* **22**, 139–180.

6. Ribi, E., Haskins, W. T., Landy, M., and Milner, K. C. (1961) Preparation and host-reactive properties of endotoxin with low content of nitrogen and lipid. *J. Exp. Med.* **114**, 647–663.
7. Galanos, C., Luderitz, O., and Westphal, O. (1969) A new method for the extraction of R lipopolysaccharides *Eur. J. Biochem.* **9**, 245–249.
8. Roberts, N. A., Gray, G. W., and Wilkinson, S. G. (1967) Release of lipopolysaccharide during the preparation of cell walls of *Pseudomonas aeruginosa. Biochim. Biophys. Acta* **135**, 1068–1071.
9. Leive, L. (1965) Release of lipopolysaccharide by EDTA treatment of *E. coli. Biochem. Biophys. Res. Commun.* **21**, 290–296.
10. Goebel, W. F., Binkley, F., and Perlman, E. (1945) Studies on the Flexner group of dysentery bacilli. I. The specific antigens of *Shigella paradysenteriae. J. Exp. Med.* **81**, 315–330.
11. Westphal, O. and Jann, K. (1965) Bacterial lipopolysaccharides. Extraction with phenol water and further applications of the procedure, in *Methods in Carbohydrate Chemistry* (Whistler, R. L., ed.), Academic Press, New York, pp. 83–91.
12. Darveau, R. P. and Hancock, R. E. W. (1983) Procedure for isolation of bacterial lipopolysaccharides from both smooth and rough *Pseudomonas aeruginosa* and *Salmonella typhimurium* strains. *J. Bacteriol.* **155**, 831–838.
13. Johnson, K. G., Perry, M. B., and McDonald, J. J. (1976) Studies of the cellular and free lipopolysaccharides from *Neisseria canis* and *N. subflava. Can. J. Microbiol.* **189**, 189–196.
14. Hitchcock, P. J. and Brown, T. M. (1983) Morphological heterogeneity among Salmonella lipopolysaccharide chemotypes in silver-stained polyacrylamide gels. *J. Bacteriol.* **154**, 269–277.
15. Westphal, O., Jann, K., and Himmelspach, K. (1983) Chemistry and immunochemistry of bacterial lipopolysaccharides as cell wall antigens and endotoxins. *Prog. Allergy* **33**, 9–39.
16. Johnson, K. G. and Perry, M. B. (1976) Improved techniques for the preparation of bacterial lipopolysaccharides. *Can. J. Microbiol.* **22**, 29–34.
17. Erwin, A. L., Munford, R. S., and Group TBPFS. (1989) Comparison of lipopolysacharides from Brazilian purpuric fever isolates and conjunctivitis isolates of *Haemophilus influenzae* biogroup aegyptius. *J. Clin. Microbiol.* **27**, 762–767.
18. Inzana, T. J. (1983) Electrophoretic heterogeneity and interstrain variation of the lipopolysaccharide of Haemophilus influenzae. *J. Infect. Dis.* **148**, 492–499.
19. Hitchcock, P. J. (1984) Analyses of gonococcal lipopolysaccharide in whole-cell lysates by sodium dodecyl sulfate-polyacrylamide gel electrophoresis: stable association of lipopolysaccharide with the major outer membrane protein (protein I) of *Neisseria gonorrhoeae. Infect. Immun.* **46**, 202–212.
20. Inzana, T. J. and Pichichero, M. E. (1984) Lipopolysaccharide subtypes of *Haemophilus influenzae* type b from an outbreak of invasive disease. *J. Clin. Microbiol.* **20**, 145–150.

21. Tsai, C.-M. and Frasch, C. E. (1982) A sensitive silver stain for detecting lipopolysaccharides in polyacrylamide gels. *Anal. Biochem.* **119**, 115–119.
22. Blake, M. S., Johnston, K. H., Russell-Jones, G. J., and Gotschlich, E. C. (1984) A rapid, sensitive method for detection of alkaline phosphatase-conjugated anti-antibody on Western blots. *Anal. Biochem.* **136**, 175–179.
23. Hitchcock, P. J. and Brown, T. M. (1983) Morphological heterogeneity among *Salmonella* lipopolysaccharide chemotypes in silver-stained polyacrylamide gels. *J. Bacteriol.* **154**, 269–277.
24. Kulshin, V. A., Zahringer, U., Lindner, B., et al. (1992) Structural characterization of the lipid A component of pathogenic *Neisseria meningitidis*. *J. Bacteriol.* **174**, 1793–1800.
25. El Hamidi, A., Tirsoaga, A., Novikov, A., Hussein, A., and Caroff, M. (2005) Microextraction of bacterial lipid A: easy and rapid method for mass spectrometric characterization. *J. Lipid Res.* **46**, 1773–1778.
26. Yi, E. C. and Hackett, M. (2000) Rapid isolation method for lipopolysaccharide and lipid A from Gram-negative bacteria. *Analyst* **125**, 651–656.

2

Proteomic Analysis of Proteins Secreted by *Streptococcus pyogenes*

Michelle A. Chaussee, Emily J. McDowell, and Michael S. Chaussee

Summary

Streptococcus pyogenes secretes various proteins to the extracellular environment. During infection, these proteins interact with human macromolecules and contribute to pathogenesis. We describe a proteomic approach routinely used in our laboratory to characterize culture supernatant proteins using small-format two-dimensional gel electrophoresis. Proteins are collected after overnight growth of the bacteria in broth media. Compounds that inhibit isoelectric focusing, such as salts, are removed by enzymatic treatment and precipitation with trichloroacetic acid and acetone. Following resuspension in denaturing solution, the proteins are separated by isoelectric focusing using a 7-cm immobilized strip with a pH gradient of 4–7. Subsequently, proteins are further separated with sodium dodecyl sulfate–polyacrylamide gel electrophoresis (SDS–PAGE) and stained with SYPRO Ruby. The small-gel format requires less time, reagents, and smaller culture volumes compared with large-format approaches, while still resolving and detecting a large proportion of the exoprotein fraction.

Key Words: Proteomics; exoprotein; *Streptococcus pyogenes*; two-dimensional gel electrophoresis.

1. Introduction

Streptococcus pyogenes secretes many proteins that interact with human macromolecules *(1,2)*. Not surprisingly, many are important determinants of host–pathogen interactions and contribute to virulence. In vitro, the secreted proteins can be readily isolated from culture supernatants. Extracellular proteins include both those with type II signal secretion signals and (for reasons

From: *Methods in Molecular Biology, vol. 431: Bacterial Pathogenesis*
Edited by: F. DeLeo and M. Otto © Humana Press, Totowa, NJ

that remain unknown) proteins that lack signal sequences, such as glycolytic proteins *(3–5)*.

The relatively small number of culture supernatant proteins (CSPs), compared with proteins in the cytoplasm, simplifies proteomic analysis. However, several aspects can make two-dimensional gel electrophoresis (2-DE) of CSPs challenging *(6)*. These include (1) the presence of basic CSPs, which can become insoluble during isoelectric focusing, (2) the presence of hyaluronic acid, which can interfere with isoelectric focusing, (3) the presence of peptides in complex media, such as Todd–Hewitt broth, and (4) the abundance of specific exoproteins such as the cysteine protease SpeB, which can mask less abundant proteins *(7)*.

Analysis typically involves three steps. First, exoproteins are isolated from bacterial cultures. The aim of this process is to retain as many proteins as possible while eliminating components that inhibit isoelectric focusing. To do so, proteins are enzymatically treated and precipitated to remove salts and other non-protein components. Second, the proteins are resolved by one- or two-dimensional gel electrophoresis (*see* **Fig. 1**). Third, proteins are visualized and the composition and quantities of protein spots determined. The protocol described below uses a mini-gel format that is relatively fast and requires small culture volumes compared with large-format two-dimensional gel electrophoresis.

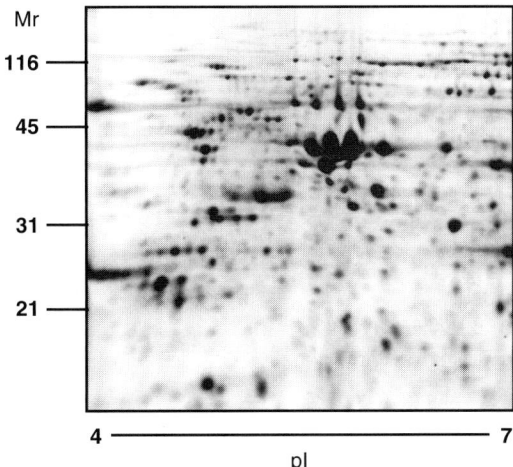

Fig. 1. Two-dimensional gel electrophoresis of culture supernatant proteins. Proteins were isolated from wild-type *Streptococcus pyogenes* strain NZ131 after 18 h of culture in 40 mL of chemically defined medium. Hundred micrograms of protein was separated with a 7-cm immobilized gradient strip (pH 4–7) (IPG; GE Biosciences) and SDS–PAGE. Proteins were stained with SYPRO Ruby.

2. Materials

2.1. Isolation of CSPs

1. 100% trichloroacetic acid (TCA) (*see* **Note 1**).
2. Acetone (*see* **Note 1**).
3. Absolute ethanol.
4. 25-mm, 0.2-µm syringe filter.
5. 10-mL luer-lock syringe.
6. Broth medium (Todd–Hewitt or chemically defined media [*see* **Note 2**]).
7. 10× DNase buffer: 100 mM Tris–HCl (pH 7.5), 25 mM MgCl$_2$, 5 mM CaCl$_2$.
8. RNase A.
9. DNase.

2.2. Isoelectric Focusing

1. IPGphor (GE Healthcare Life Sciences, Piscataway, NJ, USA).
2. 7-cm Immobiline DryStrips (pH 4–7) (GE Healthcare Life Sciences).
3. 7-cm IPG ceramic strip holder (GE Healthcare Life Sciences).
4. Dry-strip cover fluid.
5. IPG buffer (pH 4–7) (GE Healthcare Life Sciences).

2.3. Casting Polyacrylamide Gels

1. 1.5 M Tris–HCl (pH 8.8).
2. 10% (w/v) ammonium persulfate (*see* **Note 3**).
3. 10% (w/v) sodium dodecyl sulfate (SDS).
4. Acrylamide solution: acrylamide:bis-acrylamide (37.5:1). Acrylamide is a potent neurotoxin. Wear gloves and protective clothing when using solutions containing acrylamide (*see* **Note 4**).
5. Mini-PROTEAN 3 large glass plates with 1.0-mm spacers (Bio-Rad Laboratories, Hercules, CA, USA).
6. Mini-PROTEAN 3 Multi-Casting Chamber (Bio-Rad Laboratories).
7. Mini-PROTEAN 3 short glass plates (Bio-Rad Laboratories).
8. N,N,N´,N´-tetramethylethylenediamine (TEMED).
9. Water-saturated butanol: Mix 50 mL of *n*-, *i*-, or *t*-butanol and 10 mL of deionized H$_2$O. Combine in a bottle and shake. Use the top phase to overlay gels. Store at room temperature.

2.4. Second-Dimension Gel Electrophoresis

1. Mini-PROTEAN 3 Dodeca Cell (Bio-Rad Laboratories).
2. MultiTemp III Thermostatic Circulator (GE Healthcare Life Sciences) (*see* **Note 5**).
3. Solubilization and rehydration solution (SR solution): 8 M urea, 2 M thiourea *(8)*, and 4% (w/v) CHAPS. Add dry reagents and fill to volume with distilled

H$_2$O. Stir until all reagents are dissolved, which may take 1 h or longer. Do not heat the solution (*see* **Note 6**). Dispense 1-mL aliquots into microcentrifuge tubes and store at $-20\,°C$. Just prior to use, thaw the SR solution and add 75 mM dithiothreitol (DTT) (*see* **Note 7**).

4. 1× SDS electrophoresis buffer: 25 mM Trizma base, 0.192 M glycine, 0.1% (w/v) SDS (*see* **Note 8**).
5. Agarose sealing solution: 1× SDS electrophoresis buffer, 0.5% (w/v) agarose NA (GE Healthcare Life Sciences), and a trace of bromophenol blue (*see* **Note 9**).
6. SDS equilibration buffer: 50 mM Tris–HCl, 6 M urea, 2% (w/v) SDS, 30% (v/v) glycerol, and a trace of bromophenol blue. Dispense 10-mL aliquots of the solution into tubes and store at $-20\,°C$ until needed. Just before use, thaw and add DTT (100 mM final concentration).
7. Broad-range SDS–polyacrylamide gel electrophoresis (SDS–PAGE) molecular weight standards (Bio-Rad Laboratories).
8. Sample Application Pieces (GE Healthcare Life Sciences).

2.5. Protein Detection, Quantitation, and Analysis

1. SYPRO Ruby protein stain (Bio-Rad Laboratories).
2. Fixer and destain solution: 10% (v/v) methanol and 6% (v/v) glacial acetic acid.
3. 2-D Quant Kit (GE Healthcare Life Sciences).
4. Typhoon 9410 imager (GE Healthcare Life Sciences).

3. Methods
3.1. Isolation of CSPs

1. Culture *S. pyogenes* in 40 mL of broth media in 50-mL polypropylene centrifuge tubes at $37\,°C$ for 18 h (*see* **Note 10**).
2. Centrifuge at $3220 \times g$ for 10 min at $4\,°C$ to pellet the bacteria.
3. Pass the supernatant fluid through a 0.2-μm syringe filter into a sterile 50-mL polypropylene centrifuge tube. Proceed to next step or store the supernatant fluid at $-80\,°C$ (*see* **Note 11**).
4. Incubate freshly filtered unfrozen supernatants at $-20\,°C$ for 30 min (*see* **Note 12**). If the supernatant has been frozen, thaw it at room temperature and proceed to next step when the solution is completely thawed but ice cold.
5. Keep samples on ice during **steps 6–41**.
6. Add 100% ice-cold TCA to a final concentration of 10% (v/v) and invert the tube several times to mix.
7. Add ice-cold acetone to a final concentration of 5% (v/v).
8. Invert tube several times to mix (*see* **Note 13**).
9. Incubate the tube at $-20\,°C$ for 1 h inverting several times every 20 min.
10. Centrifuge at $6000 \times g$ for 20 min at $4\,°C$.
11. Discard supernatant fluid in a hazardous waste container.
12. Add 3 mL of ice-cold ethanol to the pellet and resuspend the pellet by vortexing.

13. Transfer 750 µL of the solution into four 1.5-mL microcentrifuge tubes (*see* **Note 14**).
14. Vortex well and incubate at –20 °C for 30 min.
15. Centrifuge at 25,000 × *g* for 5 min at 4 °C.
16. Remove the supernatant fluid and discard it in a hazardous waste container.
17. Add 1.5 mL of ice-cold ethanol to each tube.
18. Vortex well. The pellet will disperse but may not completely dissolve.
19. Incubate at –20 °C for 30 min.
20. Vortex well and centrifuge at 25,000 × *g* for 5 min at 4 °C.
21. Pour off supernatant fluid. Invert tubes and let any remaining ethanol evaporate.
22. Resuspend the pellet with 250 µL of 0.04 *M* Tris–HCl (pH 7.5).
23. Vortex well and combine the solutions into one microcentrifuge tube.
24. Add 24,000 U/mL of DNase, 75,000 U/mL of RNase, and 10 µL of 10× DNase buffer (*see* **Note 15**). Incubate for 2 h at 4 °C.
25. Add 100% cold TCA to a final concentration of 10% (v/v) and invert the tube several times to mix.
26. Add ice-cold acetone to a final concentration of 5% (v/v) and invert the tube several times to mix (*see* **Note 13**).
27. Incubate at –20 °C for 1 h. Invert the tube several times every 20 min.
28. Centrifuge at 25,000 × *g* for 20 min at 4 °C.
29. Discard supernatant fluid.
30. Add 1.5 mL of ice-cold ethanol to the pellet and resuspend the pellet by vortexing.
31. Vortex well and incubate at –20 °C for 30 min.
32. Centrifuge at 25,000 × *g* for 5 min at 4 °C.
33. Remove supernatant fluid and discard in a hazardous waste container.
34. Add 1.5 mL of ice-cold ethanol to the tube.
35. Vortex well. The pellet will disperse but may not completely dissolve.
36. Incubate at –20 °C for 30 min.
37. Centrifuge at 25,000 × *g* for 5 min at 4 °C.
38. Pour off supernatant fluid. Invert tubes and let any remaining ethanol evaporate.
39. Resuspend the pellet in SR solution containing 75 m*M* DTT (*see* **Note 16**).
40. Centrifuge samples at 25,000 × *g* for 5 min at 4 °C and transfer the supernatant fluid to a clean microcentrifuge tube. Proceed to the next step or store the sample at 4 °C until the protein concentration can be determined.
41. Determine the protein concentration by using the 2-D quant kit (GE Healthcare Life Sciences) (*see* **Note 17**).
42. Store sample at 4 °C.

3.2. Isoelectric Focusing

1. Remove protein samples from 4 °C storage and equilibrate to room temperature.
2. Vortex well. Dilute the samples to a concentration of 0.8 mg/mL in a final volume of 125 µL with SR solution containing 75 m*M* DTT. Add 1.875 µL of IPG buffer (pH 4–7) (*see* **Note 18**). Vortex well and mix samples on a platform rocker at room temperature for at least 1 h.

3. Centrifuge samples at 25,000 × *g* for 10 min at 4 °C. Transfer the supernatant fluid to a clean microcentrifuge tube.
4. Slowly pipet 125 µL (*see* **Note 19**) of the sample to the center of a 7-cm ceramic strip holder (*see* **Note 20**). Remove any bubbles.
5. Remove the protective plastic cover from the IPG strip. Position the IPG strip with the gel side down and the anodic end of the strip directed toward the anodic end of the strip holder. Lower the strip into the sample starting with the anodic end (*see* **Note 21**). Finally, lower the cathodic end of the IPG strip into the channel, making sure that the gel contacts the strip holder electrodes at each end. Be careful not to trap bubbles under the IPG strip.
6. To minimize evaporation and urea crystallization, apply IPG strip cover fluid.
7. Place the cover on the strip holder and place the ceramic holder on the IPGphor.
8. Rehydrate and focus the strips according to the following parameters:

   a. Rehydration for 12 h at 20 °C (*see* **Note 22**).
   b. Isoelectric focusing (IEF) parameters: 20 °C, 50 µA per strip.
   c. Step 1: 500 V for 500 V/h (step and hold).
   d. Step 2: 1000 V for 1000 V/h (step and hold).
   e. Step 3: 8000 V for 32,000 V/h (step and hold).

9. Use forceps to remove strip from the ceramic strip holder and place IPG strips into separate 15-mL polypropylene centrifuge tubes with the support film toward the tube wall (*see* **Note 23**).
10. Proceed to **Subheading 3.4.2** or store the IPG strips at –80°C.

3.3. Casting Polyacrylamide Gels

1. Assemble the Mini-PROTEAN 3 gel casting chamber (*see* **Note 24**).
2. Prepare a 12% acrylamide gel solution. For six gels (8 × 7.3 × 0.001 cm), mix 13.5 mL of 40% acrylamide/bis-acrylamide stock solution, 18.6 mL of deionized water, 12 mL of 1.5 *M* Tris–HCl (pH 8.8), and 0.45 mL of 10% SDS. Mix well. Just before pouring gels, add 0.45 mL of 10% ammonium persulfate and 12.6 µL TEMED and mix well (*see* **Note 25**).
3. Immediately pour the acrylamide solution into the casting chamber, filling it until the solution is approximately 3 mm from the top of the short plate.
4. Gently overlay each gel with water-saturated butanol.
5. Allow the gels to polymerize 45–60 min.
6. Just before electrophoresis, disassemble the casting chamber.
7. Rinse the gels with distilled water to remove any butanol.

3.4. Second-Dimension Gel Electrophoresis

3.4.1. Preparation for Electrophoresis

1. Place 3.4–4.4 L of 1× SDS electrophoresis buffer in the Mini-PROTEAN 3 Dodeca Cell chamber and pre-cool the buffer by setting the MultiTemp III Thermostatic Circulator to 10 °C (*see* **Note 26**).

2. Cut a sample application piece in half and place each half in a separate tube. Use half of a sample application piece for each gel being run.
3. Add 5 µL of the SDS–PAGE molecular weight standard solution to a sample application piece. Add two drops of molten agarose sealing solution to the sample application piece (*see* **Note 27**).

3.4.2. Equilibration of IPG strips

1. Remove strips from freezer.
2. Add 5 mL of the equilibration buffer containing 100 m*M* DTT to each tube and place on a rocking platform for 15 min.

3.4.3. Running Second Dimension

1. Remove the strip from the screw cap tube and dip it in 1× SDS electrophoresis buffer.
2. Position the strip between the gel plates with the plastic backing against one of the glass plates.
3. With a thin plastic ruler, or similar instrument, gently push the IPG strip down so the entire lower edge of the IPG strip is in contact with the top surface of the slab gel (*see* **Note 28**).
4. Apply a sample application piece containing molecular weight markers at the far end of the strip where there is no IPG gel. Make sure the sample application piece touches the top surface of the second dimension gel.
5. Gently overlay the IPG strip and sample application piece with agarose sealing solution. Use care not to introduce bubbles while applying the agarose sealing solution.
6. Place gels in the Mini-PROTEAN 3 Dodeca Cell electrophoresis chamber.
7. Set the power supply for the following parameters:

    a. 50 V for 15 min.
    b. 100 V for 2–3 h.

8. Monitor the migration of the bromophenol blue dye and turn off the power supply when the dye is a few millimeters above the bottom of the gel.
9. Remove gel cassettes from chamber, and using a plastic wedge, carefully open the cassette (*see* **Note 29**).

3.5. Protein Detection, Quantitation, and Analysis

1. Place the gels in fixing solution (10% methanol, 6% glacial acetic acid) for 30 min.
2. Pour off fixing solution.
3. Add 30 mL of SYPRO Ruby stain. Place the gels on a rocking platform and stain overnight in a foil covered or lightproof plastic container (*see* **Note 30**).
4. The next day, pour off the SYPRO Ruby stain.

5. Add 30 mL of destain (10% methanol, 6% glacial acetic acid) and place the gels on a rocking platform. Allow the gels to destain for 4–24 h (*see* **Note 31**).
6. Pour off the destain.
7. Add distilled water to the gels (*see* **Note 32**).
8. Image the gel with a Typhoon 9410 instrument using the 610BP 30 filter and 457 (blue) laser.

4. Notes

1. Use TCA and acetone in a ventilated fume hood and wear suitable protective gear and clothing.
2. Chemically defined media are preferred because they lack the peptides present in complex media. If complex media are used, adjustments must be made empirically in the amount of protein to load because the peptides will alter protein determinations.
3. Store at 4 °C and prepare fresh weekly.
4. It is safer to purchase liquid solutions of acrylamide:bis-acrylamide rather than purchasing the powdered form of the reagents because mixing the dry reagents may create dust that is harmful if inhaled.
5. To prevent gels from overheating during electrophoresis, one needs a circulating cooler when running the Mini-PROTEAN 3 Dodeca.
6. Heating solutions containing urea can cause protein carbamylation (*9*).
7. Up to 100 m*M* DTT can be added, if desired.
8. SDS electrophoresis buffer can be made as a 10× solution (250 m*M* Trizma base, 1.92 *M* glycine, 1.0% (w/v) SDS) and diluted to 1× as needed. Our laboratory makes a carboy of 10 L of 10× buffer and we store it at room temperature.
9. The addition of just a few grains of bromophenol blue is sufficient. This will provide a dye marker to monitor electrophoresis.
10. Relatively few proteins are detected in supernatant fluids of exponential phase cultures.
11. The CSPs can now be stored at –80 °C for up to 1 week. –80 °C storage is for convenience only and does not improve protein recovery.
12. Supernatants should be ice-cold but not frozen.
13. Solution may become cloudy.
14. If sample volume is more than 3 mL, dispense 750 µL into the appropriate amount of 1.5-mL tubes until the entire sample has been divided.
15. Adding 10 µL of the 10× DNase buffer assumes there is a final volume of 1.0 mL. If the volume is different, add the appropriate amount of 10× DNase buffer.
16. Resuspension volume will range from 100 to 350 µL depending on the size of the pellet. Start off with a small volume and vortex well. Continue adding SR solution until all of the pellet is resuspended. It may be necessary to rock the samples overnight to get the best resuspension of the precipitated protein. Using the smallest volume of SR solution necessary to dissolve the pellet will give a

high concentration of protein. This is important during isoelectric focusing as 125 μL is the maximum volume of sample that can be separated with 7-cm strips.

17. Usually 15–25 μL of sample will result in optical density values that in the middle of the standard curve.

18. IPG buffer is supplied in ranges that match the range of the IPG strip being used. The concentration of IPG buffer can vary from 0.5 to 2.0%, depending on the sample.

19. To ensure complete sample uptake, do not exceed the 125 μL volume limit for the strips.

20. Handle the ceramic holders with care, as they are brittle, and with gloves to avoid contamination. The holder must be clean and completely dry before use.

21. To help coat the entire strip, gently lift and lower the strip and slide it back and forth along the surface of the solution, tilting the strip holder slightly as needed to ensure complete and even hydration.

22. Rehydrate for a minimum of 10 h, but overnight is best *(10)*.

23. If not proceeding to the equilibration step, strips must be frozen immediately after **step 3** is finished.

24. Make sure to start with clean glass plates and chamber equipment. Using lint-free wipes, clean the glass plates with methanol before assembling chamber.

25. This recipe will make six gels and can be scaled up or down depending on the number of gels being run.

26. SDS electrophoresis buffer temperature should be maintained at approximately 20 °C. Setting the MultiTemp III at 10 °C will usually keep the buffer in the recommended temperature range.

27. Adding molten agarose to the sample application piece containing the SDS–PAGE molecular weight standard solution will prevent the markers from diffusing out of the sample application piece.

28. Make certain that no air bubbles are trapped between the IPG strip and the gel surface or between the gel backing and the glass plate.

29. Before placing the gel in fixer, locate the acidic end of the IPG strip, and make a diagonal cut on the acidic side of the gel at the very top to determine the proper orientation when imaging the gels.

30. SYPRO Ruby is light sensitive. Always store the gels in a lightproof container.

31. Destaining overnight minimizes background and produces better gel images.

32. Store the gels in water in a covered plastic container at 4 °C.

References

1. Cunningham, M. W. (2000) Pathogenesis of group A streptococcal infections. *Clin. Microbiol. Rev.* **13,** 470–511.

2. Sumby, P., Barbian, K. D., Gardner, D. J., Whitney, A. R., Welty, D. M., Long, R. D., Bailey, J. R., Parnell, M. J., Hoe, N. P., Adams, G. G., DeLeo, F. R.,

and Musser, J. M. (2005) Extracellular deoxyribonuclease made by group A *Streptococcus* assists pathogenesis by enhancing evasion of the innate immune response. *Proc. Natl. Acad. Sci. U.S.A.* **102,** 1679–1684.

3. Pancholi, V. and Fischetti, V. A. (1992) A major surface protein on group A streptococci is a glyceraldehyde-3-phosphate-dehydrogenase with multiple binding activity. *J. Exp. Med.* **176,** 415–426.

4. Pancholi, V. and Fischetti, V. A. (1993) Glyceraldehyde-3-phosphate dehydrogenase on the surface of group A streptococci is also an ADP-ribosylating enzyme. *Proc. Natl. Acad. Sci. U.S.A.* **90,** 8154–8158.

5. Pancholi, V. and Fischetti, V. A. (1998) Alpha-enolase, a novel strong plasmin(ogen) binding protein on the surface of pathogenic streptococci. *J. Biol. Chem.* **273,** 14503–14515.

6. Rabilloud, T. (2000) Two-dimensional gel electrophoresis in proteomics: old, old fashioned, but it still climbs up the mountains. *Proteomics* **2,** 3–10.

7. Lei, B., Mackie, S., Lukomski, S., and Musser, J. M. (2000) Identification and immunogenicity of group A *Streptococcus* culture supernatant proteins. *Infect. Immun.* **68,** 6807–6818.

8. Rabilloud, T. (1998) Use of thiourea to increase the solubility of membrane proteins in two-dimensional electrophoresis. *Electrophoresis* **19,**758–760.

9. Gorg, A., Weiss, W., and Dunn, M. W. (2004) Current two-dimensional electrophoresis technology for proteomics. *Proteomics* **4,** 3665–3685.

10. Sanchez, J. C., Rouge, V., Pisteur, M., Ravier, F., Tonella, L., Moosmayer, M., Wilkins, M. R., and Hochstrasser, D. F. (1997) Improved and simplified in-gel sample application using reswelling of dry immobilized pH gradients. *Electrophoresis* **18,** 324–327.

3

Proteomic Analysis to Investigate Regulatory Networks in *Staphylococcus aureus*

Susanne Engelmann and Michael Hecker

Summary

The analysis of the expression of virulence genes and the elucidation of metabolic and regulatory pathways of *Staphylococcus aureus* provide us with important information about the interaction between the pathogen and its host, mechanisms by which this organism causes diseases, and the resistance to antibiotics. In order to investigate regulatory networks of *S. aureus*, we analyze the cytoplasmic and extracellular proteome by using two-dimensional (2D) gel analyses combined with matrix-assisted laser ionization–time-of-flight mass spectrometry (MALDI–TOF MS). Gel-based proteomics is an extremely valuable tool in microbial physiology that can, in combination with various visualization and quantitation software packages, very rapidly provide comparative and quantitative data for multi-sample comparison.

Key Words: 2D PAGE; protein expression profiling; DIGE labeling; [^{35}S]-L-methionine pulse labeling; virulence gene expression; dual-channel imaging technique.

1. Introduction

The genome sequence information provides the basis for functional genomic approaches such as DNA chip technologies and proteome analyses. The first genome sequences of two *Staphylococcus aureus* strains, N315 and Mu50, were published in 2001 *(1)*. To date, the complete genome sequences of nine *S. aureus* strains have become available in the databases (www.tigr.org; www.ncbi.nlm.nih.gov). The number of open reading frames varies from 2600 to 2700 between these strains. Despite this wealth of genome information, the function of ~1000 of the proteins encoded by these genes is still unknown.

From: *Methods in Molecular Biology, vol. 431: Bacterial Pathogenesis*
Edited by: F. DeLeo and M. Otto © Humana Press, Totowa, NJ

Whereas the genome sequence only provides the blueprint of life, the proteome brings this genome sequence to fruition. Protein expression profiling not only reveals the overall pattern of protein expression under various environmental conditions, but also provides information on expression level, post-translational modifications, and subcellular localization of individual proteins. We are using functional genomics techniques to focus on two main areas: (1) a basic under-standing of cell metabolism and stress/starvation responses of *S. aureus* under different conditions, and (2) a comprehensive analysis of the function and structure of virulence-associated regulons.

The first step in the physiological proteomics analysis of *S. aureus* is the estab-lishment of a comprehensive two-dimensional (2D) gel proteome map that should cover most of the cytoplasmic proteins expressed under specific circumstances (*see* **Fig. 1**). For physiological studies, the highly sensitive two-dimensional polyacrylamide gel electrophoresis (2D PAGE) originally invented by O'Farrel *(2)* and Klose *(3)* is still the state of the art. With our standard 2D gel system (pH 4–7), we identified 473 proteins that cover about 40% of the cytoplasmic proteome of *S. aureus* predicted for this proteomic window *(4)*. The proteins are associated with various cellular functions ranging from the transcriptional and translational machinery, citrate cycle, glycolysis, and fermentation pathways to biosynthetic pathways of nucleotides, fatty acids, and cell wall components *(4,5)*.

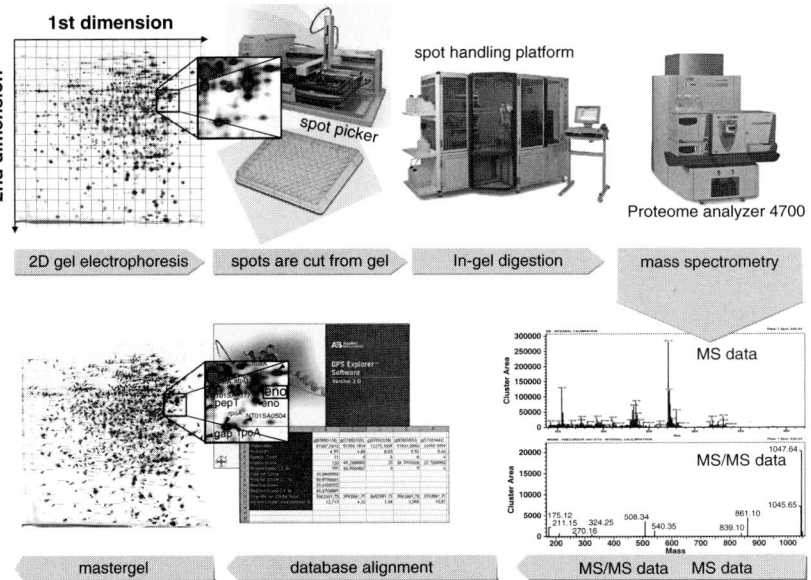

Fig. 1. Schematic representation of protein identification by mass spectrometry and preparation of reference 2D maps.

With this comprehensive proteome map, we have one essential tool for a better understanding of cell physiology of the human pathogen, *S. aureus*.

Extracellular proteins constitute a reservoir of virulence factors and have important roles in the pathogenicity of bacteria. Preliminary genome data show that the various strains encode very different sets of virulence factors. In addition to this genomic variability, differences in the activity of virulence-associated regulators are believed to be responsible for variations in the expression of some virulence factors in clinical isolates. A comprehensive analysis of the extracellular proteome of *S. aureus* can show whether individual virulence genes are (1) expressed and secreted, and if yes, (2) in which quantities, and (3) under which environmental conditions. The expression of virulence genes is regulated in a coordinated fashion during the growth cycle by a complex network of regulators. The production of extracellular proteins takes place mainly during the late exponential and post-exponential phase at high cell densities, and at the same time, the synthesis of surface-associated proteins is down-regulated *(6)*. The best characterized regulators of virulence gene expression to date are Agr (*accessory gene regulator*) and SarA (*staphylococcal accessory regulator*) *(7)*. We have compared the 2D pattern of extracellular protein extracts of wild-type cells with the 2D patterns of the respective regulatory mutants in order to identify proteins whose synthesis is influenced by these regulatory genes *(6,8)*.

The secretome map is also a valuable tool for evaluating the fitness and virulence of clinical isolates from patients with different diseases. The comparison of the extracellular protein patterns revealed considerable hetero-geneity between these strains, which might explain their different virulence potential. To date, at least 90 of the proteins identified in the supernatant of various *S. aureus* strains showed signal sequences typical for Sec-translocated proteins. As expected, many of these proteins were already known to play a role in the virulence of *S. aureus* *(6,8–10)*.

2. Materials
2.1. Cell Culture, Cell Lysis, and Protein Preparation

1. Tryptone soy broth (TSB) (Oxoid, Wesel, Germany), autoclaved for 15 min.
2. Luria–Bertani broth (LB) (Invitrogen, Karlsruhe, Germany) pH 7.5, autoclaved for 20 min.
3. Basic medium (2×): Dissolve 3.54 g of Na_2HPO_4, 2.72 g of KH_2PO_4, 0.4 g of $MgSO_4$ $7H_2O$, 1 g of NH_4Cl, and 1 g of NaCl in 1 L deionized water. Autoclave for 20 min.
4. 1 *M* citrate: Prepare in deionized water and autoclave for 20 min.
5. 20% (w/v) glucose: Prepare in deionized water and autoclave for 10 min.

6. MOPS buffer (25×): Dissolve 209.26 g of MOPS in deionized water and fill to 1 L. Adjust the pH to 7.0 and autoclave buffer for 10 min.

7. L-Amino acid mixtures (50×): Prepare mixtures of one to four different amino acids: (1) alanine, valine, leucine, isoleucine; (2) aspartate, glutamate; (3) serine, threonine, cysteine; (4) arginine, lysine, proline; (5) phenylalanine, tryptophane, histidine; (6) glycine. With the exception of glycine, dissolve 0.8 g of each amino acid in 100 mL of deionized water. In case of glycine, use 8 g. It is important to dissolve the amino acids of one mixture successively. Sterilize mixtures by filtration (0.45 μm).

8. Vitamins (1000×): Prepare all vitamin solutions separately—0.036 mM cyanocobalmine, 0.29 mM p-aminobenzoate, 0.04 mM biotin, 0.81 mM nicotinic acid, 0.21 mM Ca-D-pantothenic acid, 0.62 mM pyridoxamine hydrochloride, 0.29 mM thiaminium dichloride, and 0.26 mM riboflavin. Dissolve the vitamins in deionized water and sterilize by filtration (0.45 μm).

9. Trace elements (1000×): All trace element solutions are prepared separately— 0.51 mM $ZnCl_2$, 0.5 mM $MnCl_2$, 0.097 mM H_3BO_3, 1.46 mM $CoCl_2$, 0.015 mM $CuCl_2$, 0.1 mM $NiCl_2$, and 0.148 mM Na_2MoO_4. Dissolve trace elements in deionized water and sterilize by filtration (0.45 μm).

10. 0.5 mM $FeCl_2$: Dissolved in 2N HCl.

11. 2N NaOH in deionized water.

12. TE Buffer: 10 mM Tris, 1 mM EDTA, pH 7.5, autoclaved.

13. Glass beads (0.10–0.11 mm, Braun, Melsungen, Germany).

14. Ribolyser (Thermo Electron Cooperation, Waltham, USA).

15. 100% (w/v) trichloroacetic acid (TCA) solution: Dissolve TCA in deionized water and prepare fresh.

16. 8 M Urea/2 M Thiourea solution: Dissolve 2.4 g of urea and 0.76 g of thiourea in 1 mL of water (deionized water additionally purified by using the water purification system "Synergy" from Millipore, Schwalbach, Germany) and bring to 5 mL with the same water. Store aliquots at −20 °C.

17. Stop solution: 15 mg of L-methionine and 10 mg of chloramphenicol dissolved in 10 mL 0.1 M Tris–HCl (pH 7.5).

2.2. Difference Gel Electrophoresis Labeling of Proteins

1. 99.8% anhydrous dimethylformamide (DMF).

2. 10 mM L-lysine in deionized water.

3. CyDye Difference Gel Electrophoresis (DIGE) Fluors for Ettan DIGE (GE-Healthcare, Little Chalfont, UK): The dye tubes (Cy2, Cy3, Cy5) should be allowed to warm to room temperature for 5 min. Add 10 μL DMF (freshly opened) to each dye tube and mix. The tubes now contain 1 mM Cy2, Cy3, or Cy5 dye in DMF and should be vortexed vigorously for 30 s. Afterward, centrifuge the tubes for 30 s at 12,000 × g. The fluorochrome can now be used. Unused CyDye stock solutions should be returned to −20 °C as soon as possible and stored in the dark.

4. pH indicator strips (pH 4.5–10.0).

5. 50 mM NaOH in deionized water.

2.3. Two-Dimensional Polyacrylamide Gel Electrophoresis

2.3.1. First Dimension

1. Rehydration solution (10×): 100 mg of CHAPS (Roth, Karlsruhe, Germany) and 30 mg of DTT (MP Biomedicals, Illkirch, France) in 8 *M* urea/2 *M* thiourea solution (*see* **step 16**, **Subheading 2.1.**). Subsequently, add 52 μL Pharmalyte (pH 3–10) (GE-Healthcare). Store aliquots at –20 °C.
2. 18-cm IPG strips: linear pH range of 4–7 for cytoplasmic proteins and 3–10 for extracellular proteins (GE-Healthcare).
3. Rehydration chamber.
4. Multiphor II unit (GE-Healthcare).
5. Electrode strips (GE-Healthcare).

2.3.2. Second Dimension

1. 1.5 *M* Tris–HCl, pH 8.8 (adjusted at 12 °C). Store at room temperature (*see* **Note 1**).
2. 10% (w/v) sodium dodecyl sulfate (SDS) in deionized water. Store at room temperature.
3. 10% (w/v) ammonium persulfate (APS) dissolved in deionized water. Store at 4 °C no longer than 1 week.
4. 0.5 *M* Tris–HCl, pH 6.8 (adjusted at 12 °C). Store at room temperature (*see* **Note 1**).
5. Upper buffer (4×): 0.4% (w/v) SDS dissolved in 0.5 *M* Tris–HCl, pH 6.8.
6. Equilibration solution (100 mL): Dissolve 36 g of urea in a mixture of 10 mL of 0.5 *M* Tris–HCl (pH 6.8), 30 mL of glycerol, and 40 mL of 10% SDS. For solution A, add 175 mg of DTT (MP Biomedicals) to 50 mL equlibration solution. For solution B, add 2.25 g of iodoacetamide (toxic!) and a trace of bromphenol blue to 50 mL equilibration solution.
7. Running buffer (10×): Dissolve 150 g of Tris base, 720 g of glycine, and 50 g of SDS in deionized water. When all components are completely in solution, add deionized water to 5 L.
8. Protean® plus Dodeca™ cell gel system, power supply, glass plates (Biorad-Laborities, Hercules, USA).

2.4. Protein Detection

1. The Coomassie staining solution is prepared as follows: add 40 mL of 85% *o*-phosphoric acid, 40 g of $(NH_4)_2SO_4$, and 0.48 g of Coomassie Blue G-250 to 360 mL deionized water. Stir solution and add 100 mL methanol.
2. Fixing solution for silver staining: 50% (v/v) ethanol and 12 % (v/v) acetic acid. Prior to use, add 500 μL of 37% formaldehyde to 1 L fixing solution.
3. Washing solution: 50% (v/v) ethanol in deionized water.
4. Sodium thiosulfate solution: Add 0.2 g of $Na_2S_2O_3 \cdot 5H_2O$ to 1 L deionized water and dissolve by stirring.
5. Silver nitrate solution: Add 2 g of $AgNO_3$ to 1 L deionized water and dissolve by stirring. Prior to use, add 750 μL of 37% formaldehyde to 1 L silver nitrate solution.

6. Developing solution: Dissolve 30 g of K_2CO_3 and 4 mg of $Na_2S_2O_3 \cdot 5H_2O$ in 1 L deionized water. Prior to use, add 500 µL 37% formaldehyde to 1 L developing solution.

7. Stop solution: 1% (w/v) glycine dissolved in deionized water.

8. Storage Phosphor Screens 20 cm × 25 cm (Molecular Dynamics, Krefeld, Germany).

9. Storm 840 Phosphor Imager (Molecular Dynamics).

10. Typhoon laser scanner (9200, 9210, 9400, or 9410).

11. Scanner X finity ultra (Quato Graphic, Braunschweig, Germany).

2.5. In-Gel Digestion

1. Spot cutter (Proteome Work™) with a picker head of 2 mm (GE-Healthcare).

2. Ettan Spot Handling Workstation (GE-Healthcare).

3. 50 mM ammonium bicarbonate/50% (v/v) methanol.

4. 75% (v/v) acetonitrile.

5. Trypsin solution: 1 mg/mL mg trypsin dissolved (Promega, Madison, WI, USA) in 20 mM ammonium carbonate (freshly prepared).

6. 50% (v/v) acetonitrile/0.1% (w/v) trifluoroacetic acid (TFA).

7. Matrix solution 50% (v/v) acetonitrile/0.5% TFA saturated with α-cyano-4-hydroxycinnamic acid (CHCA).

8. 0.5% (w/v) TFA/50% (v/v) acetonitrile.

2.6. Protein Identification by Mass Spectrometry

1. Proteome-Analyzer 4700 (Applied Biosystems, Foster City, CA, USA).

2. 4700 Explorer™ software.

3. Genome sequences of *S. aureus* in FASTA format (www.ncbi.nlm.nih.gov).

3. Methods

In order to investigate the physiology and virulence of *S. aureus*, we sought to analyze the cytoplasmic and extracellular proteome of *S. aureus* by using 2D gel analysis combined with MALDI–TOF MS/MS (*see* **Fig. 1** and **Note 2**). For low-complexity organisms such as bacteria, the majority of cellular and extracellular proteins can be visualized within the main analytical window of pHs ranging from 4–7 (cytoplasmic proteins) or 3–10 (extracellular proteins); thus, gel-based proteomics will remain an extremely valuable tool, because it economically permits the comparative and quantitative protein profiling in multi-sample comparisons. Protein master gels showing most of the proteins predicted for the respective cell compartment (e.g., cytosol, supernatant) (*see* **Fig. 1**) offer the chance to study the regulation of the synthesis and/or the amount of all these proteins under various conditions and in different mutant strains. In this way, the 2D protein pattern of cells grown under various growth restricting conditions was compared with those of exponentially growing cells. All proteins induced by a single stimulus belong to a stimulon (*see* **Fig. 2** and Color Plate 1, following p. 46). The next step is to dissect the stimulons into their individual regulation groups, the regulons (*see* **Fig. 2**). Comparison of the protein expression profile of the wild type with mutants lacking specific regulatory proteins under inducing conditions is currently the best way to define the regulon structure. Furthermore, the allocation of as yet unknown proteins to stimulons and regulons of already known function is a simple but convincing approach to an initial prediction of their function.

The synthesis of proteins under different conditions can be analyzed by incorporation of [^{35}S]-L-methionine into proteins that were newly synthesized during

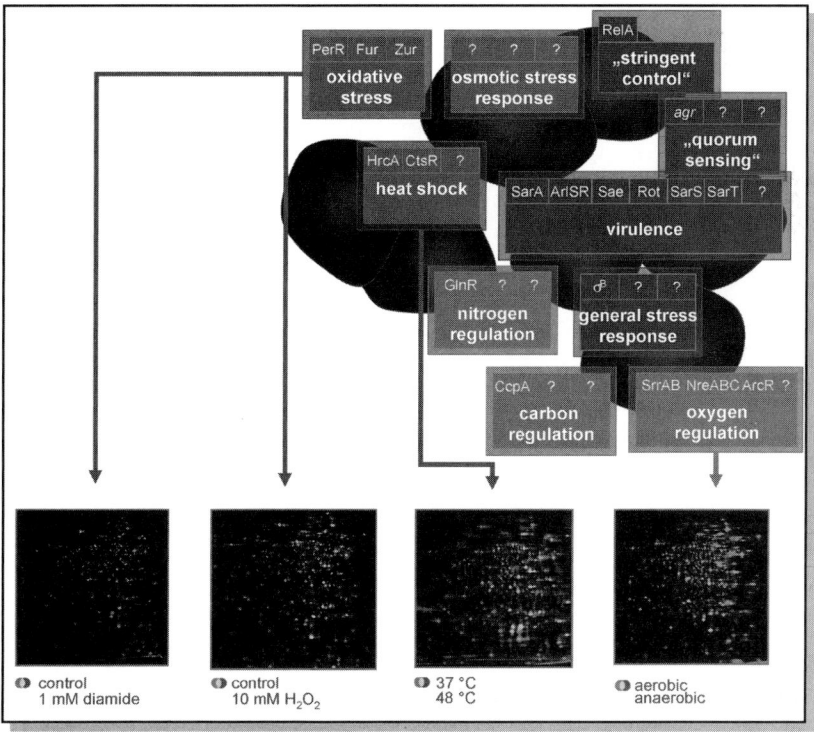

Fig. 2. Proteomic signatures of different stress/starvation conditions in *Staphylococcus aureus*. Comparison of the protein synthesis profile of exponentially growing cells (green) with that of stressed *S. aureus* cells (red) reveals changes in protein synthesis that are particular for the certain stress stimuli. Cells were cultivated in synthetic medium and exposed to the respective stimulus at an optical density at 500 nm (OD_{500}) of 0.5. Protein synthesis was analyzed by [^{35}S]-L-methionine labeling (5 min pulse) under control conditions and 10 min after imposition to stress. All proteins induced by one stimulus belong to a stimulon. (*See* Color Plate 1, following p. 46.)

a defined period (5 min) (*see* **Fig. 3**). The resulting autoradiograms reflect the instantaneous synthesis rates of individual proteins at the time of labeling and can be identified by using the respective protein master gel. The silver or colloidal Coomassie-stained protein pattern represents the actual level of proteins accumulated within or outside the cell until the time of harvesting (*see* **Fig. 3**).

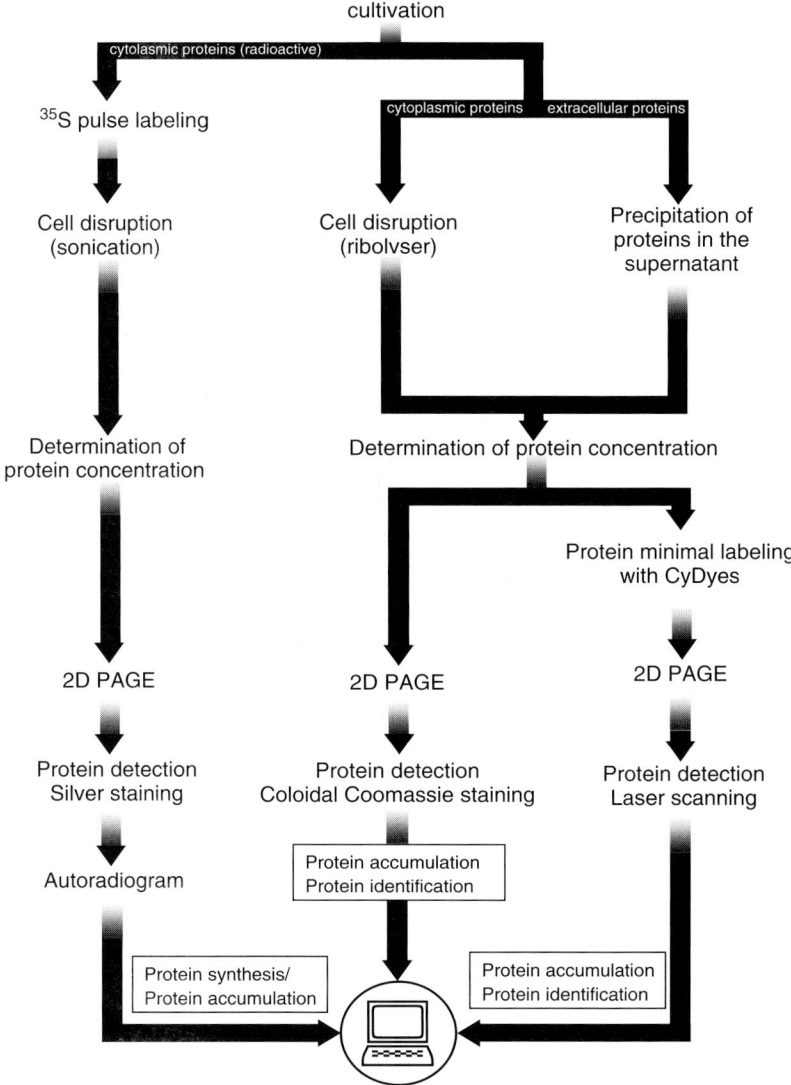

Fig. 3. Strategies of 2D gel experiments for protein identification and for analyzing protein accumulation and synthesis.

By using the 2D fluorescence DIGE (Ettan DIGE) technique, three different protein extracts are labeled with CyDye fluors (Cy2, Cy3, and Cy5, respectively) that are subsequently separated on the same 2D gel (*see* **Fig. 3**). Thus, the process of detecting, identifying, and quantifying proteins is simplified. Each of the three protein extracts is labeled with one CyDye. After labeling, the three samples are mixed together and run on the same gel. In this way, the same protein labeled with the different CyDye will migrate to the same position on the 2D gel, which helps to limit errors due to experimental variation (*see* **Fig. 4** and Color Plate 2, following p. 46). Each of the individual samples can be visualized independently by selecting the appropriate excitation and emission wavelength to scan for each CyDye (*see* **Fig. 4**).

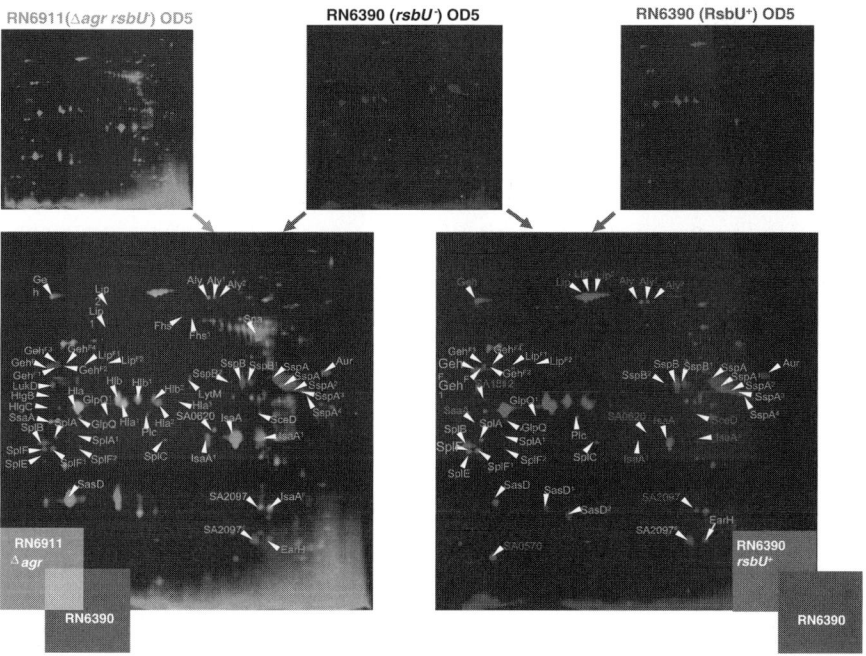

Fig. 4. Difference gel electrophoresis (DIGE) labeling technique to illustrate differences in protein patterns of three protein extracts. Extracellular protein patterns of wild-type RN6390 (blue), the isogenic *agr* mutant (green), and RN6390 complemented with *rsbU* of *Staphylococcus aureus* COL (red). Prior to separation by 2D gel electrophoresis technique, protein extracts of the respective strains were labeled with CyDye fluorochromes [Cy2 RN6390, Cy3 RN6911, Cy5 RN6390 (RsbU+)]. A mixture of 50 µg of each protein extract was separated on one gel. The proteins were detected by using a Typhoon [Cy2: Blue2-Laser (488 nm), Cy3: Green-Laser (532 nm), Cy5: the Red-Laser (633 nm)]. (*See* Color Plate 2, following p. 46.)

3.1. Bacterial Growth Conditions

1. For radioactive pulse-labeling experiments, cultivate bacterial strains in 50 mL of synthetic medium without methionine *(11)*. For 100 mL synthetic medium, add 4 mL of 25× MOPS buffer, 14 μL of 1 *M* citrate, 740 μL of 20% glucose, 2 mL of each amino acid mixture, 100 μL of each vitamin solution, 100 μL of each trace element solution, 150 μL of $FeCl_2$, and 150 μL of 2*N* NaOH to 50 mL of 2× basic medium. Afterward, bring the volume of the medium to 100 mL with sterilized, deionized water. Inoculate the synthetic medium with exponentially growing cells of the appropriate *S. aureus* strain (grown overnight to start a fresh culture) in the same medium to an initial optical density at 500 nm (OD_{500}) of 0.07–0.1. Cultivate cells with vigorous agitation at 37 °C.
2. For preparation of unlabeled cytoplasmic and extracellular protein extracts, inoculate 100 to 1500 mL of complex medium (TSB or LB) or synthetic medium with exponentially growing cells of the appropriate *S. aureus* strain to an initial OD_{540} of 0.05. Grow cells with vigorous agitation at 37 °C.
3. For stress kinetic experiments, inoculate 100 mL of medium with exponentially growing cells of the appropriate *S. aureus* strain to an initial OD_{540} of 0.05. At an OD of 0.5–0.7, transfer 10–50 mL of cell culture to new, preheated Erlenmeyer flasks and expose to stress conditions. Harvest cells at fixed time intervals after imposition of stress. As a control, harvest 20–50 mL from the untreated culture immediately before and at the end of the stress experiment.

3.2. Protein Preparation

3.2.1. Cytoplasmic Proteins

1. For preparation of cell extracts at different ODs ($OD_{540} = 0.5$ and 10) or different time points after imposition of stress, separate cells of 50-mL cultures from the supernatant by centrifugation at 7000 × *g* for 10 min at 4 °C. (Simultaneously, the supernatant can be used for preparation of extracellular proteins [*see* **Subheading 3.2.2.**].)
2. Wash cell pellets with 1 mL of ice-cold TE buffer, resuspend in 1 mL TE buffer, and transfer into screw top tubes containing 500 μL of glass beads.
3. Disrupt cells by homogenization with glass beads using the Rybolyser (Thermo Electron Corporation) for 30 s at 6.5 m/s.
4. Centrifuge lysate at 21,000 × *g* for 25 min at 4 °C to remove cell debris and then transfer supernatant to a new tube.
5. Centrifuge at 21,000 × *g* for 45 min (all at 4 °C) to remove insoluble and aggregated proteins, which disturb the isoelectric focusing (IEF) of the proteins.
6. Determine protein concentration by using Roti-Nanoquant (Roth) (*see* **Note 3**). Store the protein solution at –20 °C.

3.2.2. Extracellular Proteins

1. Separate the supernatant of a 50-mL culture from cells by centrifugation (10 min, 7000 × *g*, 4 °C) and transfer into a new tube. (Optional: centrifugation step can be repeated if there are cells remaining in the supernatant.)

2. Add 5 mL of freshly prepared 100% (w/v) TCA (end concentration 10% (v/v)) to the supernatant and precipitate the proteins overnight at 4 °C.
3. Centrifuge precipitate for 1 h and decant the supernatant very carefully. Wash the precipitate with 20 mL of 100% (v/v) ice-cold ethanol, centrifuge at $8000 \times g$ for 10 min at 4 °C, and decant supernatant. Repeat the washing procedure twice.
4. Afterward, remove the pellet from the tube, mix with 10 mL of 100% ice-cold ethanol, transferred to a 15-mL Falcon tube, and centrifuge at $8000 \times g$ for 10 min at 4 °C. The supernatant should be decanted very carefully. Repeat this washing procedure at least six times. One of these steps should be done with 70% (v/v) ice-cold ethanol instead of 100% ice-cold ethanol. After decanting the supernatant, dry the protein pellet at room temperature.
5. Dissolve the dried protein pellet in 8 M urea/2 M thiourea and mix by shaking for 20 min at room temperature. Centrifuge the solution for 15 min at $21,000 \times g$ and transfer the supernatant into a new microtube.
6. Determine protein concentration by using ROTI-Nanoquant (Roth) (*see* **Note 3**). Store protein solution at –20 °C.

3.2.3. Preparation of Cytoplasmic Protein Extracts Labeled With [^{35}S]-L-Methionine (Pulse-Labeling Reaction)

1. Grow cells in synthetic medium at 37 °C to an OD of 0.5.
2. Transfer 10 mL of the culture volume to a new Erlenmeyer flask and add 10 μL of (100 μCi) L-[^{35}S]-methionine to the culture. Stop the labeling reaction after 5 min by adding 1 mL of stop solution and by transferring the Erlenmeyer flask to ice.
3. Pellet cells by centrifugation at $8000 \times g$ for 5 min at 4 °C.
4. Wash cell pellets twice with 1 mL of ice-cold TE buffer and resuspend in 400 μL of TE buffer.
5. For cell lysis, add 10 μL of lysostaphin solution (10 mg/mL) to the cell suspension.
6. After incubation on ice for 10 min, disrupt cells by sonication. For this step, place a beaker of ice water around the sample tube to keep it cold. Sonicate the samples for 1 min (0.5/s, low) followed by a 1-min cooling break. Repeat this process three times. Sonication is complete when the solution appears noticeably less cloudy than the starting solution.
7. After sonication, centrifuge the sample at $21,000 \times g$ for 10 min at 4 °C. Transfer the supernatant to a new tube and repeat the centrifugation process for 30 min.
8. Determine protein concentration by using ROTI-Nanoquant (Roth) (*see* **Note 3**). Store the protein solution at –20 °C.

3.3. CyDye DIGE Minimal Labeling (for Extracellular Proteins)

1. The recommended concentration of protein is between 5 and 10 mg/mL.
2. The pH of the protein solution should be between pH 8.0 and 9.0. Check by spotting 1 μL on a pH indicator strip. If the pH is lower than 8.0, then the pH

will need to be adjusted before labeling. The pH can be increased to pH 8.5 by the careful addition of 50 m*M* NaOH.

3. For labeling experiments, prepare 400 μ*M* CyDye working solution by adding one volume of CyDye stock solution to 1.5 volumes of high-grade DMF (*see* **Note 4**).

4. It is recommended that 50 μg of protein is labeled with 400 pmol of CyDye. If labeling more than 50 μg of protein, the same fluorochrome-to-protein ratio must be used.

5. Add a volume of sample equivalent to 50 μg of protein and 1 μL of diluted CyDye to a microfuge tube, mix, and centrifuge briefly. Afterward, incubate the sample on ice for 30 min in the dark.

6. Stop the labeling reaction by adding 1 μL of 10 m*M* lysine to the sample that has incubated on ice for 10 min in the dark. Labeled protein extracts can now be stored at –80 °C.

7. The protein samples that are going to be separated on the same gel should be pooled now.

8. Follow the protocol for rehydration of IPG strips (*see* **Subheading 3.4.1.**).

3.4. 2D Gel Electrophoresis

3.4.1. Isoelectric Focusing

1. Make up protein samples [80–100 μg of radioactively labeled proteins, 3 × 50 μg of DIGE-labeled proteins (*see* **Note 5**), or 350–600 μg of unlabeled proteins for Colloidal Coomassie staining) (*see* **Note 6**)] to 360 μL with 8 *M* urea/2 *M* thiourea. Subsequently, add 40 μL of 10× rehydration buffer and mix the solution by shaking at room temperature for 30 min. Centrifuge the rehydration mix for 5 min at 21,000 × *g* (20 °C) to remove insoluble proteins.

2. Dispense equally the supernatant in one slot of the rehydration chamber. Position the IPG strip with the gel side down and lower the strip onto the solution. To help coat the entire IPG strip, gently lift and lower the strip and slide it back and forth along the surface of the solution. Avoid large bubbles under the IPG strip. Rehydration occurs overnight for at least 15 h (no longer than 24 h).

3. Perform IEF with the Multiphor II unit. The dry-strip aligner and the electrodes should be cleaned with soft tissues and all cables should be connected. Set the temperature of the thermostatic circulator to 20 °C. Place two 110-cm electrode strips on a clean flat surface and soak with distilled water. Remove excess water by blotting with filter paper (*see* **Note 7**).

4. To remove rehydrated IPG strips from the slots, grab the ends of the strip with tweezers and lift it out of the tray. Any contact with the gel surface should be avoided. Position the strips in the dry-strip aligner in adjacent grooves and align all strips so that the anodic gel edges are lined up. Place the electrode strips across the cathodic and anodic ends of the aligned IPG strips. The electrode strips must at least partially contact the gel surface of each IPG strip. When each electrode is aligned with the electrode strip, press down on each to contact the electrode strips. Cover the IPG strips with mineral oil.

5. Perform IEF by using the following voltage profile: step 1—500 V (gradient), 2 mA, 5 W, 2 Vh; step 2—3500 V (gradient), 2 mA, 5 W, 3 kVh; step 3—3500 V, 2 mA, 5W, 23.5 kVh. As IEF proceeds, the current must be checked (low current is normal for IPG gels). If no current is flowing, check the contact between electrodes and electrode strips.
6. After IEF, proceed to the second dimension immediately (*see* **Subheading 3.4.2.**) or store the IPG strips at –20 °C in foil.

3.4.2. SDS–PAGE

Focused proteins are separated according to their molecular weight in 12.5% acrylamide and 2.6% bis-acrylamide polyacrylamide gels using the Tris–glycine system (*see* **Note 8**).

1. To prepare 12 slab gels, mix 335.3 mL of 40% acrylamide, 179 mL of 2% bis-acrylamide, 272.54 mL of 1.5 *M* Tris–HCL (pH 8.8), 11.48 mL of 10% SDS, and 300 mL of deionized water while stirring on a magnetic stirrer. Add 2.8 mL of 10% APS and 0.55 mL of TEMED to this solution and mix by stirring (*see* **Note 9**). Generation of bubbles should be avoided. Pour gels immediately by filling the gel cassette about 1 cm below the top of the glass plates. Overlay gels immediately with a thin layer (1 mL) of water-saturated n-butanol or water to minimize gel exposure time to oxygen and to create a flat gel surface. Polymerization takes at least 3 h. Each gel should be inspected and the top of surface of each gel should be straight and flat. Remove butanol or water and rinse the gel surface with deionized water. The butanol should be completely removed.
2. Prepare the stacking gel by mixing 10.8 mL of 40% acrylamide, 3.4 mL of 2% bis-acrylamide, 30 mL of upper buffer (4×), and 74 mL of deionized water. Afterward, add 0.3 mL of 10% APS and 0.05 mL of TEMED to the solution while stirring. Use an appropriate amount of the stacking gel solution to quickly rinse the top of each gel and overlay with a thin layer (1 mL) of water. The stacking gel should polymerize within 1 h.
3. Prepare equilibration buffers A and B. Place each IPG strips with the gel side up in one slot of an equilibration chamber and add 4–5 mL of equilibration buffer A to each slot. Equilibrate the IPGs for at least 15 min with gently shaking. Decant equilibration solution A and add 4–5 mL of equilibration solution B to each slot. Equilibrate the IPGs again for at least 15 min while shaking. Decant equilibration solution B and place the IPG strips on filter paper so that they rest on an edge to help drain the equilibration solution.
4. Add 1× running buffer to the gel system. Put gels in the gel system filled with 1× running buffer. The running buffer should cover the gels.
5. Place IPG strips between the plates on the surface of the stacking gel by gently pushing the IPG strip down so that the entire lower edge of the IPG strip is in contact with the top surface of the stacking gel. No air bubbles should be trapped between the IPG strips and the stacking gel.

6. Perform electrophoresis at constant power (2 W per gel) at 12 °C. A constant temperature during electrophoresis is very important for gel-to-gel reproducibility. When starting electrophoresis, the buffer temperature should be at 12 °C. Therefore, cooling has to be started before preparing the gels (*see* **step 1**, this **Subheading**).

7. After electrophoresis, remove gels from glass plates in preparation for staining. Each gel should be marked to identify the acidic end of the first dimension preparation.

3.5. Protein Detection

3.5.1. Colloidal Coomassie Staining (12)

1. Add 250 mL of Coomassie staining solution to each gel and incubate gels for 24 h on a stirrer.
2. After removing Coomassie staining solution, rinse the gels with deionized water several times.
3. Seal gels in foil after staining.
4. Scan gels with a Scanner X finity ultra (Quato Graphic) in transmission mode at a resolution of 200 dpi.
5. Store gels at 4 °C.

3.5.2. Silver Staining (13)

1. Treat the resulting 2D gels with fixing solution for silver staining (200–250 mL per gel) for 1–2 h.
2. Wash gels three times with 200–250 mL of 50% (v/v) ethanol for 20 min and pretreat with 200–250 mL of sodium thiosulfate solution for 1 min.
3. Rinse the gels three times with deionized water for 20 s.
4. Incubate gels with 200–250 mL of silver nitrate solution for 20 min.
5. After removing the silver nitrate solution, rinse the gels twice with deionized water for 20 s.
6. For developing, incubate gels in 200–250 mL of potassium carbonate solution for 2–10 min. Stop the developing reaction by incubating the gels with 200–250 mL of 1% (w/v) glycine for 20 s, rinsing with deionized water for 20 s and stopping again with 200–250 mL of 1% (w/v) glycine for 20 min.
7. Wash gels two times with deionized water for 20 min and seal in foil.
8. Scan gels with a Scanner X finity ultra (Quato Graphic) in transmission mode at a resolution of 200 dpi.
9. Store gels at 4 °C.

3.5.3. Detection of DIGE-Labeled Proteins

Cy2, Cy3, and Cy5 can be detected using a Typhoon laser scanner (9200, 9210, 9400, or 9410). The unfixed gels are scanned according to the Ettan Dige User manual (GE-Healthcare).

1. Place gels directly on the wet surface of the Typhoon.
2. For scanning, use the BLUE2-Laser (488 nm) for Cy2-labeled proteins, the green laser (532 nm) for Cy3-labeled proteins, and the red laser (633 nm) for Cy5-labeled proteins (*see* **Note 10**).
3. For quantitation, scan the gels at 100 dpi resolution.

3.5.4. Detection of L-[³⁵S]-Methionine-Labeled Proteins

1. After silver staining and scanning, place gels on Whatman paper, cover with cellophane sheets, and dry on a vacuum dryer at 75 °C for at least 2–4 h (*see* **Note 11**).
2. Expose the dried gels to "Storage Phosphor Screens" for 2 h to several days (depending on signal intensity). Scan Storage Phosphor Screens with a Storm 840 Phosphor Imager at a resolution of 200 µm and a color depth of 16 bit (65536 gray scale levels) (*see* **Note 12**).

3.6. Protein Identification by Mass Spectrometry (see Fig. 1)

1. For identification of proteins by MALDI–TOF MS, cut Coomassie-stained protein spots from gels using a spot cutter (Proteome Work™) with a picker head of 2 mm and transfer into 96-well microtiter plates.
2. Digestion of proteins with trypsin and subsequent spotting of peptide solutions onto the MALDI targets are performed automatically in the Ettan Spot Handling Workstation (GE-Healthcare) using the following standard procedure *(14)*: wash the gel pieces twice with 100 µL of 50 m*M* ammonium bicarbonate/50% (v/v) methanol for 30 min and once with 100 µL of 75% (v/v) acetonitrile for 10 min. After 17 min of drying, add 10 µL of trypsin solution containing 20 ng/µL trypsin and incubate at 37 °C for 120 min. For peptide extraction, cover gel pieces with 60 µL of 50% (v/v) acetonitrile/0.1% (w/v) TFA and incubate for 30 min at 37 °C. Transfer the supernatant containing peptides into a new microtiter plate and repeat the extraction with 40 µL of the same solution. Dry the supernatants completely at 40 °C for 220 min and dissolve the peptides in 2.2 µL of 0.5% (w/v) TFA/50% (v/v) acetonitrile.
3. Spot 0.7 µL of this solution directly onto the MALDI target. Add 0.4 µL of matrix solution and mix with the sample solution by aspirating the mixture five times. Prior to the measurement in the MALDI–TOF instrument, dry the samples on the target for 10–15 min.
4. Carry out MALDI–TOF MS analyses of spotted peptide solutions on a Proteome-Analyzer 4700 (Applied Biosystems). The spectra are recorded in a reflector mode in a mass range from 900 to 3700 Da. For one main spectrum, 25 subspectra with 100 spots per subspectrum are accumulated using a random search pattern. If the autolytical fragment of trypsin with the monoisotopic $(M+H)^+$ *m/z* at 2211.104 reaches a signal-to-noise ratio (S/N) of at least 10, an internal calibration is automatically performed using the peak for one point calibration. The peptide search tolerance is 50 ppm, but the actual standard deviation is between 10 and 20 ppm.

5. Calibration is performed manually for the less than 1% samples for which automatic calibration fails. After calibration, the peak lists are created using the "peak to mascot" script of the 4700 Explorer™ software with the following settings: mass range from 900 to 3700 Da, peak density of 50 peaked per range of 200 Da, minimal area of 100 and maximal 200 peaks per protein spot, and minimal S/N ratio of 6. The resulting peak lists are compared with organism-specific sequence databases by using the mascot search engine (Matrix Science, London, UK). Peptide mixtures that yield at least twice a Mowse score of at least 49 (depending on the size and quality of the database) and sequence coverage of at least 30% are regarded as positive identifications.

6. Proteins that fail to exceed the 30% sequence coverage cut-off are subjected to MALDI–MS/MS. MALDI–TOF–TOF analysis is performed for the three strongest peaks of the TOF spectrum. For one main spectrum, 20 subspectra with 125 shots per subspectrum are accumulated using a random search pattern. The internal calibration is automatically performed as one-point calibration if the monoisotopic arginine $(M+H)^+$ m/z at 175.119 or lysine $(M+H)^+$ m/z at 147.107 reaches an S/N of at least 5. The peak lists are created using the "peak to mascot" script of the 4700 Explorer™ software with the following settings: mass range from 60 Da to a mass that was 20 Da lower than the precursor mass, peak density of five peaks per 200 Da, minimal area of 100 and maximal 20 peaks per precursor, and a minimal S/N ratio of 5. Database searches are performed using the GPS explorer software with the organism-specific databases. Proteins with a Mowse score of at least 49 in the reflector mode that is confirmed by subsequent peptide/fragment identifications of the strongest peaks (MS/MS) are regarded as identified. MS/MS analysis is particularly useful for the identification of spots containing more than one protein.

3.7. Quantitation and Bioinformatic Approaches

The 2D gel image analysis is performed with the Delta2D Software (Decodon GmbH, Greifswald, Germany). Three different data sets of each experiment have to be analyzed in order to screen for differences in the amount or synthesis of the proteins identified on 2D gels.

The dual-channel imaging technique is an excellent tool for identifying all proteins induced or repressed by growth-restricting stimuli *(15)*. In this technique, two digitized images of 2D gels are generated and combined in alternate additive dual-color channels (*see* **Fig. 5** and Color Plate 3, following p. 46). The first image (densitogram), which shows protein levels visualized by various staining techniques, is false-colored green. The second image (autoradiograph), representing proteins synthesized and radioactively labeled during a 5-min pulse labeling with $[^{35}S]$-L-methionine, is false-colored red. When the two images are combined, proteins accumulated and synthesized in growing cells are colored yellow. However, proteins not previously accumulated in the cell but newly synthesized after the imposition of a stress or

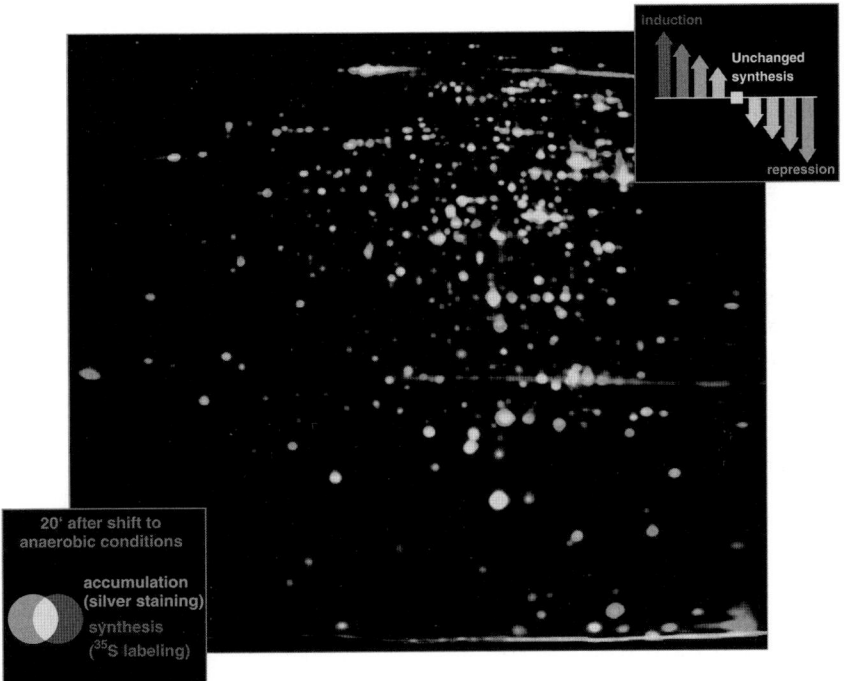

Fig. 5. Dual-channel imaging technique. Bacterial cells were grown in synthetic medium without nitrate to an optical density at 500 nm (OD_{500}) of 0.5. The proteins were labeled with [^{35}S]-L-methionine 20 min after transfer to anaerobic conditions. Protein extracts were separated on 2D gels. The resulting images from the protein synthesis ([^{35}S]-L-methionine labeling) and the protein accumulation (staining with silver nitrate) were overlaid in "dual channel images." Proteins whose synthesis was induced in response to the stressor but have not accumulated yet are colored red, repressed spots that are still present but not synthesized anymore are colored green, and spots without changed expression behavior in response to stress are shown in yellow. (*See* Color Plate 3, following p. 46.)

starvation stimulus are colored red. Looking for such red-colored proteins is a simple approach for visualizing all proteins induced by a single stimulus, thereby defining the entire stimulon. Proteins repressed by the stimulus can also be visualized by this powerful technique. Green-colored proteins that are no longer synthesized (no longer red) but still present in the cell are the candidates for repression by the stimulus. For a more detailed study, the synthesis rate of the proteins after exposure to stress can be quantified (*see* **Fig. 5**).

By using the Ettan DIGE, a pooled internal protein standard labeled with Cy2 can be incorporated on every 2D gel. Normally, the internal standard represents a mix of all protein extracts of an experiment and is used to match the protein patterns across gels, thereby negating the problem of inter-gel variation. By this technique, it is possible to plot the relative abundance of each protein on different gels against the normalized internal standard. This allows accurate quantitation of samples, with an associated increase in statistical significance. Both the DeCyder software (GE-Healthcare) and Delta2D (Decodon GmbH) can be used to automatically apply statistical tests. These tests compare the average ratio and variation within each group to the average ratio and variation in the other group to see if any changes between the groups are significant.

All proteomic data can be assembled into an adaptational network consisting of a large number of stress and starvation stimulons and regulons. An essential feature of this network is the interplay between the various regulation groups, because the individual stress and starvation stimulons are tightly connected. The visualization of such a complex protein expression pattern requires specialized software tools, such as the "color coding" tool of Delta2D. Using this tool, one can display proteins induced by more than one stimulus by a specific color code, thus providing convenient intuitive visualization of complex regulation patterns (*see* **Fig. 6** and Color Plate 4, following p. 46). By applying this technique, one can display already known and probably new overlapping regions within the regulatory network.

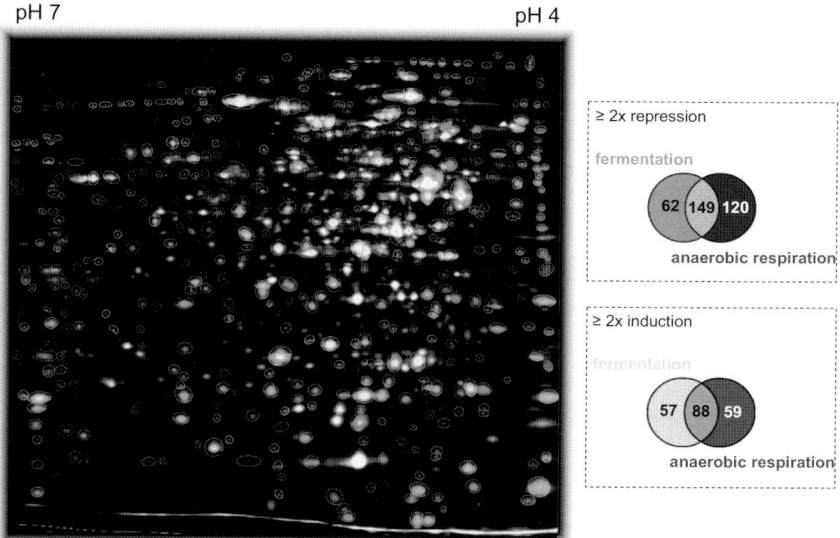

Fig. 6. Multicolor imaging of protein expression patterns of *Staphylococcus aureus* COL under anaerobic conditions in the presence and absence of nitrate. Delta2D software is used to visualize complex protein expression patterns on the 2D image in the standard pH range 4–7. The color code is represented on the right side. (*See* Color Plate 4, following p. 46.)

4. Notes

1. For highly reproducible 2D gels, the pH of these buffers should be adjusted at a fixed temperature.
2. These protocols are easily adaptable to other organisms. However, you will have to appropriately modify the culture conditions and the cell disruption protocols.
3. Ensure that at least 10 µL of each protein solution is used. Otherwise, dilute an appropriate volume of the protein extract prior to measurement. Ensure that the ratio E_{590nm}/E_{450nm} is always the same for each protein extract. Protein extracts in TE or water should be frozen at –20 °C before measuring the protein concentration.
4. Add the DMF first to the tube followed by the CyDye.
5. When separating DIGE-labeled proteins on 2D protein gels, all steps should be carried out in the dark.
6. If the volume containing the desired amount of cytoplasmic proteins exceeds 40 µL, the volume should be reduced by using a speed vac. (Caution: this step can only be applied for protein extracts dissolved in TE buffer!)
7. Important: strips must be damp, not wet!
8. These instructions assume the use of the Protean® plus Dodeca™ cell gel system (Biorad-Laborities). They are easily adaptable to other gel systems.
9. The total volume of the solution needed depends on the gel sizes, the gel thickness, and the number of the gels cast.
10. For an optimized scanning process, the gels should be scanned separately by the three different lasers.
11. Ensure that the gels are completely dried before interrupting the vacuum. The gel should have lost its softness when you press your finger on it.
12. For quantitation of protein synthesis, ensure that no protein spot is saturated.

Acknowledgments

We are very grateful to Christian Kohler, Anne-Kathrin Ziebandt, Harald Kusch, Stephan Fuchs, and Robert S. Jack for critical reading the manuscript and preparing of figures. We thank all co-workers and students for their excellent work on proteome analyses. Furthermore, we thank the Decodon GmbH for providing the Delta2D software. This work was supported by grants from the Bildungsministerium für Bildung und Forschung (031U107A/-207A; 031U213B), the DFG (GK212/3-00), the Land Mecklenburg-Vorpommern, and the Fonds der Chemischen Industrie to M.H. and S.E.

References

1. Kuroda, M., Ohta, T., Uchiyama, I., Baba, T., Yuzawa, H., Kobayashi, I., Cui, L., Oguchi, A., Aoki, K., Nagai, Y., Lian, J., Ito, T., Kanamori, M., Matsumaru, H., Maruyama, A., Murakami, H., Hosoyama, A., Mizutani-Ui, Y., Takahashi, N. K.,

Sawano, T., Inoue, R., Kaito, C., Sekimizu, K., Hirakawa, H., Kuhara, S., Goto, S., Yabuzaki, J., Kanehisa, M., Yamashita, A., Oshima, K., Furuya, K., Yoshino, C., Shiba, T., Hattori, M., Ogasawara, N., Hayashi, H., and Hiramatsu, K. (2001) Whole genome sequencing of meticillin-resistant *Staphylococcus aureus*. *Lancet* **357**, 1225–1240.

2. O'Farrell, P. H. (1975) High resolution two-dimensional electrophoresis of proteins. *J. Biol. Chem.* **250**, 4007–4021.

3. Klose, J. (1975) Protein mapping by combined isoelectric focusing and electrophoresis of mouse tissues. A novel approach to testing for induced point mutations in mammals. *Humangenetik* **26**, 231–243.

4. Kohler, C., Wolff, S., Albrecht, D., Fuchs, S., Becher, D., Büttner, K., Engelmann, S., and Hecker, M. (2005) Proteome analyses of *Staphylococcus aureus* in growing and non-growing cells: a physiological approach. *Int. J. Med. Microbiol.* **295**, 547–565.

5. Hecker, M. and Engelmann, S. (2006) Physiological proteomics of *Bacillus subtilis* and *Staphylococcus aureus*: towards a comprehensive understanding of cell physiology and pathogenicity, in *Pathogenomics* (Hacker, J. and Dobrindt, U., eds.), Wiley-VCH Verlag GmbH & Co. KGaA, Weinheim, pp. 43–68.

6. Ziebandt, A. K., Becher, D., Ohlsen, K., Hacker, J., Hecker, M., and Engelmann, S. (2004) The influence of *agr* and sigma(B) in growth phase dependent regulation of virulence factors in *Staphylococcus aureus*. *Proteomics* **4**, 3034–3047.

7. Novick, R. P. (2003) Autoinduction and signal transduction in the regulation of staphylococcal virulence. *Mol. Microbiol.* **48**, 1429–1449.

8. Ziebandt, A. K., Weber, H., Rudolph, J., Schmid, R., Höper, D., Engelmann, S., and Hecker, M. (2001) Extracellular proteins of *Staphylococcus aureus* and the role of SarA and sigma B. *Proteomics* **1**, 480–493.

9. Rogasch, K., Rühmling, V., Pané-Farré, J., Höper, D., Weinberg, C., Fuchs, S., Schmudde, M., Bröker, B. M., Wolz, C., Hecker, M., and Engelmann, S. (2006) The influence of the two component system SaeRS on global gene expression in two different *Staphylococcus aureus* strains. *J. Bacteriol.* **188**, 7742–7758.

10. Sibbald, M. J. J. B., Ziebandt, A. K., Engelmann, S., Hecker, M., de Jong, A., Harmsen, H. J. M., Raangs, G. C., Stokroos, I., Arends, J. P., Dubois, J. Y. F., and van Dijl, J. M. (2006) Mapping of pathways to staphylococcal pathogenesis by comparative secretomics. *Microbiol. Mol. Biol. Rev.* **70**, 755–788.

11. Gertz, S., Engelmann, S., Schmid, R., Ohlsen, K., Hacker, J., and Hecker M. (1999) Regulation of sigmaB-dependent transcription of *sigB* and *asp23* in two different *Staphylococcus aureus* strains. *Mol. Gen. Genet.* **261**, 558–566.

12. Candiano, G., Bruschi, M., Musante, L., Santucci, L., Ghiggeri, G. M., Carnemolla, B., Orecchia, P., Zardi, L., and Rigetti, P. G. (2004) Blue silver: a very sensitive colloidal Coomassie G-250 staining for proteome analysis. *Electrophoresis* **25**, 1327–1333.

13. Blum, H., Beier, H., and Gross, H. J. (1987) Improved silver staining of plant proteins, RNA and DNA in polyacrylamide gels. *Electrophoresis* **8**, 93–99.

14. Eymann, C., Dreisbach, A., Albrecht, D., Bernhardt, J., Becher, D., Gentner, S., Tam, L. T., Büttner, K., Buurman, G., Scharf, C., Venz, S., Völker, U., and Hecker, M. (2004) A comprehensive proteome map of growing *Bacillus subtilis* cells. *Proteomics* **4**, 2849–2876.
15. Bernhardt, J., Büttner, K., Scharf, C., and Hecker, M. (1999) Dual channel imaging of two-dimensional electropherograms in *Bacillus subtilis. Electrophoresis* **20**, 2225–2240.

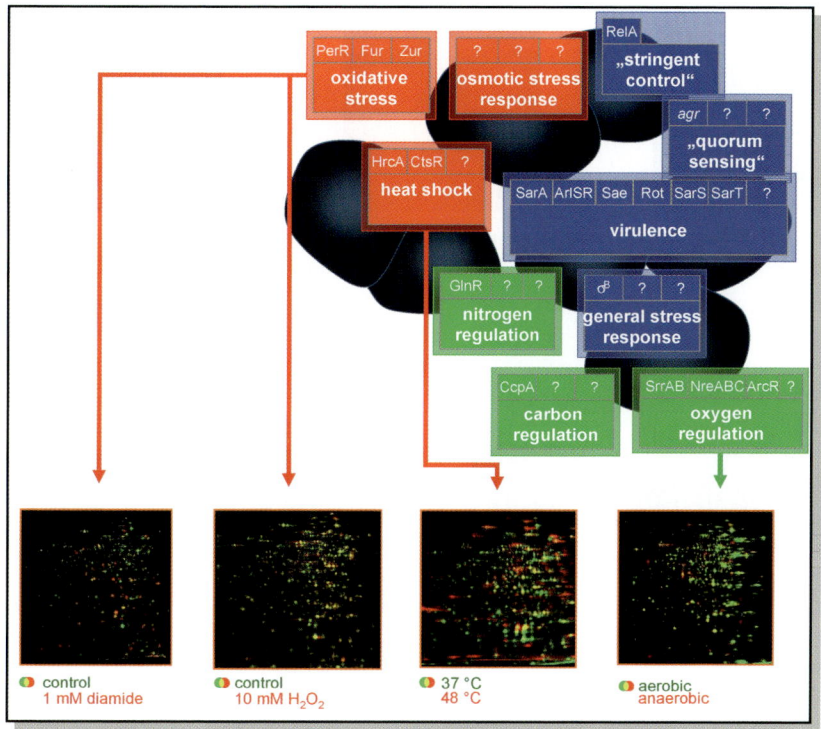

Color Plate 1. Proteomic signatures of different stress/starvation conditions in *Staphylococcus aureus*. Comparison of the protein synthesis profile of exponentially growing cells (green) with that of stressed *S. aureus* cells (red) reveals changes in protein synthesis that are particular for the certain stress stimuli. Cells were cultivated in synthetic medium and exposed to the respective stimulus at an optical density at 500 nm (OD_{500}) of 0.5. Protein synthesis was analyzed by [^{35}S]-L-methionine labeling (5 min pulse) under control conditions and 10 min after imposition to stress. All proteins induced by one stimulus belong to a stimulon (Chapter 3, Fig. 2; *see* discussion on p. 31).

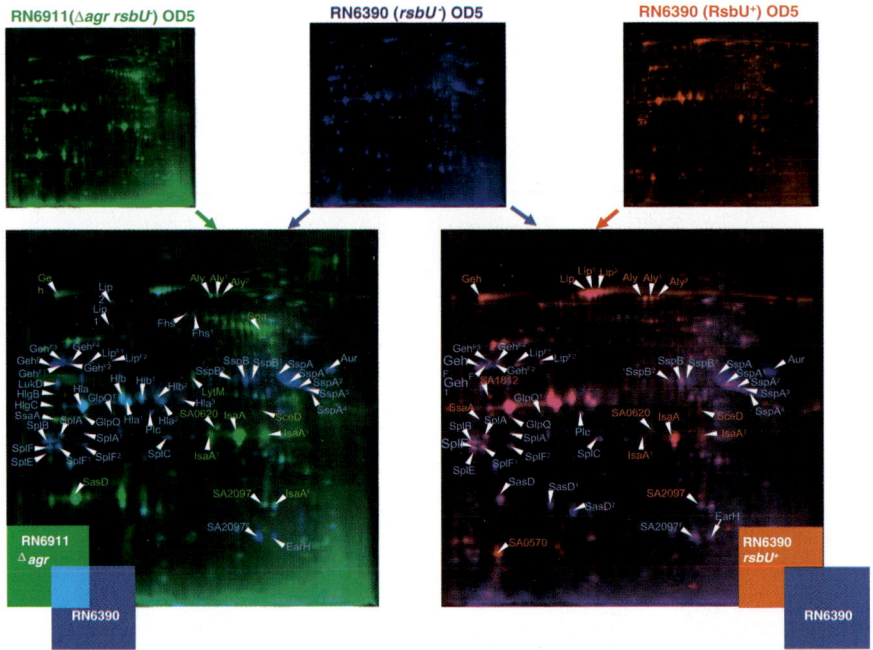

Color Plate 2. Difference gel electrophoresis (DIGE) labeling technique to illustrate differences in protein patterns of three protein extracts. Extracellular protein patterns of wild-type RN6390 (blue), the isogenic *agr* mutant (green), and RN6390 complemented with *rsbU* of *Staphylococcus aureus* COL (red). Prior to separation by 2D gel electrophoresis technique, protein extracts of the respective strains were labeled with CyDye fluorochromes [Cy2 RN6390, Cy3 RN6911, Cy5 RN6390 (RsbU$^+$)]. A mixture of 50 µg of each protein extract was separated on one gel. The proteins were detected by using a Typhoon [Cy2: Blue2-Laser (488 nm), Cy3: Green-Laser (532 nm), Cy5: the Red-Laser (633 nm)] (Chapter 3, Fig. 4; *see* discussion on p. 33).

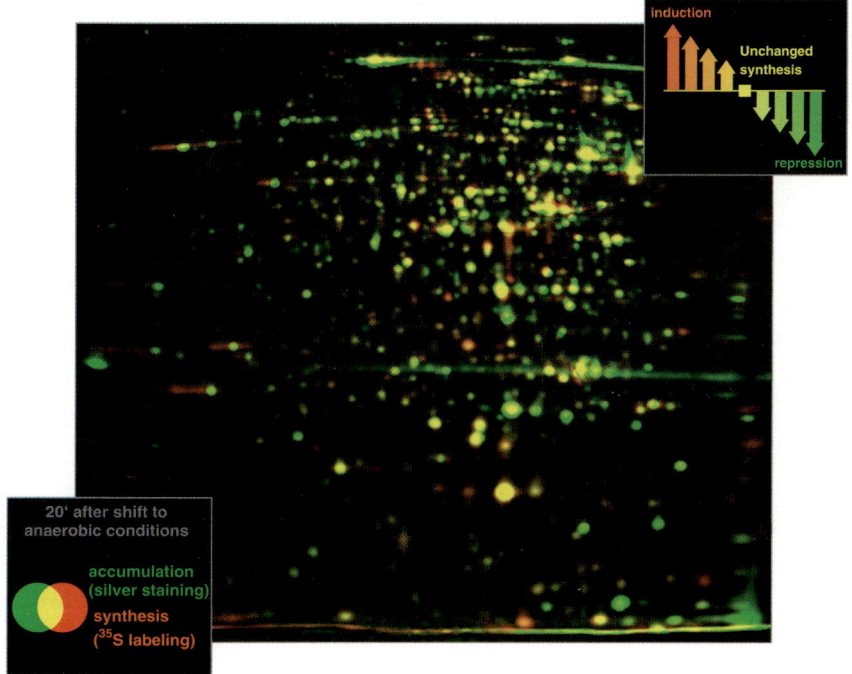

Color Plate 3. Dual-channel imaging technique. Bacterial cells were grown in synthetic medium without nitrate to an optical density at 500 nm (OD_{500}) of 0.5. The proteins were labeled with [^{35}S]-L-methionine 20 min after transfer to anaerobic conditions. Protein extracts were separated on 2D gels. The resulting images from the protein synthesis ([^{35}S]-L-methionine labeling) and the protein accumulation (staining with silver nitrate) were overlaid in "dual channel images." Proteins whose synthesis was induced in response to the stressor but have not accumulated yet are colored red, repressed spots that are still present but not synthesized anymore are colored green, and spots without changed expression behavior in response to stress are shown in yellow (Chapter 3, Fig. 5; *see* discussion on p. 40).

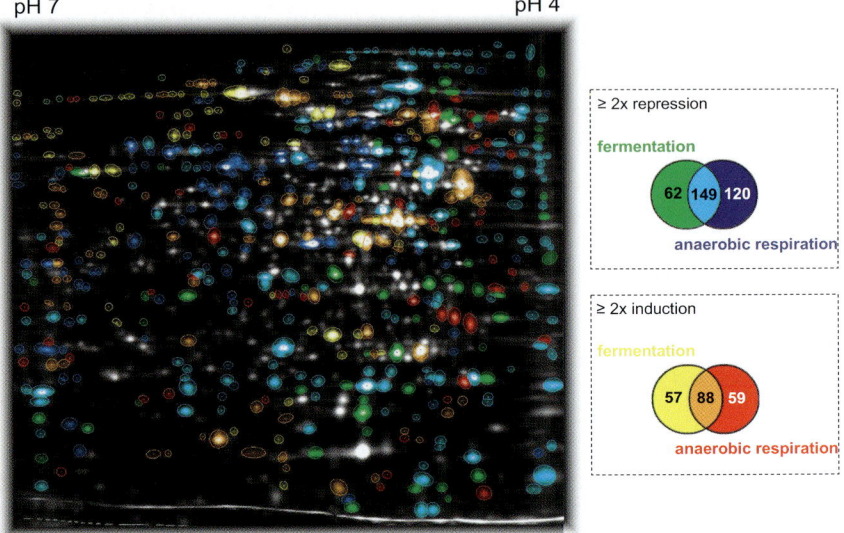

pH 7   pH 4

≥ 2x repression

fermentation

62 149 120

anaerobic respiration

≥ 2x induction

fermentation

57 88 59

anaerobic respiration

Color Plate 4. Multicolor imaging of protein expression patterns of *Staphylococcus aureus* COL under anaerobic conditions in the presence and absence of nitrate. Delta2D software is used to visualize complex protein expression patterns on the 2D image in the standard pH range 4–7. The color code is represented on the right side (Chapter 3, Fig. 6; *see* discussion on p. 42).

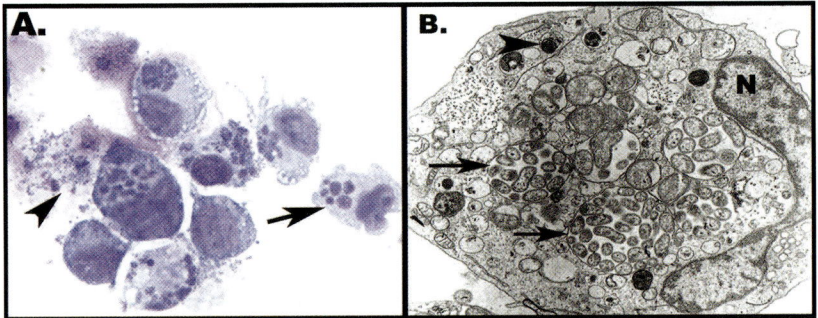

A.   B.   N

Color Plate 5. Images of *Anaplasma phagocytophilum*-infected HL60 cells. (**A**) Cytofuge smear of highly infected HL60 cells. Arrow indicates intact, multiple pathogen morulae within the HL60 cell cytoplasm (*6*). Arrowhead indicates HL60 cell lysis with abundant extracellular bacteria (1000× magnification; Hema-Quik stain). (**B**) Electron micrograph of a highly infected HL60 cell with abundant intracellular bacteria found within cytoplasmic vacuoles (morula). The N labels the cell nucleus. Arrows indicate the "reticulate," possibly replicative, stage of organism development. The arrowhead indicates the "dense cored," possibly infective, stage of organism development (Chapter 13, Fig. 1; *see* discussion on p. 162).

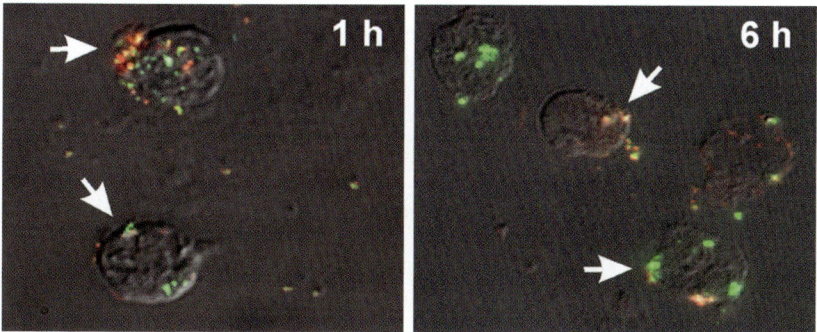

Color Plate 6. Interaction of *Anaplasma phagocytophilum* with human neutrophils. **(A)** Ingestion of *A. phagocytophilum* by human neutrophils. Neutrophils were incubated with ~5–20 *A. phagocytophilum*/neutrophil for the indicated times, and percent ingestion (number of neutrophils containing internalized *A. phagocytophilum*) was determined by fluorescence microscopy. **(B)** Micrographs illustrating ingestion of *A. phagocytophilum* by human neutrophils. Green, intracellular bacteria. Red or red-green (yellow), extracellular bacteria. Reprinted with permission from reference *(13)*. Copyright 2005 The American Association of Immunologists, Inc (Chapter 13, Fig. 2; *see* discussion on p. 167).

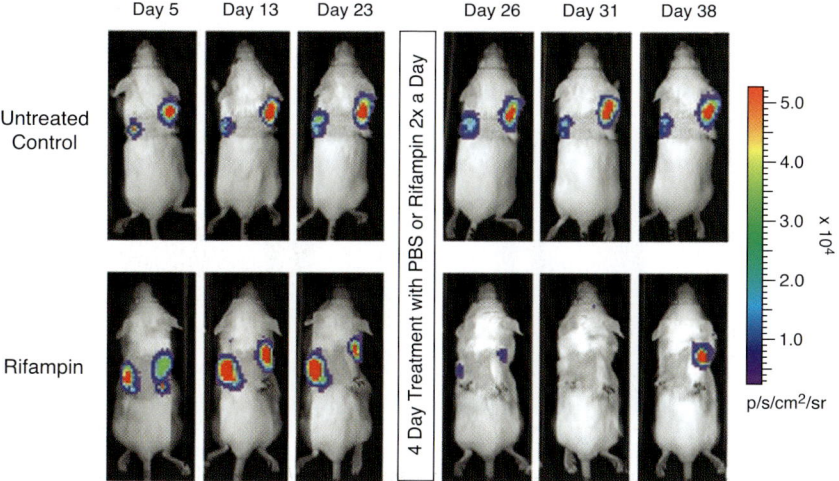

Color Plate 7. Real-time *in vivo* bioluminescence monitoring of *S. aureus* Xen-29 in the mouse model of soft-tissue biofilm infection during treatment with rifampin. Effect of 4 days of rifampin treatment (twice a day with 30 mg/kg) on a 3-week-old biofilm. A representative animal from the group receiving antibiotic or the untreated infected group is shown. Note the response to treatment including relapse that can be monitored non-invasively within the same animal throughout the study period (Chapter 18, Fig. 2; *see* discussion on p. 230).

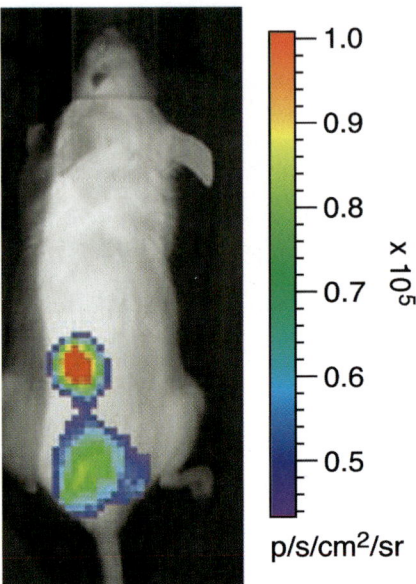

Color Plate 8. *In vivo* bioluminescent monitoring of the progression of *P. mirabilis* Xen 44 in the mouse model of catheter-induced urinary track infection (UTI) (Chapter 18, Fig. 4; *see* complete caption on p. 233 and discussion on p. 232).

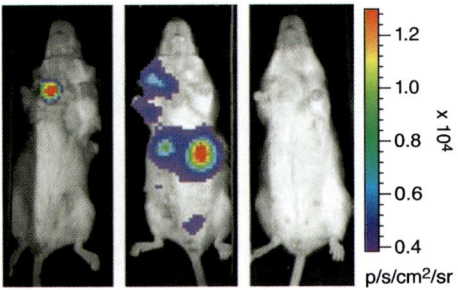

Color Plate 9. Daily imaging of mice inoculated by tail vein injection with *S. aureus* Xen 29 reveals the colonization of *S. aureus* around the indwelling catheter in jugular vein catheterized mice as early as one day after infection (left). Six days later, evidence of metastatic disease with staphylococci spreading to heart, kidney, and liver as a result of seeding of bacteria from the infected vascular catheter (center). In contrast, whole-body imaging of non-catheterized mice challenged with a similar dose of bacterial suspension showed no sign of infection around heart or any other organ, even after a week into the hematogenously induced infection (right) (Chapter 18, Fig. 5; *see* complete caption and discussion on p. 235).

4

Microarray Comparative Genomic Hybridization for the Analysis of Bacterial Population Genetics and Evolution

Caitriona M. Guinane and J. Ross Fitzgerald

Summary

Comparative genomic hybridization analyses have contributed greatly to our understanding of bacterial evolution, population genetics, and pathogenesis. Here, we describe a robust protocol for microarray-based comparison of genome content, which could be applied to any bacterial species of interest.

Key Words: Whole-genome microarrays; comparative genomics; bacterial evolution; population genetics.

1. Introduction

Comparative genomics is a rapidly expanding discipline revolutionized by the availability of increasing numbers of bacterial genome sequences. DNA microarrays that represent single or multiple bacterial genomes are a cost-effective way of comparing the genome content of large numbers of strains of the same species. Comparative genomic hybridizations (CGHs) with microarrays allow determination of the presence or absence of genes based on the level of sequence homology required for hybridization to occur (*see* **Note 1**). CGH analyses can result in broad insights into bacterial genomics, population genetics, and evolution. Moreover, whole-genome microarrays can be used to address specific issues of bacterial pathogenesis including the basis for host specificity, epidemic waves, and the emergence and re-emergence of pathogenic clones *(1)*. Importantly, the findings of such studies could be used in the rational design of disease-controlling strategies.

From: *Methods in Molecular Biology, vol. 431: Bacterial Pathogenesis*
Edited by: F. DeLeo and M. Otto © Humana Press, Totowa, NJ

Studies on the evolution and pathogenesis of an increasing number of bacterial pathogens including *Escherichia coli* O157:H7 *(2)*, *Mycobacterium* spp. *(3)*, *Helicobacter pylori* *(4)*, *Yersinia* spp. *(5)*, and *Staphylococcus aureus* *(6–9)* have been performed with microarrays, resulting in a greatly enhanced understanding of their biology.

Several different approaches to microarray construction have been developed, including oligonucleotide- or DNA-based arrays on glass slides or nylon membranes *(10)*. Here, we describe a robust CGH protocol based on the use of a PCR-generated microarray *(6)*. Detailed descriptions of genomic DNA template preparation, labeling, microarray hybridization, and data analysis are provided.

2. Materials

2.1. Genomic DNA Template Preparation (S. aureus)

1. Edge Biosystems DNA extraction kit (Edge Biosystems, Gaithersburg, MD). All components are kept at 4 °C; spheroplast buffer is stored at –20 °C.
2. 100% isopropanol (molecular grade). Chill before use.
3. 70% ethanol in sterile deionized water. Chill before use.
4. Lysostaphin (AMBI products, Lawrence, NY) lyophilized powder is resuspended in nuclease-free water (Stock; 5 mg/mL). Store at –20 °C.
5. Agilent 2100 Bioanalyser (Agilent Technologies, Palo Alto, CA) or equivalent.
6. *Hin*P1I restriction endonuclease (New England Biolabs, Ipswich, MA). Store at –20 °C.

2.2. Labeling of DNA

1. 400 ng of genomic DNA templates (test and reference strains; *see* **Note 2**).
2. Prime-a-gene labeling kit (Promega, Madison, WI):

   a. Bovine serum albumin (BSA, 10 mg/mL); nuclease free.
   b. Unlabeled dNTPs (1.5 m*M*). Store at –20 °C .
   c. DNA polymerase I large fragment (Klenow). Store at –20 °C.
   d. Labeling buffer (5×) including 26.0 A_{260} U/mL random hexadeoxyribonucleotides.

3. TE Buffer: 10 m*M* Tris–HCl, 1 m*M* EDTA, pH 7. Store at room temperature.
4. 0.2 *M* EDTA (pH 8.0). Store at room temperature.
5. Cyanine 3-dCTP (Cy3); Cyanine 5-dCTP (Cy5) (NEN-Dupont, Boston, MA). Store at –20 °C; light sensitive.
6. Sephadex® G-50 spin columns (Roche Diagnostics, Basel, Switzerland) (optional).
7. Centri-spin[20] label clean up kit (Princeton Separations, Adelphia, NJ). Store at room temperature.
8. Agilent 2100 Bioanalyser (Agilent Technologies) or equivalent to determine genomic DNA labeling efficiency.

2.3. Slide Preparation and Pre-Hybridization

1. Microarray hybridization chamber (Telechem, Sunnyvale, CA).
2. Hybridization buffer, ULTRAhyb (Ambion, Austin, TX), pre-heated to 68 °C.

2.4. Hybridization

1. 20× sodium chloride sodium citrate (SSC): 3.0 M NaCl, 0.3 M Na-citrate, pH 7.0.
2. 2× SSC, 0.1× SSC, and 0.05× SSC in nuclease-free water. Store at room temperature.
3. 0.1% (w/v) SDS. Store at room temperature.
4. 95% ethanol, molecular grade.
5. Lifterslip (Fischer Biosciences, Rockville, MD).

2.5. Scanning of Slides and Data Analysis

1. GenePix 4100A microarray scanner or equivalent and GenePix Pro 6.0 software (Axon Instruments, Foster city, CA) or similar.
2. GeneSpring software (Agilent Technologies).
3. Microarray validation system (*see* **Note 3**).

2.6. Phylogenetic Analysis

1. Cluster and TreeView programs *(11)*.

3. Methods

The major steps in the CGH protocol include genomic DNA isolation and labeling, microarray hybridization, scanning of slides and calculation of fluorescence intensity ratios, and data analysis by use of appropriate statistical methods (*see* **Fig. 1**).

3.1. Genomic DNA Isolation and Digestion

1. Perform genomic DNA isolation from *S. aureus* test and reference strains using a modification of the Edge Biosystems genomic DNA isolation kit. The modification consists of addition of lysostaphin (100 µg/mL final concentration) to bacterial cells and incubation at 37 °C for 10 min prior to DNA extraction.
2. Determine the quality of DNA on a Bioanalyser or equivalent; the $A_{260/280}$ nm absorbance ratio should be >1.8. The procedure outlined is optimal for 400 ng of staphylococcal genomic DNA.
3. Digest genomic DNA with the restriction enzyme *Hin*P1I at 37 °C overnight and purify before using as a template for labeling.

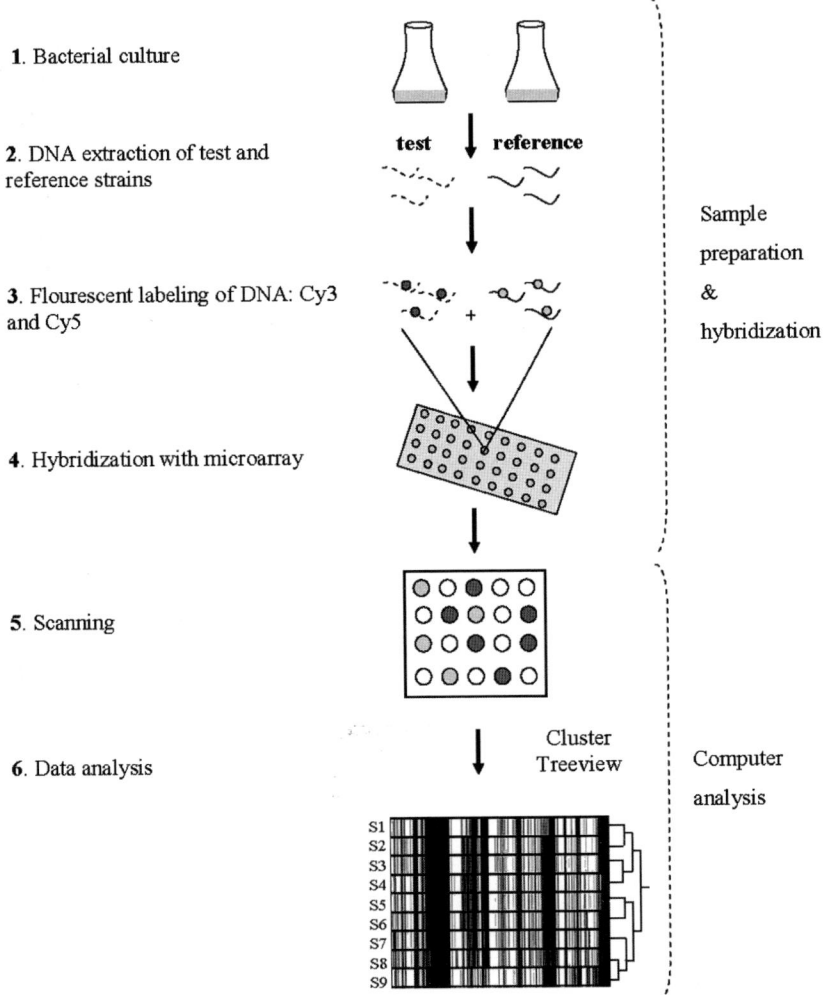

1. Bacterial culture

2. DNA extraction of test and reference strains

   test   reference

3. Flourescent labeling of DNA: Cy3 and Cy5

4. Hybridization with microarray

5. Scanning

6. Data analysis

Sample preparation & hybridization

Cluster Treeview

Computer analysis

S1
S2
S3
S4
S5
S6
S7
S8
S9

Fig. 1. Overview of the protocol for microarray-based comparative genomic hybridization. S1–S9 represent genomic profiles of the strains analyzed.

3.2. Labeling of DNA (Prime-A-Gene Labeling System; Promega Modified Protocol)

1. Thaw all components on ice. Keep the Klenow fragment at –20 °C until required.
2. Denature purified, digested genomic DNA in TE buffer at 95 °C for 2 min followed by immediate chilling on ice.
3. Carry out the labeling reaction in a volume of 50 µL including 5× buffer, unlabeled dNTPs (each at 20 µM) (*see* **Note 4**), denatured template (400 ng),

BSA (400 µg/mL), either Cy5- or Cy3-labeled dUTP (2.5 nmol), and Klenow DNA polymerase (100 U/mL) at room temperature for 4 h to overnight.

4. Terminate the reaction by incubation at 95 °C for 2 min. Chill on ice and add 20 mM EDTA.

5. Purify labeled genomic DNA using the Centri-spin 20 kit (Princeton Separations) according to the manufacturer's instructions and use immediately in hybridization experiments or store at –20 °C for later use (*see* **Note 5**).

6. Determine the efficiency of labeling with an Agilent Bioanalyser or equivalent.

3.3. Slide Preparation and Pre-Hybridization

1. Add 1.8 mL of hybridization buffer (pre-warmed to 68 °C) to the slide and incubate at 42 °C for 45 min in a hybridization chamber.

2. Wash the slide in pre-warmed (42 °C) distilled H_2O (dH_2O) for 30–45 s and follow by washing in dH_2O at room temperature.

3. Centrifuge the slide at 15 × g for 7 min. The slide is now ready for hybridization.

3.4. Hybridization and Washing

1. Add equal volumes of dH_2O and hybridization buffer (100 µL each) to the end reservoirs of the hybridization chamber.

2. Mix the labeled probes together with an equal amount of pre-warmed hybridization buffer.

3. Denature the mixture at 100 °C for 5 min and add to the edge of the lifterslip.

4. Perform hybridization at 48 °C for a minimum of 2 h and as long as overnight.

5. Following hybridization, immerse the slide in 2× SSC, 0.1% SDS (pre-warmed to 42 °C) to remove the coverslip.

6. Wash the slide twice in 0.1× SSC containing 0.1 % SDS for 10 min at room temperature, followed by two washes in 0.1× SSC for 5 min and two washes in 0.05× SSC for 2 min.

7. Rinse the slide in dH_2O, followed by 95% ethanol. Dry the slide by centrifugation at 15 × g for 7 min.

3.5. Scanning of Slides and Data Analysis

1. Scan the slides using a microarray scanner such as a GenePix 4100A instrument or equivalent with line scans performed to maximize signal and minimize background. During scanning, normalize fluorescence intensity between channels to a serial dilution of reference strain chromosomal DNA by adjusting laser power and/or photomultiplier gain. Problems that have occurred throughout the microarray experiment can become apparent upon scanning of the slides (*see* **Notes 6–8**).

2. GenePix Pro 6.0 software or similar is used to convert the pixel images into signal intensity before exporting as a flat text file to GeneSpring (Agilent Technologies). For each slide, adjust spot intensities for background and normalize to the median

intensity of the slide. The median ratio of normalized test strain fluorescence to normalized control strain fluorescence is used to identify gene differences. Genes are typically considered lacking or highly variable in the test vs. control strain below a 0.5 test/control ratio and considered present in greater copy number in the test vs. control strain above a 2.0 test/control ratio (*see* **Note 9**).

3. Use GeneSpring software (Agilent Technologies) for further statistical analysis of the data.

3.6. Phylogenetic Analysis

1. To determine the evolutionary relatedness of strains, use the Cluster program *(11)* based on the presence or absence of all open reading frames (ORFs) represented on the microarray.

2. The output can be represented in a phylogenetic tree constructed using the program TreeView (*see* **Note 10**).

4. Notes

1. ORFs defined as "absent" either are not present in the genome that is being tested or have sufficient sequence divergence that hybridization will not occur with the ORFs represented on the microarray under conditions of high stringency. For each strain, two or three hybridizations are carried out.

2. The reference strain is representative of one of the sequenced strains encoded on the microarray, and the test strain represents a strain of unknown genomic content.

3. To validate the microarray hybridization process, use an array validation system such as the Spot report oligo array validation system (Stratagene, La Jolla, CA).

4. The unlabeled dNTPs are prepared as outlined in the Promega protocol. Briefly, mix 1 µL of each non-isotopically labeled dNTP to yield 3 µL of a premix containing the three unlabeled dNTPs, each at 500 µ*M*. The volume of aqueous labeled dNTPs should not exceed 50% of the total reaction volume.

5. As suggested in the Promega protocol, unincorporated, labeled nucleotides may be removed by size-exclusion chromatography using Sephadex® G-50 spin columns or by selective precipitation of the labeled DNA. This step is unnecessary if dye incorporation is above 60%.

6. Streaks of labeled product from strong spots ("comet tails") may be observed after scanning. Steps can be taken to reduce this effect including denaturing of the slides at 95 °C for 2 min before hybridization, reducing the amount of label used in the reaction, or diluting the PCR products before printing the microarray.

7. To minimize high background fluorescence, one may increase the spacing between the spots and lower the photomultiplier tube gain (PMT) during the scanning process. Only genomic DNA of >1.8 $A_{260/280}$ ratio should be used for labeling and hybridization.

8. The occurrence of intense spots in the background of the slide can be minimized by the use of powder-free gloves and working in a dust-free environment. It is

also important to ensure that the probe has been sufficiently denatured before hybridization.

9. A global normalization of all samples in addition to individual chip normalization should be performed. The reference strain should display a 45° scatter plot whereas the test strain should display a skewed histogram toward the negative. GeneSpring can be used to construct a dendrogram and run a Pearson correlation to determine whether the replicate hybridization experiments are in agreement.

10. The Pearson correlation coefficient is used to represent the scale of a dendrogram constructed with TreeView after phylogenetic analysis with Cluster. For a pairwise comparison, a coefficient of 1 indicates absolute identity and zero indicates complete independence.

Acknowledgments

The authors are grateful to Dan Sturdevant for providing technical information.

References

1. Fitzgerald, J. R. and Musser, J. M. (2001) Evolutionary genomics of pathogenic bacteria. *Trends Microbiol.* **9,** 547–553.
2. Whittam, T. S. and Bumbaugh, A. C. (2002) Inferences from whole-genome sequences of bacterial pathogens. *Curr. Opin. Genet. Dev.* **12,** 719–725.
3. Behr, M. A., Wilson, M. A., Gill, W. P., Salamon, H., Schoolnik, G. K., Rane, S., and Small, P. M. (1999) Comparative genomics of BCG vaccines by whole-genome DNA microarray. *Science* **284,** 1520–1523.
4. Salama, N., Guillemin, K., McDaniel, T. K., Sherlock, G., Tompkins, L., and Falkow, S. (2000) A whole-genome microarray reveals genetic diversity among *Helicobacter pylori* strains. *Proc. Natl. Acad. Sci. USA* **97,** 14668–14673.
5. Hinchliffe, S. J., Isherwood, K. E., Stabler, R. A., Prentice, M. B., Rakin, A., Nichols, R. A., Oyston, P. C., Hinds, J., Titball, R. W., and Wren, B. W. (2003) Application of DNA microarrays to study the evolutionary genomics of *Yersinia pestis* and *Yersinia pseudotuberculosis*. *Genome Res.* **13,** 2018–2029.
6. Fitzgerald, J. R., Sturdevant, D. E., Mackie, S. M., Gill, S. R., and Musser, J. M. (2001) Evolutionary genomics of *Staphylococcus aureus*: insights into the origin of methicillin-resistant strains and the toxic shock syndrome epidemic. *Proc. Natl. Acad. Sci. USA* **98,** 8821–8826.
7. Witney, A. A., Marsden, G. L., Holden, M. T., Stabler, R. A., Husain, S. E., Vass, J. K., Butcher, P. D., Hinds, J., and Lindsay, J. A. (2005) Design, validation, and application of a seven-strain *Staphylococcus aureus* PCR product microarray for comparative genomics. *Appl. Environ. Microbiol.* **71,** 7504–7514.
8. Lindsay, J. A., Moore, C. E., Day, N. P., Peacock, S. J., Witney, A. A., Stabler, R. A., Husain, S. E., Butcher, P. D., and Hinds, J. (2006) Microarrays

reveal that each of the ten dominant lineages of *Staphylococcus aureus* has a unique combination of surface-associated and regulatory genes. *J. Bacteriol.* **188,** 669–676.

9. Koessler, T., Francois, P., Charbonnier, Y., Huyghe, A., Bento, M., Dharan, S., Renzi, G., Lew, D., Harbarth, S., Pittet, D., and Schrenzel, J. (2006) Use of oligoarrays for characterization of community-onset methicillin-resistant *Staphylococcus aureus. J. Clin. Microbiol.* **44,** 1040–1048.

10. Cook, K. L. and Sayler, G. S. (2003) Environmental application of array technology: promise, problems and practicalities. *Curr. Opin. Biotechnol.* **14,** 311–318.

11. Eisen, M.B., Spellman, P.T., Brown, P.O., and Botstein, D. (1998) Cluster analysis and display of genome-wide expression patterns. *Genetics* **95,** 14863–14868.

5

Detection and Inhibition of Bacterial Cell–Cell Communication

Scott A. Rice, Diane McDougald, Michael Givskov, and Staffan Kjelleberg

Summary

Bacteria communicate with other members of their community through the secretion and perception of small chemical cues or signals. The recognition of a signal normally leads to the expression of a large suite of genes, which in some bacteria are involved in the regulation of virulence factors, and as a result, these signaling compounds are key regulatory factors in many disease processes. Thus, it is of interest when studying pathogens to understand the mechanisms used to control the expression of virulence genes so that strategies might be devised for the control of those pathogens. Clearly, the ability to interfere with this process of signaling represents a novel approach for the treatment of bacterial infections. There is a broad range of compounds that bacteria can use for signaling purposes, including fatty acids, peptides, N-acylated homoserine lactones, and the signals collectively called autoinducer 2 (AI-2). This chapter will focus on the latter two signaling systems as they are present in a range of medically relevant bacteria, and here we describe assays for determining whether an organism produces a particular signal and assays that can be used to identify inhibitors of the signaling cascade. Lastly, the signal detection and inhibition assays will be directly linked to the expression of virulence factors of specific pathogens.

Key Words: Quorum sensing; cell–cell signaling; acylated homoserine lactone; autoinducer 2; virulence factor expression; global regulation; inhibitors.

1. Introduction

It is now recognized that bacteria regulate high-density phenotypes, such as biofilm formation, and phenotypes that are important for a number of pathogenic

From: *Methods in Molecular Biology, vol. 431: Bacterial Pathogenesis*
Edited by: F. DeLeo and M. Otto © Humana Press, Totowa, NJ

bacteria, through signal-based regulatory systems *(1)*. In general, these systems function via extracellular signals that bacteria use to assess their local population density. When these signals reach sufficient concentrations (e.g., at high cell densities), a feedback regulatory loop is induced, resulting in a very rapid expression of phenotypes in the population of cells. The best studied of these regulatory loops are the acylated homoserine lactone (AHL) system in Gram-negative bacteria and the autoinducer 2 (AI-2) system, first identified in the genus *Vibrio* and subsequently found in a broad range of Gram-negative and Gram-positive bacteria.

AHL-mediated density-dependent signaling (or quorum sensing) occurs when a threshold concentration of the AHLs (produced by the product of the *synthase I* gene) accumulates to a sufficient level to interact with receptor proteins (produced from the products of the *R* gene), which then act as transcriptional regulators by binding to the promoter of target genes *(1)*. *Pseudomonas aeruginosa* is a particularly aggressive and problematic pathogen because it produces a number of virulence factors (extracellular proteases, toxins, siderophores), many of which are regulated by AHL-mediated quorum-sensing pathways *(1)*, and thus *P. aeruginosa* has become the model bacterium for AHL-mediated gene regulation. Several virulence factors in *P. aeruginosa* are AHL-regulated, including elastase, pyocyanin, rhamnolipids, alkaline protease, and exotoxin A (reviewed in **ref.** *(2)*). Importantly, the colonization of surfaces and biofilm formation are also regulated by the AHL system in *P. aeruginosa*. For example, AHL-deficient mutants of *P. aeruginosa* form aberrant biofilms *(3)* and are more easily removed from surfaces. Moreover, AHL-deficient strains of *P. aeruginosa* are not able to establish chronic lung infections in mouse models of cystic fibrosis infection *(4)* or ocular disease *(5)*, which was presumably due to reduced biofilm formation and virulence factor expression. These findings have been replicated with AHL mutants in other pathogenic bacteria including *Burkholderia cepacia* *(6)* and *Serratia marcescens* (*liquefaciens*) *(7)*. AHL systems in *S. marcescens* control surface colonization, including attachment, surface motility and biofilm development *(7)*, and expression of other virulence factors [e.g., proteases *(8)*]. We also have recently demonstrated that the AHL regulatory system in *S. marcescens* controls genes essential for attachment and invasion into corneal epithelial cells (unpublished data).

The AI-2 quorum-sensing system is found in a variety of Gram-negative and Gram-positive bacteria and has been shown to regulate a number of phenotypes, including virulence factor expression and biofilm formation in *Streptococci*, virulence factor expression in enterohemorrhagic *Escherichia coli* (EHEC), and virulence factor expression and biofilm formation in *Vibrio vulnificus* and *Vibrio cholerae* *(9,10)* (for a review, see **ref.** *(11)*). There are at least 12 pathogenic *Vibrio* species recognized to cause human illness. Among the non-cholera *Vibrio* species are *Vibrio parahaemolyticus*, which is responsible for gastroenteritis and wound infections, and *Vibrio alginolyticus*, which causes wound infections. The most commonly isolated species is *V. vulnificus*, which is of great medical importance due to the high mortality rate associated with infections *(12)*. Because of the introduction of antibiotics for the treatment of cholera, the occurrence of drug-resistant *V. cholerae* strains is being reported with increasing frequency. Ninety-eight percent of clinical O1 isolates showed resistance to one or more antibiotics, whereas 50% of the environmental isolates showed resistance to one to three drugs, and one clinical and one environmental isolate showed resistant to seven antibiotics *(13)*.

The trend of acquisition of resistance to antibiotics has increased for many pathogenic bacterial species. Yet, the number of therapeutically useful compounds that emerge is continuously decreasing, and only one new class of antibiotic has been introduced in the last 40 years. Only the development of new classes of antimicrobials with novel mechanisms of action can address the drug resistance problem we face today. Schmidt *(14)* states "a prioritized goal in the search for new targets is to identify and inactivate disease-causing functions instead of focusing on killing. Since these factors have no vital functions for microbes, there is less pressure to develop resistance, which would ultimately reduce global selective pressure and prevent the acceleration and dissemination of resistance." Thus, quorum-sensing systems such as the AHL and AI-2 regulatory systems represent novel targets for the control of biofilm formation and virulence factor production by bacteria.

2. Materials

2.1. Detection and Inhibition of N-Acylated Homoserine Lactone Quorum Sensing

2.1.1. ABt Medium *(15)*

1. Solution A—in 100 mL of distilled water (dH_2O), 4 g of $(NH_4)_2SO_4$, 6 g of $Na_2HPO_4 \cdot 2H_2O$, 3 g of KH_2PO_4, 3 g of NaCl. Autoclave and store at room temperature.
2. 1 M $MgCl_2$.
3. 0.1 M $CaCl_2$.
4. 0.01 M $FeCl_2$ (make fresh as Fe oxidizes rapidly into insoluble hydroxides).
5. Thiamine, 1 mg/mL.
6. Solution B—to 1 L of sterile dH_2O, add 1 mL of 1 M $MgCl_2$, 1 mL of 0.1 M $CaCl_2$, 1 mL of 0.01 M $FeCl_2$, and 2.5 mL of thiamine at 1 mg/mL. Store at room temperature.
7. 20% glucose, filter sterilized. Store at room temperature.
8. 20% casamino acids, filter sterilized. Store at room temperature.

9. To prepare the ABt medium, add one part of Solution A to nine parts of Solution B. Add glucose and casamino acids to a final concentration of 0.5%. Store at room temperature away from direct sunlight (*see* **Note 1**).

2.1.2. Signals and Antagonists

1. Stock solution of *N*-3-oxo-hexanoyl homoserine lactone (OHHL) (Sigma, St. Louis, MO, USA), 10 μM dissolved in dimethylsulfoxide (DMSO). Store at −20°C.
2. High-performance liquid chromatography (HPLC) grade dichloromethane (DCM) (*see* **Note 2**).
3. Stock solution of sample extract, resuspended to 5 mg/mL in 100% ethanol (*see* **Note 3**).

2.2. Inhibition of the AHL-Regulated Phenotype of Elastase Production

1. Elastin-Congo Red: Elastin-Congo Red buffer, 0.1 *M* Tris–HCl (pH 7.2), 1 m*M* CaCl$_2$, Elastin-Congo Red (Sigma), 20 mg of powder per reaction tube containing 1 mL of Elastin-Congo buffer. The powder is insoluble in the buffer, and hence, each reaction tube has to be measured separately.

2.3. Detection and Inhibition of AI-2 Quorum Sensing

1. Autoinducer bioassay (AB) medium (according to **ref.** *(16)*): Solution A—0.3 *M* NaCl, 0.05 *M* MgSO$_4$, 0.2% vitamin-free casamino acids (Difco, Franklin Lakes, NJ, USA). Make 1 L of solution A by adding the reagents above to the indicated final concentrations in dH$_2$O. Adjust pH to 7.5 with KOH, autoclave, and cool. Add to solution A, 10 mL of 1 *M* potassium phosphate buffer (pH 7.0), 10 mL of 0.1 *M* L-arginine, 20 mL of glycerol, 1 mL of riboflavin (10 µg/mL) and 1 mL of thiamine (1 mg/mL).
2. Hemolysin buffer: 10 m*M* Tris–HCl (pH 7.2), 0.15 *M* NaCl, 10 m*M* MgCl$_2$, 10 m*M* CaCl$_2$, 0.2% (w/v) gelatin. Add the reagents above to the indicated final concentration in dH$_2$O. Sterilize by autoclaving.
3. Lauria Broth 20 (according to **ref.** *(17)*): To 1 L of dH$_2$O, add 10 g of tryptone, 5 g of yeast extract, and 20 g of NaCl. Autoclave.

3. Methods
3.1. AHL Signal, Inducer, or Antagonist Extraction and Preparation

In many cases, one may be interested in recovering either quorum-sensing signals from particular bacteria or potential quorum-sensing inhibitors (QSIs) from eukaryotic organisms or from microorganisms. For the isolation of QSIs from eukaryotes, whole materials are normally extracted for active compounds, whereas the isolation of signals or antagonists from bacteria is performed on supernatants of spent cultures. This section describes both procedures.

3.1.1. Extraction of Whole Materials or Supernatants

1. For whole materials, place material to be extracted in fresh DCM and shake (*see* **Note 4**).
2. Transfer the material to fresh DCM and repeat for a total of three extractions (skip to **step 6**).
3. For the extraction of spent supernatants of microorganisms, remove the cells by centrifugation and decanting off the supernatant (*see* **Note 5**).
4. Add the DCM in a 2:1 ratio to the supernatant in a separatory funnel, shake well, and let the mixture separate into two phases.
5. Collect the lower solvent fraction and add fresh DCM to the aqueous phase, shake, and reseparate. Repeat for a total of three extractions.
6. Combine the solvent fractions from the three extractions.
7. Evaporate off the solvent using a rotary evaporator.
8. When dry, resuspend the extract in a small (e.g., 1 mL) volume of a suitable organic solvent such as ethyl acetate or DCM and transfer to a clean vial that has been weighed.
9. Evaporate off the solvent, determine the amount of extracted material by reweighing the dried vial, and resuspend to a concentration of 5 mg/mL in ethanol or DMSO (*see* **Note 3**).

3.1.2. Minimum Inhibitory Concentration (MIC) Determination for Antagonists

Prior to testing of extracts, particularly for extracts of whole organisms, it is useful to determine whether the extracted material affects the growth of the monitor strains. This information can then be used to set the maximum non-inhibitory concentration to be used in the signal induction or inhibition assays.

1. Prepare an overnight culture of the assay strain, *E. coli* MT102 pJBA132 (*luxR*-P_{luxI}-RBSII-*gfp*(ASV)-T_0-T_1) (*18*) by growing the organism in 2 mL of ABt medium, with 6 µg/mL tetracycline at 37°C with shaking (200 rpm).
2. The following morning, add 100 µL of fresh medium to the wells of a 96-well microtiter plate. To the wells of the top row, add 200 µL total.
3. Add the extracted material to the first row to a maximum of 200 µg/mL.
4. Perform doubling dilutions of the material from the top row to the second to last row, leaving the bottom row without extract as a control row.
5. Dilute the overnight culture 1:100 into fresh medium, with or without antibiotic, respectively, and aliquot 100 µL into the wells of the 96-well microtiter plate containing the extracts for testing (*see* **Note 6**).
6. Incubate at 37°C and take regular readings for up to 24 h to determine growth yield and to calculate the MIC values.

3.1.3. AHL Induction/Inhibition Detection

For testing for the presence of inducers or repressors of AHL-based quorum sensing, the same *E. coli* monitor strain can be used. To detect inducers, one can use directly the monitor strain, which does not produce its own AHL signal. Whereas this monitor strain works well for most AHL molecules, it is not sensitive to C4-AHLs and another strain must be used, and the method for testing for the C4-AHLs is described elsewhere *(19)*. To test for inhibition of signaling, one must use the monitor strain in the presence of both the test substance and exogenously added AHL, which in this case is the 3-oxo-6C-AHL.

1. Prepare an overnight culture of the assay strain, *E. coli* MT102 pJBA132 (*luxR*-P_{luxI}-RBSII-*gfp*(ASV)-T_0-T_1) *(18)*, by growing the organism in 2 mL of ABt medium, with 6 µg/mL tetracycline at 37°C with shaking (200 rpm).

2a. To test for induction of the AHL-mediated quorum-sensing monitor, the following morning, add 100 µL of fresh ABt medium to the wells of a 96-well microtiter plate. To the wells of the top row, add 200 µL total.

2b. To test for the presence of AHL inhibitors in the extract, the following morning, add 100 µL of fresh ABt medium with 3-oxo-C6-AHL added to a final concentration of 20 nM to the wells of a 96-well microtiter plate. To the wells of the top row, add 200 µL total volume.

3. Add the extracted material to the first row with the maximum concentration being equal to twice the MIC as determined above or 200 µg/mL, which ever is lower.

4a. When testing for the presence of AHL inducers, perform doubling dilutions of the material from the top row to the second to last row, leaving the bottom row without extract as a control row. To the last row, add 3-oxo-C6-AHL added to a final concentration of 20 nM as a positive control for induction.

4b. When testing for the presence of AHL inhibitors, perform doubling dilutions of the material from the top row to the second to last row, leaving the bottom row without extract as a control row. This last row will serve as a control for maximum induction in the absence of inhibitor.

5. Dilute the overnight culture 1:100 into fresh medium, with or without antibiotic, respectively, and aliquot 100 µL into the wells of the 96-well microtiter plate containing the extracts for testing (*see* **Note 6**).

6. Incubate at 37°C and take regular readings for up to 24 h to determine growth yield (optical density at 600 nm—OD_{600}) and quorum-sensing-regulated Gfp expression (excitation 475 nm, emission 515 nm) to calculate quorum-sensing induction and inhibition.

7. Normalize the Gfp expression by dividing the Gfp expression by the OD_{600} to correct for any differences in cell number (*see* **Note 7**). The inhibition can also be visualized using fluorescence microscopy, as shown in **Fig. 1**.

3.2. Inhibition of AHL Quorum-Sensing Phenotypes

To verify that a particular extract or compound is active as an inducer or inhibitor of AHL-based quorum sensing in a relevant pathogen, it is important

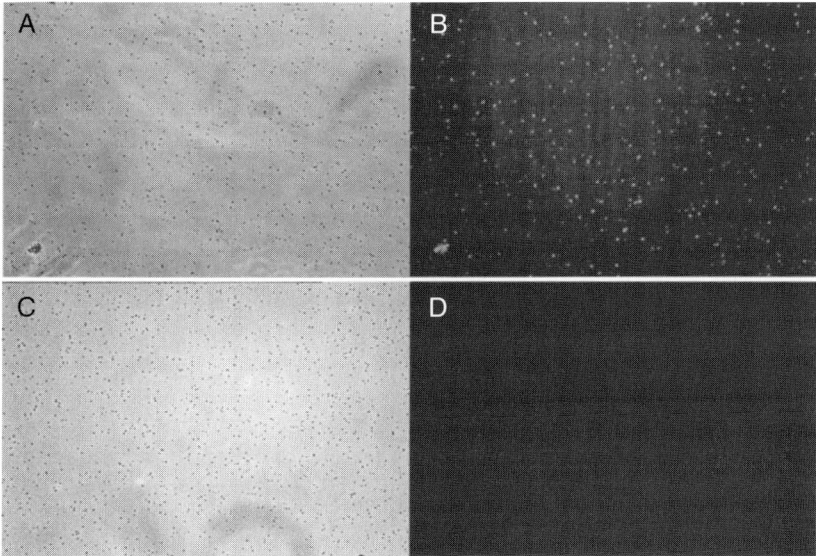

Fig. 1. Quorum-sensing-mediated expression of green fluorescent protein. Light (**A** and **C**) and fluorescent (**B** and **D**) images of the *gfp*-quorum-sensing reporter under inducing conditions (**A** and **B**) or non-inducing conditions (**C** and **D**). Magnification ×400.

to validate the bioassay results with a phenotype relevant to the particular pathogen. For this purpose, the expression of elastase, the product of the gene *lasB* of *P. aeruginosa*, will be exemplified. This protease has been selected because elastase expression has been demonstrated to be regulated by quorum sensing in *P. aeruginosa* and the assay is simple to perform. The assay can be performed using either a wild-type strain of *P. aeruginosa* or a quorum sensing mutant strain that does not produce its own signal [e.g., a *lasI/rhlI* mutant strain *(20, 21)*]. Presented below is a protocol using a wild-type strain, which is optimum for testing for inhibitors because the elastase production is expressed from native promoters and by endogenously produced signal.

3.2.1. MIC Determination

1. Prepare an overnight culture of *P. aeruginosa* by growing the organism in 2 mL of ABt medium at 37°C with shaking (200 rpm).
2. The following morning, add 100 µL of fresh medium to the wells of a 96-well microtiter plate. To the wells of the top row, add 200 µL total volume.
3. Add the extracted material to the first row to a maximum of 200 µg/mL.
4. Perform doubling dilutions of the material from the top row to the second to last row, leaving the bottom row without extract as a control row.

5. Dilute the overnight culture 1:100 into fresh medium and aliquot 100 µL into the wells of the 96-well microtiter plate containing the extracts for testing.
6. Incubate at 37°C and take regular readings for up to 24 h to determine growth yield and to calculate the MIC values.

3.2.2. The Induction or Inhibition of Elastase Activity in

P. aeruginosa *Performed as Described in ref. (22) with Some Modifications*

1. Grow the *P. aeruginosa* in ABt at 37°C with shaking (200 rpm).
2. The following morning, dilute the culture 1:100 into fresh medium, either add extract at half the MIC dose, as determined above or 200 µg/mL (*see* **Note 8**), or add an equal volume of solvent to the control culture and incubate at 37°C with shaking (200 rpm).
3. At hourly intervals, collect 1.5 mL of supernatant and also determine the OD_{600}. Filter out the cells from the 1.5-mL aliquot using a 0.2-µm filter (Pall Acrodisc, 0.2-µm Supor Membrane, Pall Life Sciences, East Hills, New York, USA) and store at –20°C (*see* **Note 9**).
4. Once samples have been collected, thaw on ice. Whilst thawing, prepare tubes with 1 mL of Elastin-Congo red buffer and 20 mg of Elastin-Congo red.
5. Add 100 µL of sterile supernatant to the Elastin-Congo red tubes and incubate at 37°C for 18 h with agitation.
6. Stop the reaction by the addition of EDTA to a final concentration of 12 mM. Centrifuge at 16,000 × g to remove insoluble Elastin-Congo red.
7. Collect the supernatant and measure the OD_{495}. A blank, with Elastin-Congo red and buffer, but without sample, should be used to subtract out the background. Results should be normalized (*see* **Note 10**) to correct for differences in cell density. **Fig. 2** presents representative data on elastase repression for a QSI at two concentrations.

3.3. Detection of AI-2 Activity or Inhibition

AI-2 activity is commonly detected by the use of AI-2 bioassay developed over a decade ago. *Vibrio harveyi* BB170 *(23)* (sensor 1$^-$, sensor 2$^+$) has a null mutation in the gene for sensor 1 and thus is a bioluminescent reporter strain for AI-2 activity. *V. harveyi* BB152 *(23)* (autoinducer 1$^-$, autoinducer 2$^+$) has a null mutation in the gene for autoinducer 1 synthase (AI-1) and produces only AI-2, and thus serves as a positive control for assays of AI-2 production. For the generation of cell-free supernatants, *V. harveyi* strains were grown in AB medium *(16)* overnight at 30°C with shaking at 200 rpm and cell-free supernatants prepared as below.

3.3.1. Collection of AI-2 Signal

1. For generation of cell-free supernatants, test cultures are inoculated at a 1:100 dilution and the cultures grown to late stationary phase (*see* **Notes 11** and **12**).

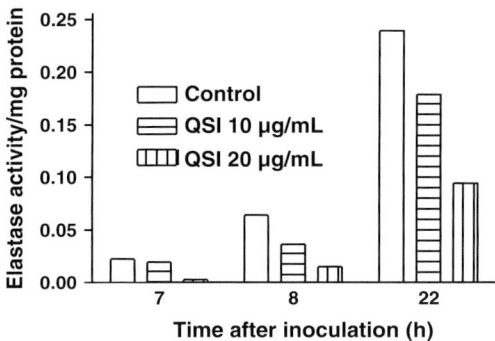

Fig. 2. Elastase activity of *Pseudomonas aeruginosa* in the presence and absence of a quorum-sensing inhibitor (QSI). Samples were collected at 7, 8, and 22 h post inoculation in the presence of a quorum-sensing antagonist at 10 and 20 µg/mL. Elastase activity was determined on cell-free supernatants from the different time points and was normalized based on total protein concentration of the supernatants.

Growth is followed spectrophotometrically and aliquots removed at various time points for the preparation of cell-free supernatants.

2. Cell-free supernatants are prepared by centrifugation of culture aliquots for 5 min at 6000 × *g* followed by filtration of the supernatant through a 0.22-µm filter (Pall acrodisc supor syringe filter) and collection of the supernatant in a sterile 1.5-mL Eppendorf tube.

3. Cell-free supernatants can be frozen and stored at –20°C until used.

3.3.2. Detection of AI-2 Activity/Inhibition

To test for the presence of AI-2 activity, grow the *V. harveyi* reporter strain BB170 for 16 h at 30°C with shaking in AB medium.

1. Dilute cells 1:5000 into 30°C pre-warmed AB medium.
2. Prepare microtiter plates by adding 90 µL of the diluted suspension to wells of 96-well microtiter plates and then add 10 µL of either medium (for negative controls) or supernatants to be tested.
3. Incubate the microtiter plates at 30°C with shaking at 175 rpm. Hourly determinations of the total luminescence as well as cell density are quantified.
4. Report activity as the percentage of activity obtained from *V. harveyi* BB152 cell-free supernatant.
5. To test for inhibition of AI-2-mediated cell–cell signaling, add the diluted monitor strain to the plates as indicated in **steps 1–3**, where the supernatant added is from the positive control strain, *V. harveyi* BB152. Add the antagonist from the stock solution, as extracted in **Subheading 3.1.1**. Again, it is advisable to first perform MIC determinations of the inhibitor extract as described in

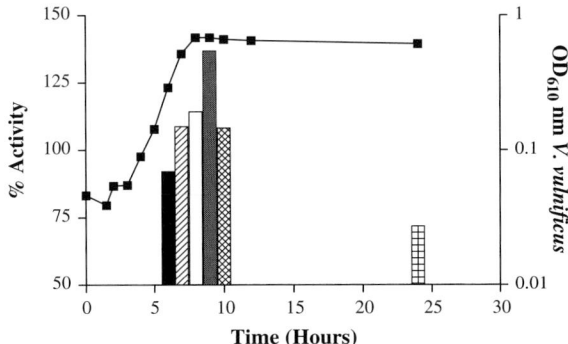

Fig. 3. Production of autoinducer 2 (AI-2) signal by *Vibrio vulnificus* during different phases of growth. Supernatant samples from *V. vulnificus* were collected at different time points, and the percent AI-2 activity compared with that of positive control was determined (left axis). The growth of the culture, as determined by optical density at 610 nm (OD_{610}), is plotted on the right axis and is used to show that AI-2 production peaks during early stationary phase and decreases after 24 h.

Subheading 3.1.2 to avoid effects on growth. Representative data are shown in **Fig. 3**, which show the timing of AI-2 production during the growth of *V. vulnificus*.

3.4. Inhibition of Hemolysin Assay with Signal Antagonist

To test for inhibition of hemolysin activity by signal antagonists, routinely grow the test cultures overnight. The next morning, 10 mL of fresh medium is inoculated with a 1:100 dilution of the overnight culture in a sidearm flask and incubated at 37°C with shaking (200 rpm) and growth measured at OD_{610}. Antagonists are added to the cultures during log phase (OD_{610} is approximately 0.3). DMSO is added to control samples to account for the solvent volume added, and growth should be followed spectrophotometrically to ensure the concentrations of antagonist compounds used are not growth inhibitory. Cell-free supernatants should be collected across the growth curve.

1. Wash 4 mL of sheep blood three times with five volumes (20 mL) of PBS (pH 7.2). Centrifuge at $1140 \times g$ for 5 min and resuspend in five volumes (20 mL) of PBS.
2. Incubate aliquots of cell-free supernatants (200 µL), 600 µL of hemolysin buffer, and 200 µL washed red blood cells at 37°C for 30 min.
3. Remove intact red blood cells by centrifugation and determine the OD_{450} of the supernatant. Results for the effects of an AI-2 inhibitor are shown in **Fig. 4**.

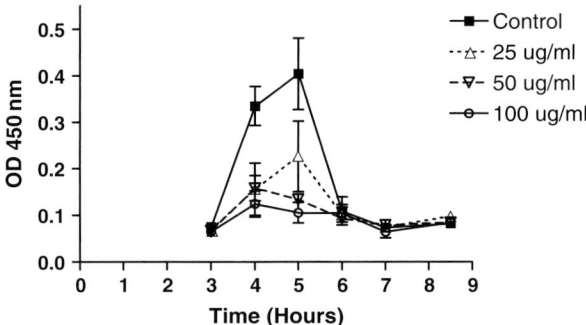

Fig. 4. Inhibition of the autoinducer 2 (AI-2)-mediated production of hemolysin in *Vibrio vulnificus*. Hemolysin production was monitored using sheep red blood cells in the absence, control, or presence of an AI-2 inhibitor at three concentrations. Samples were taken at different times during the growth phase. The inhibitor was used at non-growth inhibitory concentrations.

4. Notes

1. ABt medium can be stored at room temperature, but should not be more than 1 month old when using.
2. It is important to use high-quality solvents or to distill them because many commercial grade solvents contain plasticizers that are concentrated during the extraction procedure. These concentrated compounds can be toxic to the indicator strains or can interfere with the interpretation of results.
3. It is preferable to use a solvent that is miscible in aqueous media, such as ethanol or DMSO for the bioassay testing. However, if the extract does not dissolve well in ethanol, it is possible to evaporate off the ethanol and resuspend the sample in an organic solvent and then dilute into ethanol or DMSO prior to adding to the bioassay medium. DMSO is not a volatile solvent and hence if the sample does not dissolve in DMSO, it is not easy to remove the DMSO to try another solvent.
4. Some materials, such as plants, can lyse in organic solvents. This is only an issue if the experimental design intends to separate surface active or presented compounds from intracellular compounds. If the former is the case, then different organic solvents and extraction times should be tested to identify conditions that do not lyse the tissues.
5. It is best to grow the organisms in defined minimal medium because some components of complex media interfere with the AHL assays and can give false-positive results *(24)*. Typically, 100 mL of culture supernatant is sufficient for detection of signals or antagonists.
6. At this point, the maximum concentration has been halved due to the addition of the cells.

7. This method will not identify the chemical species that induces or inhibits the quorum-sensing monitor strain. There are other methods that can aid in this, such as thin layer chromatography or gas chromatography–mass spectroscopy methods (described in **ref.***(19)*).
8. The timing of addition relative to sampling is partly dependent on the stability of the signal or antagonist in the presence of bacteria. Therefore, if no effect is observed with a single dose, it may be necessary to repeat the dose during growth or to time the addition of the extract more closely with the onset of elastase production.
9. Filtration typically results in the loss of some sample volume, so a slight excess is collected.
10. Normalization for cell density can be done by dividing the OD_{495} by either the OD_{600} or by protein concentration.
11. It is important to note that 5 g/L of glucose has been shown to inhibit bioluminescence in *V. harveyi (25)*; thus care must be taken to ensure the growth medium used for the test strains does not contain high levels glucose. The cell-free supernatant can be made glucose depleted by growing *V. harveyi* strain MM30 (AI-2$^-$) in the supernatant prior to testing *(26)*. By contrast, low levels of glucose may actually increase bioluminescence and lead to false positives.
12. It is important that the cell-free supernatants be neutral in pH as acidity can interfere with the assay.

Acknowledgments

This work has been supported by the National Health and Medical Research Council of Australia and funding from Biosignal Ltd, Australia.

References

1. Fuqua, C. and Greenberg, E. P. (2002) Listening in on bacteria: acyl-homoserine lactone signalling. *Nat. Rev. Mol. Cell Biol.* **3,** 685–695.
2. Venturi, V. (2006) Regulation of quorum sensing in *Pseudomonas. FEMS Microbiol. Rev.* **30,** 274–291.
3. Davies, D. G., Parsek, M. R., Pearson, J. P., Iglewski, B. H., Costerton, J. W., and Greenberg, E. P. (1998) The involvement of cell-cell signals in the development of a bacterial biofilm. *Science* **280,** 295–298.
4. Wu, H., Song, Z., Hentzer, M., Andersen, J. B., Molin, S., Givskov, M., and Hoiby, N. (2004) Synthetic furanones inhibit quorum-sensing and enhance bacterial clearance in *Pseudomonas aeruginosa* lung infection in mice. *J. Antimicrob. Chemother.* **53,** 1054–1061.
5. Zhu, H., Bandara, R., Conibear, T., Thuruthyil, S. J., Rice, S. A., Kjelleberg, S., Givskov, M., and Willcox, M. D. P. (2004) *Pseudomonas aeruginosa* with *lasI* quorum-sensing deficiency is avirulent during corneal infection. *Invest. Ophthalmol. Vis. Sci.* **45,** 1897–1903.

6. Huber, B., Riedel, K., Hentzer, M., Heydorn, A., Gotschlich, A., Givskov, M., Molin, S., and Eberl, L. (2001) The *cep* quorum-sensing system of *Burkholderia cepacia* H111 controls biofilm formation and swarming motility. *Microbiology* **147,** 2517–2528.

7. Labbate, M., Queck, S. Y., Koh, K. S., Rice, S. A., Givskov, M., and Kjelleberg, S. (2004) Quorum sensing controlled biofilm development in *Serratia liquefaciens* MG1. *J. Bacteriol.* **186,** 692–698.

8. Riedel, K., Ohnesorg, T., Krogfelt, K. A., Hansen, T. S., Omori, K., Givskov, M., and Eberl, L. (2001) N-Acyl-L-homoserine lactone-mediated regulation of the Lip secretion system in *Serratia liquefaciens* MG1. *J. Bacteriol.* **183,** 1805–1809.

9. McDougald, D., Rice, S. A., and Kjelleberg, S. (2001) SmcR-dependent regulation of adaptive phenotypes in *Vibrio vulnificus. J. Bacteriol.* **183,** 758–762.

10. Sperandio, V., Torres, A. G., Giron, J. A., and Kaper, J. B. (2001) Quorum sensing is a global regulatory mechanism in enterohemorrhagic *Escherichia coli* O157:H7. *J. Bacteriol.* **183,** 5187–5197.

11. Federle, M. J. and Bassler, B. L. (2003) Interspecies communication in bacteria. *J. Clin. Invest.* **112,** 1291–1299.

12. Linkous, D. A. and Oliver, J. D. (1999) Pathogenesis of *Vibrio vulnificus. FEMS Microbiol. Lett.* **174,** 207–214.

13. Campos, L. C., Zahner, V., Avelar, K. E. S., Alves, R. M., Pereira, D. S. G., Vital Brazil, J. M., Freitas, F. S., Salles, C. A., and Karaolis, D. K. R. (2004) Genetic diversity and antibiotic resistance of clinical and environmental *Vibrio cholerae* suggests that many serogroups are reservoirs of resistance. *Epidemiol. Infect.* **132,** 985–992.

14. Schmidt, F. R. (2004) The challenge of multidrug resistance: actual strategies in the development of novel antibacterials. *Appl. Microbiol. Biotechnol.* **63,** 335–343.

15. Clark, D. J. and Maaloe, O. (1967) DNA replication and the division cycle in *Escherichia coli. J. Mol. Biol.* **23,** 99–112.

16. Greenberg, E. P., Hastings, J. W., and Ulitzer, S. (1979) Induction of luciferase synthesis in *Beneckea harveyi* by other marine bacteria. *Arch. Microbiol.* **120,** 87–91.

17. Bertani, G. (1951) Studies on lysogenesis. *J. Bacteriol.* **62,** 293–300.

18. Andersen, J. B., Heydorn, A., Hentzer, M., Eberl, L., Geisenberger, O., Christensen, B. B., Molin, S., and Givskov, M. (2001) *gfp*-based *N*-acyl homoserine lactone sensor systems for detection of bacterial communication. *Appl. Environ. Microbiol.* **67,** 575–585.

19. Rice, S. A., Kjelleberg, S., Givskov, M., De Boer, W., and Chernin, L. (2004) In situ detection of bacterial quorum sensing signal molecules, *in Molecular Microbial Ecology Manual* (Kowalchuck, G. A., Ed.), Dordrecht/Boston/London, pp. 1629–1650, Kluwer Academic Publishers.

20. Beatson, S. A., Whitchurch, C. B., Semmler, A. B. T., and Mattick, J. S. (2002) Quorum sensing is not required for twitching motility in *Pseudomonas aeruginosa. J. Bacteriol.* **184,** 3598–3604.

21. Pesci, E. C., Pearson, J. P., Seed, P. C., and Iglewski, B. H. (1997) Regulation of *las* and *rhl* quorum sensing in *Pseudomonas aeruginosa. J. Bacteriol.* **179,** 3127–3132.

22. Pearson, J. P., Pesci, E. C., and Iglewski, B. (1997) Roles of *Pseudomonas aeruginosa las* and *rhl* quorum-sensing systems in control of elastase and rhamnolipid biosynthesis genes. *J. Bacteriol.* **179,** 5756–5767.

23. Bassler, B. and Silverman, M. R. (Eds.) (1995) Two component Signal Transduction. Intercellular communication in marine *Vibrio* species: density-dependent regulation of the expression of bioluminescence, ASM Press, Washington, DC, pp. 431–444.

24. Holden, M. T. G., Chahabra, S. R., deNys, R., Stead, P., Bainton, N. J., Hill, P. J., Manefield, M., Kumar, N., Labbate, M., England, D., Rice, S. A., Givskov, M., Salmond, G., Stewart, G. S. A. B., Bycroft, B. W., Kjelleberg, S., and Williams, P. (1999) Quorum sensing cross talk: isolation and chemical characterisation of cyclic dipeptides from *Pseudomonas aeruginosa* and other Gram-negative bacteria. *Mol. Microbiol.* **33,** 1254–1266.

25. Nealson, K. H., Eberhard, A., and Hastings, J. W. (1972) Catabolite repression of bacterial bioluminescence: functional implications. *Proc. Natl. Acad. Sci. USA* **69,** 1073–1076.

26. Turovskiy, Y. and Chikindas, M. L. (2006) Autoinducer-2 bioassay is a qualitative, not quantitative method influenced by glucose. *J. Microbiol. Methods* **66,** 497–503.

6

A System for Site-Specific Genetic Manipulation of the Relapsing Fever Spirochete *Borrelia hermsii*

James M. Battisti, Sandra J. Raffel, and Tom G. Schwan

Summary

The lack of a system for genetic manipulation has hindered studies on the molecular pathogenesis of relapsing fever *Borrelia*. The focus of this chapter is to describe selectable markers, manipulation strategies, and methods to electro-transform and clone wild-type infectious *Borrelia hermsii*. Preliminary studies suggest that the variable tick protein (Vtp) of *B. hermsii* is involved in tick-to-mammal transmission. To address this hypothesis, we have developed a system for genetic manipulation and have constructed clones of a Vtp mutant and an isogenic reconstituted strain. The methods described here are applicable for the inactivation of other loci in *B. hermsii* and should be adaptable for other species of relapsing fever spirochetes.

Key Words: *Borrelia hermsii*; tick-borne relapsing fever; genetic manipulation; transformation; mutagenesis; variable tick protein (Vtp).

1. Introduction

The most rigorous method to assess the contribution of a specific gene product in the virulence of a particular pathogen is by testing molecular Koch's postulates *(1)*, wherein isogenic strains (wild type, mutant, reconstituted mutant) are constructed and compared in experimental animals. Relapsing fever is a recurrent febrile illness caused by spirochetes in the genus *Borrelia*. Tick-borne relapsing fever (TBRF) in the USA is caused primarily by *Borrelia hermsii* and *Borrelia turicatae*, which are maintained in enzootic cycles between mammals and *Ornithodoros* ticks *(2)*. Outbreaks of TBRF are generally localized and sporadic. Louse-borne relapsing fever (LBRF) is caused by *Borrelia recurrentis*,

From: *Methods in Molecular Biology, vol. 431: Bacterial Pathogenesis*
Edited by: F. DeLeo and M. Otto © Humana Press, Totowa, NJ

which is transmitted among humans by their body lice (*Pediculus humanus corporis*). LBRF often occurs in epidemics where crowded and unsanitary conditions promote body louse infestations and their spread among people *(3,4)*.

The variable tick protein (Vtp = Vsp33) is a major outer-surface lipoprotein produced by *B. hermsii* during its persistent infection in tick salivary glands *(5,6)*. A number of studies suggest that Vtp, and its ortholog OspC in the Lyme disease spirochetes *Borrelia burgdorferi* sensu lato, are involved in tick transmission of the spirochetes or initial infection of mammals *(6–10)*. To study the role of Vtp in the transmission of *B. hermsii,* we developed a system for site-specific mutagenesis and genetic reconstitution. This chapter describes methods for the construction of isogenic strains of *B. hermsii* in an infectious background. The example used is the inactivation of the *vtp* gene; however, the method has been applied to other loci and the techniques described should be applicable to any gene of interest.

2. Materials
2.1. Spirochete Culture and Transformation Components

1. Modified Barbour–Stoenner–Kelly medium (mBSK) (*see* **Note 1**).
2. Electroporation Solution (EPS): 0.27 *M* sucrose (Fisher, Fair Lawn, NJ, Cat. BP220-1), 15% (v/v) glycerol (MP Biomedicals, Solon, OH, Cat. 800688) solution in dH$_2$O. Filter-sterilize and store at 4°C.
3. Kanamycin sulfate (1 mg/mL, Sigma-Aldrich, St. Louis, MO, Cat. K-1377) and gentamicin sulfate (200 µg/mL, Sigma, Cat. G3632). Prepare in dH$_2$O and sterilize by passage through a 0.22-µm filter. Store at 4°C.
4. Gene Pulser (BioRad, Hercules, CA) or other electroporation device capable of delivering a pulse of 2.5 kV, 25 µF, and 200 Ω.
5. 0.2-cm electroporation cuvettes (Harvard Apparatus, Holliston, MA, Cat. 45-0125).
6. 3 *M* sodium acetate (Fisher, Cat. BP334-1), adjust pH to 5.0 with HCl.
7. 100% ethanol (Aaper, Shelbyville, KY).
8. Reverse-osmosis/deionized water (resistivity ~18.0 MΩ/cm). This water is used for all solutions and media, and will be referred to simply as dH$_2$O in this text.
9. 25% glycerol (MP Biomedicals) (v/v) in dH$_2$O. Sterilize by autoclave.
10. 96-well round-bottom tissue culture plates (Nunc, Rochester, NY).

2.2. Nucleic Acids

2.2.1. Genomic Sequence (see Note 2)

1. Endogenous promoters: *B. hermsii* flagellin protein (*flaB-P*) and flagellar rod protein (*flgB-P*).
2. Target sequence: variable tick protein gene (*vtp*) described here, but any gene of interest can be targeted.
3. Autonomous replication sequences (ARS).

2.2.2. Plasmids (see Note 3, GenBank Accession Numbers in Bold)

1. pTABhFlgB-Kan (source of kanamycin selectable cassette) **(EF488744)**.
2. pTABhFlaB-Gent (source of gentamicin selectable cassette) **(EF488745)**.
3. pBhSV-2 (*B. hermsii* shuttle vector) **(EF488742)**.
4. pOK12 (plasmid backbone for construction of inactivation and reconstitution plasmids) *(11)* **(EF488748)**.
5. pOKvtp (plasmid containing the 5193-bp *vtp* locus, kan-r) **(EF488743)**.
6. pOKvtpKO (*vtp* inactivation plasmid, kan-r) **(EF488747)**.
7. pOKvtpRECON (*vtp* reconstitution plasmid, gent-r) **(EF488746)**.

2.2.3. Primers (Listed 5′–3′, Restriction Endonuclease Tags Are Underlined, Invitrogen, Carlsbad, CA)

1. 5′ BhflgB+NgoMIV ATGCCGGCGTTAAAGAAAATTGAAATAAACTTG.
2. 3′ BhflgB+NdeI ATTCATATGAACACCCTCTATATCAC.
3. 5′ BhflaB+NgoMIV TAGCCGGCAATTCCTAATCAGAAAAATGTG.
4. 3′ BhflaB+NdeI TCATATGTCATTTCCTCCGTG.
5. VTP F2+AvrII AATCCTAGGGCAAATACTATGCTCTTGATGG.
6. VTP B3+AscI ATTGGCGCGCCTCTTTAATGCCTTATCAC.
7. pOK12F1+AscI TGGCGCGCCTCGCCCTTCCCAACAGTTG.
8. pOK12B1+AvrII ACCTAGGGCGTATTGGAGCTTTCGCG.
9. vtpF1+ScaI AAAGTACTTAACCGCTGCTAAAGATGCAGTAG.
10. vtpB1+SpeI AAACTAGTCACCCTTTTTAAGCTCTGGGC.
11. ReconF+XmaI GTGCCCGGGGATTATAAGATTTAACAC.
12. ReconR+SpeI TTACTAGTAATATCCTAAGTCGATGAACAG.
13. Gent5′+XmaI ATTCGCCCGGGCCGGCAATTCCTAATCAGAAAAATGTGG.
14. Gent3′+SpeI AAACTAGTCTCGGCTTGAACGAATTGTTAGG.
15. BhOriF+NotI AATGCGGCCGCTAAGTCCAAAAGCGTCCC.
16. BhOriR+SphI ATTGCATGCTAAAATCTTCTTGCCCGC.
17. LONG VTP F1 AGAGAGATATTTAGAATTAAAAGAAAAGC.
18. LONG VTP B56 GTCAAGAAGAGCATTTGTCACAC.

2.3. Nucleic Acid Cloning Materials

1. GeneAmp PCR Core Kit (Roche Molecular Systems, Branchburg, NJ).
2. Expand Long Template PCR System Kit (Roche Diagnostics, Indianapolis, IN) (*see* **Note 4**).
3. TOPO TA Cloning Kit (Invitrogen).
4. QIAquick PCR Purification Kit and the QIAquick Gel Extraction Kit (Qiagen, Valencia, CA).
5. QIAprep Spin and Plasmid Maxi Kits (Qiagen).
6. Restriction enzymes and T4 DNA ligase (New England Biolabs, Beverly, MA).
7. BlueView Nucleic Acid Stain (Sigma-Aldrich) (*see* **Note 5**).
8. QIAtissue Kit (Qiagen).

9. OneShot Top10 *Escherichia coli* (Invitrogen).
10. Luria-Bertani (LB) Medium and a 37°C incubator for culture of *E. coli*.

3. Methods

The following describes the construction of genetic manipulation plasmids and methods for electro-transformation that were utilized to generate a *B. hermsii vtp* mutant and a reconstituted isogenic strain. This system can be customized for mutagenesis and reconstitution of other genes by designing endonuclease-tagged primers specific to the particular target locus. In silico construction of plasmids and confirmation of the restriction profile by restriction fragment length polymorphism (RFLP) are recommended prior to synthesis of the primers.

Complementation refers to a genetic manipulation event wherein a wild-type copy of the mutant gene is introduced, leaving the original mutation intact. This can be achieved by inserting a wild-type copy of the mutant gene (and its promoter region) into the *B. hermsii* shuttle vector pBhSV-2. When the shuttle vector is used for restoration of a mutant genotype, this autonomously replicating plasmid may artificially increase the relative number of gene copies and could result in abnormal regulation and synthesis of the gene of interest. Alternatively, genetic reconstitution involves the replacement of the mutagenized gene with wild-type copy at the wild-type locus.

3.1. Construction of Genetic Manipulation Plasmids

3.1.1. Cloning of the Target Locus into Vector pOK12

1. Identify specific target gene of interest within the genomic sequence.
2. Design primers to amplify the pOK12 amplicon containing the p15A origin + *kan-r* cassette and the target gene with 900–2000 bp of flanking DNA on each side. The primers should include restriction endonuclease sites (tags) that are not found in the target locus or within pOK12. Choose restriction enzymes that have a compatible buffering system (*see* **Note 6** for primer design and construction details relevant to pOKvtp, and **Fig. 1** for an illustration of this plasmid).
3. Prepare genomic DNA (*B. hermsii* DAH 2E7 in this example) using the QIAtissue kit (Qiagen) per the manufacturer's instructions. Other methods of preparing genomic DNA are suitable. As differences in the genomic sequence between strains is likely, it is most prudent to prepare genomic DNA from the spirochetal strain that is intended to be the host strain for genetic manipulation.
4. Amplify the target locus and the pOK12 amplicon using Expand Long Template PCR System Kit (Roche Diagnostics) per the manufacturer's instructions and analyze products by agarose gel electrophoresis for correct size (*see* **Note 7**). Adjust the annealing temperature and extension time to reduce non-specific background products that could interfere with the cloning.

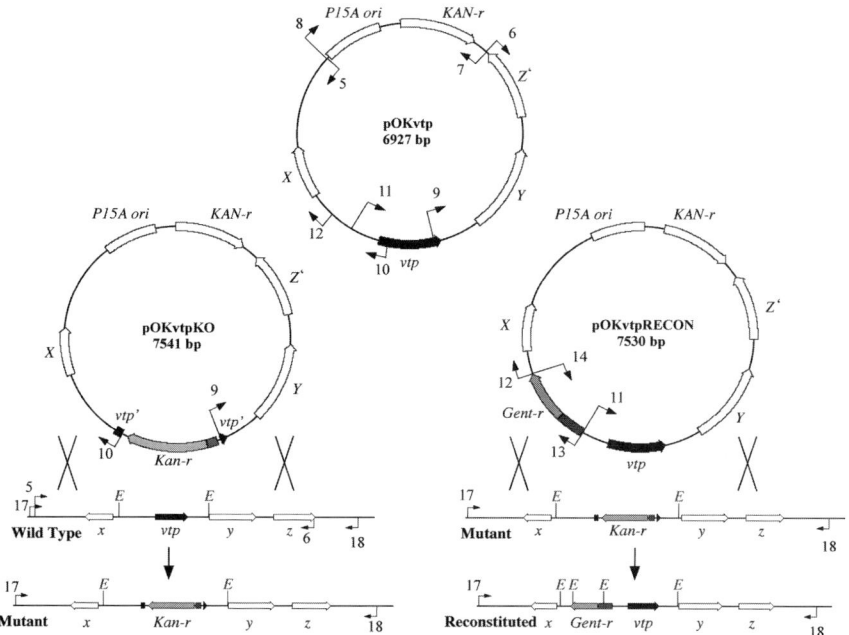

Fig. 1. The *target locus* plasmid, inactivation and reconstitution constructs, and genomic structures of developed strains. pOKvtp is the *target locus* plasmid and was generated prior to the inactivation and reconstitution constructs. Primer-binding sites, indicated by small arrows, are numbered corresponding to the list in **Subheading 2.2.2**. BLAST *(13)* analysis with amino acid sequences predicted by open reading frame (ORF) Finder (NCBI) shows that truncated *orfZ′* is most similar to *Borrelia burgdorferi* thymidylate synthase (67% identical, accession NC_001857.1), where *orfX* and *orfY* encode conserved hypothetical proteins. The amplicon generated with primers 9 and 10 from pOKvtp was fused to the *SpeI/EcoRV* fragment of pTABhFlgB-Kan, resulting in pOKvtpKO, the *vtp* inactivation construct. Electro-transformation of pOKvtpKO resulted in a double-crossover event that mutagenized *vtp* without affecting the flanking ORFs of this locus. The structure of the wild-type and mutant loci are illustrated and were verified by sequencing both strands of the amplicon generated from primers (17 and 18) flanking the cloned target locus. *EcoRV* restriction sites are indicated (*E*) and correspond to the restriction fragment length polymorphism (RFLP) shown in **Fig. 3**. Fusion of amplicons derived from pOKvtp (using primers 11 and 12) and pTABhFlaB-Gent (using primers 13 and 14) resulted in the genetic reconstitution construct, pOKvt-pRECON. Electro-transformation of the *vtp* mutant with pOKvtpRECON resulted in a double homologous recombination event that replaced the mutated *vtp* with a wild-type copy and simultaneously inserted the *PflaB::Gent-r* cassette at an intergenic location just upstream from *vtp*. The structure of the reconstituted locus is illustrated and was verified by sequencing both strands of the amplicon generated from primers (17 and 18) flanking the cloned target locus. Details relevant to the construction of these plasmids are provided in the text (*see* **Notes 6, 8**, and **9**).

5. Clean the PCR reaction using the QIAquick PCR Purification Kit (Qiagen) to remove primers and polymerase, and prepare the DNA fragment for endonuclease digestion.

6. Analyze part of the PCR reaction product by RFLP (or sequencing) (*see* **Note 7**) and include restriction enzymes that are used in the primer tags to ensure that there are no internal sites.

7. Digest the fragment with restriction endonucleases corresponding to the sites used in the primer tags.

8. Clean the reaction product to remove the enzymes used for the restriction digests with the QIAquick PCR Purification Kit (Qiagen). Gel purification of the fragment following digestion is not necessary if only one band is produced. If multiple bands are produced, then gel purification of the fragment *is* required for a specific cloning strategy. Use BlueView stain rather than ethidium bromide (EtBr) for visualization and isolation of DNA fragments (*see* **Note 5**).

9. Ligate the target locus with T4 DNA ligase to the restricted and cleaned pOK12 amplicon overnight at 16°C.

10. Use a portion of the ligation to transform OneShot Top10 *E. coli* per the manufacturer's instructions (Invitrogen), and select clones on LB plates supplemented with 50 μg/mL kanamycin sulfate at 37°C.

11. Analyze kanamycin-resistant clones by PCR, RFLP, and sequence analysis for pOK target locus.

12. Store cultures indefinitely in 20% glycerol at –80°C, or for no more than a couple of weeks at 4°C. The plasmid containing the *vtp* target locus is termed pOKvtp (*see* **Note 6** and **Fig. 1.**)

3.1.2. Construction of Inactivation and Reconstitution Plasmids

1. Design inverted endonuclease-tagged primer sets specific to the pOK-*target locus* such that PCR will generate an amplicon of the entire pOK-*target locus* plasmid, with a *deletion* where the resistance cassette is intended to be placed. For the *inactivation construct*, a major intragenic portion of the target gene is deleted and replaced with the *flgB*-promoted kan cassette, such that a double-crossover homologous recombination event will result in allelic replacement of the target gene with the resistance cassette (i.e., *flgB-kan*) (*see* **Note 8** for construction details relevant to the *vtp* inactivation plasmid, pOKvtpKO, and **Fig. 1** for an illustration of this plasmid). For the *reconstitution construct*, this *deletion* should be intergenic, where a double-crossover homologous recombination event will result in the insertion of a resistance cassette (i.e., *flaB-gent*) at a nearby location (*see* **Note 9** for construction details relevant to the *vtp* reconstitution plasmid, pOKvtpRECON, and **Fig. 1** for an illustration of this plasmid). Construction of the inactivation and reconstitution plasmids is achieved by inverse-PCR-mediated cloning utilizing the pOK-*target locus* plasmid and the resistance cassettes in pTABhFlgB-Kan and pTABhFlaB-Gent. Primers should include restriction tags that are complementary to the restriction endonuclease sites flanking the particular

resistance cassette. The resistance cassette utilized in the inactivation plasmid must be different from the cassette used for the reconstitution plasmid.

2. Prepare pOK-*target locus* plasmid from *E. coli* using QIAprep Spin Kit (Qiagen), or another plasmid DNA isolation method.

3. Generate amplicons from the pOK-*target locus* plasmid with the inactivation or reconstitution endonuclease-tagged primer sets using the Expand Long Template PCR System Kit (Roche Diagnostics) per the manufacturer's instructions.

4. Analyze products by agarose gel electrophoresis (*see* **Note 7**). Adjust the annealing temperature and extension time if needed to reduce non-specific background products that could interfere with the cloning.

5. Analyze, clean, and digest the amplicons as described in **steps 5–8, Subheading 3.1.1**.

6. Ligate the amplicons to digested resistance cassettes with T4 DNA ligase at 16°C overnight, and transform into TOP10 *E. coli* as described in **step 10, Subheading 3.1.1**.

7. For cloning the Kan-r cassette, selection is accomplished on LB plates supplemented with 50 μg/mL kanamycin sulfate. Use gentamicin sulfate at a final concentration of 10 μg/mL for cloning the Gent-r cassette in *E. coli*. Analyze clones by PCR, RFLP, and sequence analysis.

3.2. Electro-Transformation (see Note 10)

3.2.1. Preparation of Electrocompetent B. hermsii DAH 2E7 (see Note 11).

1. Grow infectious *B. hermsii* in 500 mL mBSK-c at 34°C to a density of approximately $5–7 \times 10^7$ spirochetes/mL.

2. Wash the spirochetes three times in EPS (RT), decreasing the volume after each centrifugation. First centrifuge two 250-mL cultures at $11,000 \times g$ for 15 min at 4°C. Wash and resuspend pellets (gentle vortexing) with 50 mL EPS, centrifuge for 15 min, combine pellets, resuspend in 5 mL EPS, centrifuge for 15 min, and resuspend cells in 600 μL EPS. The purpose of these washes is to remove ions contained in mBSK-c that can cause arcing during electroporation.

3. Place electrocompetent cells on ice and electroporate immediately.

3.2.2. Preparation of Genetic Manipulation Plasmid

1. Grow a 125–500 mL culture of the *E. coli* strain harboring the plasmid of interest.

2. Purify plasmid DNA using the QIAprep Plasmid Maxi Kits (Qiagen) per the manufacturer's instructions.

3. Precipitate and wash plasmid DNA to concentrate and remove salts (*see* **Note 12**).

4. Resuspend the pellet in dH$_2$O to a final concentration of 5 μg/μL.

5. Place concentrated plasmid DNA on ice for use the same day. Alternatively, store at −20°C for several days.

3.2.3. Electroporation and Selection

1. Place the concentrated DNA, electrocompetent cells, and cuvettes on ice.
2. Add 100 μL of electrocompetent cells (approximately 5×10^9 spirochetes) and 5 μL (25 μg) of plasmid DNA in a microfuge tube, mix gently, and transfer to the bottom of a 0.2-cm electroporation cuvette. Avoid introducing air bubbles in the sample that can cause arcing. Gently tap the cuvette to ensure that the sample goes to the bottom.
3. Electroporate with a Gene Pulser (BioRad), or other electroporator capable of delivering a similar pulse (2.5 kV, 25 μF, and 200Ω) (*see* **Note 13**).
4. Immediately after delivering the pulse, transfer the contents of the cuvette into 5 mL mBSK-c and incubate overnight at 34°C.
5. Transfer the 5 mL overnight culture into 45 mL mBSK-c containing the appropriate antibiotic (200 μg/mL kanamycin or 40 μg/mL gentamicin).
6. Incubate the culture at 34°C and monitor daily by dark-field microscopy for the presence of motile spirochetes. Typically 3–7 days of cultivation is required. Once detected, cryopreserve a portion of the liquid culture by combining with an equal volume of 25% glycerol and freeze at –80°C.
7. Isolate individual clones of antibiotic resistant spirochetes by limiting dilution in 96-well tissue culture plates of mBSK-c (supplemented with the appropriate antibiotic) (*see* **Note 14**).
8. Inoculate clones in mBSK-c (supplemented with the appropriate antibiotic), grow to mid-exponential phase, and cryopreserve at –80°C by combining with an equal volume of 25% glycerol.

4. Notes

1. The mBSK-c formulation is based on fortified Kelly's media *(12)*. The following recipe is for a 2× concentrate: 10× CMRL-1066 (US Biological, Swampscott, MA, C5900-05) 19.4 g/L; *N*-acetylglucosamine (Sigma A-8625) 0.8 g/L; sodium citrate (Sigma C-7254) 1.4 g/L; sodium pyruvate (Sigma P-5280) 1.6 g/L; sodium bicarbonate (Sigma S-5761) 4.4 g/L; HEPES (Sigma H-9136) 12 g/L; BSA-Fraction V (Serologicals, Norcross, GA, 81-003-6) 100 g/L; Neopeptone (Difco, MI 0119-01) 10 g/L; Yeastolate (Difco, Detroit, MI 5577-15-5) 4 g/L; D-Glucose (Sigma G-7021) 12 g/L. Adjust pH to 7.5 with NaOH. Sterilize with a 0.22-μm filter. Store 500 ml of filter-sterilized 2× concentrate in 1-L Pyrex bottles at –20°C. To obtain 1 L of complete mBSK (mBSK-c), thaw a frozen 500-mL aliquot and add 120 mL of non-hemolyzed rabbit serum (Sigma-Aldrich or PelFreez, Rogers, AR, Cat. 31125) and 380 mL sterile dH₂O. This results in a final concentration of rabbit serum of 12%. mBSK-c is dispensed into sterile polycarbonate culture tubes or pyrex bottles and used immediately or frozen at –20°C.
2. In our unpublished genome sequence database for *B. hermsii*, we identified DNA sequences of the promoter for the flagellar rod protein (*flgB-P*), the promoter for the flagellin protein (*flaB-P*), and the variable tick protein gene (*vtp*). Sequences

that confer autonomous replication (ARS) were identified using BLAST *(13)* to search for a genomic fragment that was similar to the open reading frames (ORFs) of the *B. burgdorferi* shuttle vector pBSV2 *(14)*.

3. The selectable markers in plasmids pTABhFlgB-Kan and pTABhFlaB-Gent (*see* **Fig. 2**) have been employed for genetic manipulation of three genes in infectious *B. hermsii*. The following describes the construction of these plasmids. *Construction of pTABhflgB-Kan*: A sequence contig containing the *B. hermsii* DAH flagellar rod promoter, *flgB-P*, was identified (*see* **Note 2**). Oligonucleotides 5′ BhflgB+NgoMIV and 3′ BhflgB+NdeI were designed to generate a 135-bp PCR amplicon containing *flgB-P* from genomic DNA of DAH clone 2E7. The specified restriction endonuclease sites were included in

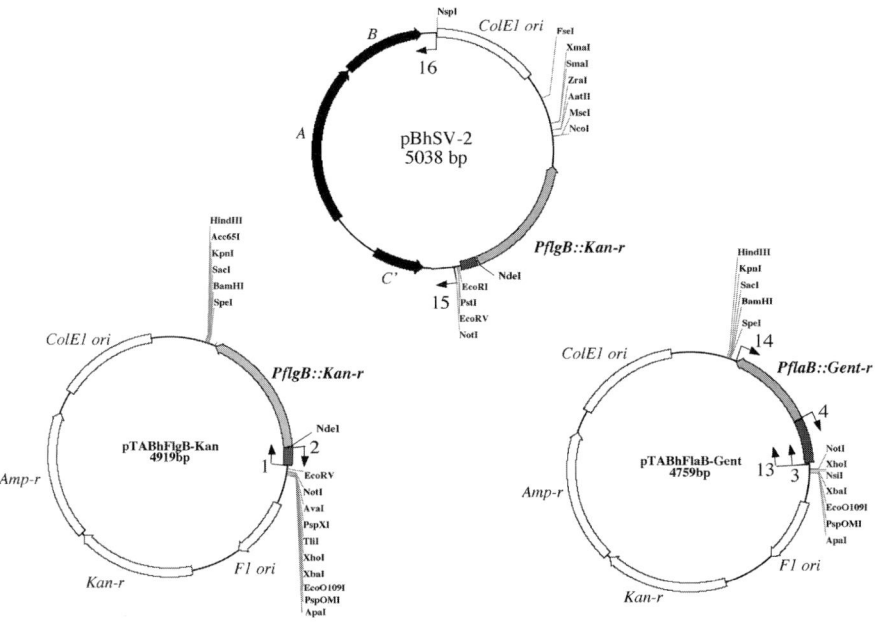

Fig. 2. The *Borrelia hermsii* shuttle vector and selectable marker plasmids. Plasmids pTABhFlgB-Kan and pTABhFlaB-Gent contain the selectable markers used in the *B. hermsii* genetic manipulation constructs. The *vtp* inactivation plasmid, pOKvtpKO (*see* **Fig. 1**), utilizes the *PflgB::kan-r* selectable marker from pTABhFlgB-Kan. The *PflaB::gent-r* cassette in pTABhFlaB-Gent was used in the *vtp* genetic reconstitution plasmid, pOKvtpRECON (*see* **Fig. 1**). The *B. hermsii* shuttle vector, pBhSV-2, was constructed following the design of pBSV2 *(14)*. Restriction endonuclease sites are shown. Primer-binding sites are specified by small arrows and correspond to sequences listed in **Subheading 2.2.2**. The source of open reading frames is indicated by shading: antibiotic resistance gene (light gray), *B. hermsii* promoter (dark gray), autonomous replication sequence (ARS) (black), and cloning vector (white). Details relevant to the construction of these plasmids are provided in the text (*see* **Note 3**).

the primers and were utilized for subsequent gene fusions. *NdeI* allows for the inframe fusion with the ATG start site. This product was cloned into pCR2.1-TOPO (Invitrogen) resulting in pCR2.1+BhFlgB-P, which was then sequenced using M13 Reverse and M13 Forward (–20) primers (Invitrogen) to confirm the sequence and orientation of the insert. pTAkan-A *(15)* was used as the source of the Tn903-derived kanamycin phosphotransferase gene, *kan-r*. The 925-bp *NdeI/BamHI* fragment of pTAkan-A containing *kan-r* was ligated to the *NdeI/BamHI*-digested pCR2.1+BhFlgB-P, resulting in the 4919-bp plasmid, pTABhFlgBP-Kan (*see* **Fig. 2**). Following transformation of *E. coli* (as in **step 10, Subheading 3.1.1.**), this plasmid construct was verified by RFLP, PCR, and sequence analysis. The *B. hermsii flgB*-promoted *kan-r* was used as the selectable marker for site-specific inactivation of *vtp* (*see* **Fig. 1**). *Construction of pTABhflaB-Gent*: A fragment containing the DNA sequence for the *B. hermsii* DAH flagellin promoter *flaB-P* was identified (*see* **Note 2**). Oligonucleotides 5´ BhflaB+*NgoMIV* and 3´ BhflaB+*NdeI* were designed to PCR-amplify a 301-bp fragment containing *flaB-P*. This product was cleaned (QIAquick), cloned into pCR2.1-TOPO resulting in pCR2.1+BhFlaB-P, and sequenced to confirm the integrity and orientation of the insert. pTAGmA *(16)* was used as the source of the *aacC1* gene, which encodes a gentamicin acetyltransferase, here termed *gent-r*. The 600-bp *NdeI/BamHI* fragment of pTAGmA containing *gent-r* was fused to the *NdeI/BamHI* digested pCR2.1+BhflaB-P resulting in the 4759-bp plasmid construct, pTABhflaB-Gent. Following transformation of *E. coli* (as in **step 10, Subheading 3.1.1.**), this plasmid construct was verified by RFLP, PCR, and sequence analysis. The *B. hermsii flaB*-promoted *gent-r* is flanked by a number of common restriction endonuclease sites and was used as the selectable marker for genetic reconstitution of *vtp* (*see* **Fig. 2**). *Construction of pBhSV-2:* This plasmid was not directly used in the manipulation of *vtp*. However, as autonomously replicating plasmids are much more efficient at transformation, this plasmid was frequently employed as a positive control for electro-transformation of *B. hermsii*. The *B. burgdorferi* shuttle vector, pBSV2 *(14)* was modified first by replacing the *NotI/NdeI* fragment containing the *B. burgdorferi flaB-P* with the 162-bp *NotI/NdeI* fragment of pTABhflgB-Kan containing *B. hermsii flgB-P*, resulting in pBhSV-1. All attempts to transform *B. hermsii* by electroporation with pBSV2 and pBhSV-1 failed to generate kanamycin-resistant spirochetes; thus another construct was made. Plasmid pBhSV-1 was then modified by replacing the three *B. burgdorferi* ORFs shown to confer autonomous replication, with the *B. hermsii* ARS loci (*see* **Note 2**). First, primers BhOriF+NotI and BhOriR+SphI were used to amplify 2643-bp DNA from *B. hermsii* DAH 2E7 containing the ARS. The *NotI* and *SphI* primer tags were used to fuse the ARS amplicon to the 2384-bp *NotI/NspI* fragment from pBhSV-1, resulting in pBhSV-2. Transformation of *E. coli*, selection, and analysis were done as described in **steps 10–12, Subheading 3.1.1**. Introduction of pBhSV-2 into *B. hermsii* by electro-transformation (*see* **Subheading 3.2.**) generated kanamycin-resistant spirochetes. Plasmid rescue *(14)* was used to verify that the plasmid replicated autonomously. The genomic source of the ARS in pBhSV-2 is presently unclear. This ARS is not the chromosomal origin. BLAST analysis *(13)* demonstrates that this ARS is similar to a number of *B. hermsii* and *B. burgdorferi* plasmid sequences. Searches were performed with the entire ARS sequence as well as with individual ORFs A, B, and C´ [identified using ORF finder (NCBI) (*see* **Fig. 2**)]. Although a contiguous sequence containing this ARS was not found in GenBank, *orfA* and *orfB* are most similar to members of the *B. burgdorferi* paralogous families 57 and 50, respectively *(17,18)*. This ARS also contains one truncated ORF *orfC´*, which is 67% identical to the carboxy-terminal portion of *B. hermsii* variable major protein VlpC54silD *(19)*.

4. On the basis of the size of the amplicon, two different PCR *(20)* systems were utilized for the construction of plasmids. Amplicons under 1000 bp were generated using a GeneAmp Core Kit (Roche Molecular Systems), and amplicons over 1000 bp were produced using an Expand Long Template PCR System Kit (Roche Diagnostics), per the manufacturer's instructions.

5. BlueView Nucleic Acid Stain (Sigma-Aldrich) was used to visualize DNA fragments over 1000 bp in agarose gels for excision and purification. This stain is less sensitive than EtBr and requires a minimum of several micrograms of DNA per well for visualization, but does not damage the samples. EtBr is more efficient for visualization but is carcinogenic, requires ultraviolet light, and prolonged exposure to UV light results in DNA damage that significantly reduces cloning efficiency. Personal protective devices including UV-resistant eyeglasses, gloves, and coat are required when working with EtBr.

6. *Construction of pOKvtp*: Primers VTP F2+*Avr*II and VTP B3+*Asc*I were designed to PCR amplify a 5193-bp fragment from *B. hermsii* containing *vtp* with approximately 2 kbp of upstream and downstream flanking sequence. The pOK12 amplicon consists of a 1720-bp portion of pOK12 *(11)* that includes the P15A origin of replication and a kanamycin resistance cassette, and was generated by the Expand PCR Kit with primers pOK12F1+AscI and pOK12B1+AvrII. Perform the cleanings and restriction digest (*see* **steps 5–8, Subheading 3.1.1.**) above to prepare the *vtp* and pOK12 amplicons for ligation. Following transformation of TOP10 *E. coli*, and selection on LB plates supplemented with 50 µg/mL kanamycin, clones were isolated and the construct was verified by PCR, RFLP, and sequence analysis. The *vtp* inactivation and reconstitution plasmids were generated by modifying pOKvtp.

7. RFLP is the analysis of DNA by restriction digestion followed by agarose gel electrophoresis to display the number and size of restriction fragments. An example of RFLP is shown in **Fig. 3**.

8. *Construction of the* vtp *inactivation plasmid*: Primers vtpF1+ScaI and vtpB1+SpeI generate 6470 bp from pOKvtp that includes all of pOKvtp excluding a 456-bp internal region of the *vtp* gene (*see* **Fig. 1**, primers 9 and 10). Following *Spe*I and *Sca*I digestion, this amplicon was ligated to the 1066-bp *Spe*I/*Eco*RV fragment of pTABhFlgB-Kan containing the *B. hermsii*

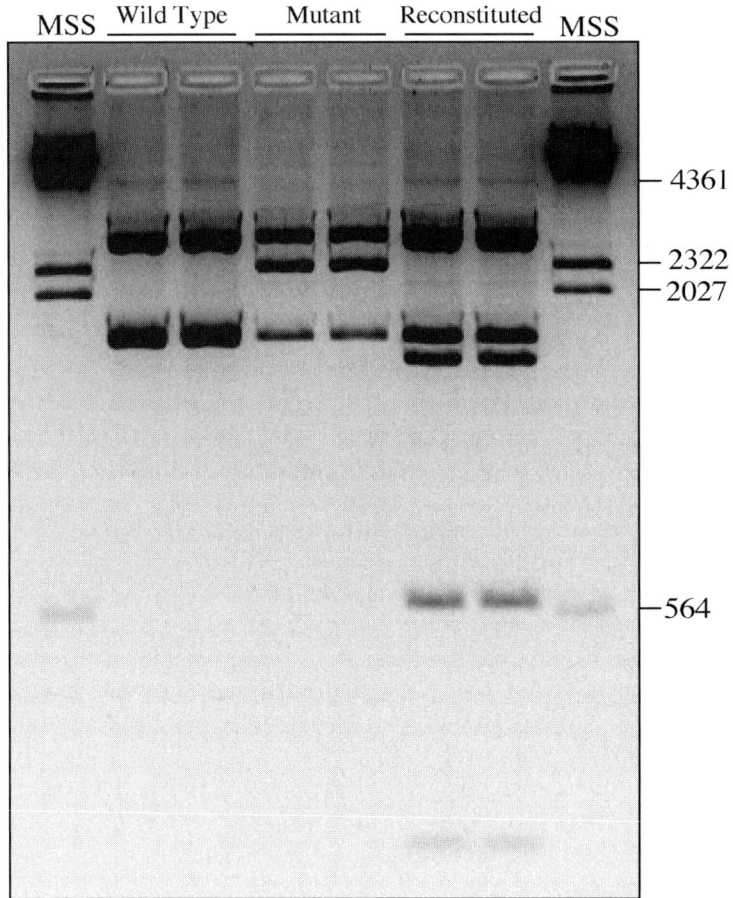

Fig. 3. Restriction fragment length polymorphism (RFLP) analysis of wild-type, *vtp* mutant, and reconstituted strains. This ethidium bromide-stained agarose gel verifies the genomic structure of the *Borrelia hermsii vtp* mutant and isogenic reconstituted strain relative to wild type. Amplicons were generated from wild-type, mutant, and reconstituted clones using primers 17 and 18 (*see* **Fig. 1**). These products were digested with *Eco*RV and the resulting restriction fragments were separated by agarose gel electrophoresis. The banding pattern (*f*ragment *l*ength *p*olymorphism) of each strain (loaded in duplicate) corresponds with the expected results: wild type (2795, 1675, 1644 bp), mutant (2795, 2289, 1644 bp), and reconstituted (2795, 1644, 1499, 582, 202 bp). Molecular size standards (MSS) are indicated to the right in base pairs.

flgB-promoted kanamycin resistance cassette (*see* **Fig. 2**). Following transformation of TOP10 *E. coli*, clones were isolated on LB plates supplemented with 50 µg/mL kanamycin. The resulting 7541-bp construct, pOKvtpKO, was verified by PCR, RFLP, and sequence analysis and used for site-specific *vtp* inactivation.

9. *Construction of the* vtp *reconstitution plasmid* : Oligonucleotides ReconF+XmaI and ReconR+SpeI (*see* **Fig. 1**, primers 11 and 12) were used to amplify 6675 bp of pOKvtp that includes all of pOKvtp with the exception of a 251-bp fragment that is 282 bp upstream of the start codon of *vtp*. Oligonucleotides Gent5´+XmaI and Gent3´+SpeI (*see* **Fig. 2**, primers 13 and 14) were used to generate an 862-bp fragment from pTABhflaB-Gent that contains the *B. hermsii flaB*-promoted gentamicin resistance cassette. The pOKvtp and pTABhflaB-Gent amplicons were digested with *Xma*I and *Spe*I, cleaned, and ligated, resulting in the plasmid pOKvtpRECON (*see* **Fig. 1**). Following transformation of TOP10 *E. coli*, clones were isolated on LB plates supplemented with 10 µg/mL gentamicin, and the construct was verified by sequence analysis.

10. All procedures are done in a laminar flow hood using aerosol barrier pipette tips and sterile technique. Methods used to transform *B. hermsii* by electroporation are similar to those procedures previously described for *B. burgdorferi* (**15, 21–23**). Spirochetes were cultivated and cloned by limiting dilution in liquid mBSK-c.

11. *B. hermsii* DAH is a wild-type non-clonal isolate that originated from blood of a human with relapsing fever (**24**). A clonal population of this isolate, designated DAH 2E7, was made by limiting dilution in liquid medium, and was infectious in RML mice by intraperitoneal injection (*see* **Fig. 4**).

12. Ethanol precipitation is used to purify and concentrate the genetic manipulation plasmids prior to electroporation. The following procedure is performed in micro-centrifuge tubes. Reagents should be prepared with sterile deionized

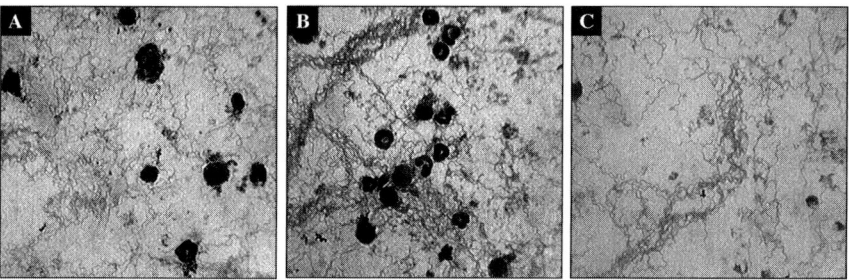

Fig. 4. Virulence of strains by mouse infection. Following in vitro manipulation, strains were assessed for virulence by intraperitoneal injection of mice and subsequent microscopic detection of Giemsa-stained spirochetes in blood. Injection of equal numbers of (**A**) electrocompetent *Borrelia hermsii* DAH 2E7, (**B**) electroporated without a manipulation construct or antibiotics and cloned by limiting dilution, and (**C**) the *vtp* mutant all produced comparable spirochetemias at 65 h post infection. Images of infected blood were taken at 400×.

H_2O to limit the ions present in the sample that could result in arcing during electroporation.

   a. Determine the volume and concentration of the DNA sample.
   b. Add 1/10 volume of 3 *M* sodium acetate (pH 5.0).
   c. Add 2.5 volumes of cold 100% ethanol.
   d. Mix and incubate at 4°C for 1 h or overnight at –80°C.
   e. Pellet the precipitated DNA by centrifugation (4°C, 16,000 × *g*, 15 min).
   f. Remove supernatant.
   g Wash the pellet one to three times with 1 mL cold 70% ethanol.
   h Aspirate supernatant and dry sample in speed vacuum or laminar flow hood.
   i. Thoroughly resuspend the pellet in dH_2O with a pipettor, such that the final concentration is approximately 5 µg/µL.
   j. Store concentrated plasmid DNA at 4°C for a few hours or at –20°C for several days.

13. The efficiency of electroporation can be estimated by monitoring the time constant (TC) associated with each pulse. First, electrocompetent spirochetes should be electroporated without additional plasmid DNA. The TC generated from the pulse of spirochetes alone should be recorded and should be between 4.0 and 4.7 ms when using 2.5 kV, 25 µF, and 200 Ω. If the spirochetes alone generate an arc, or the TC is 2–3 ms, the cells must be washed again. Addition of plasmid should not alter the TC significantly from the spirochete-alone control. If addition of plasmid DNA results in arcing, either the plasmid was not de-salted properly or it is too concentrated. First try using less plasmid. If arcing still occurs, the plasmid must be re-precipitated and washed more thoroughly. pBhSV-2 is a very useful positive control plasmid for monitoring the efficiency of transformation. The standard method for quantifying transformation efficiency (number of transformants/µg DNA) is not applicable with liquid culture and cloning by limiting dilution. One modification that increases the efficiency of transformation is to linearize the inactivation plasmid DNA prior to precipitation and electro-transformation.

14. Counting the spirochetes with a dark-field microscope helps to approximate the concentration of spirochetes and the number of 10-fold serial dilutions required. Following limiting dilution, incubate at 34°C and monitor the red-to-yellow color change provided by the phenol red pH indicator in mBSK-c. The time required for color change is shorter in the more concentrated wells and is apparent in approximately 3–5 days. Do not use color change to conclude the presence or absence of growth in a particular well. Analyze individual wells by dark-field microscopy and Poisson distribution statistics to determine clonality.

Acknowledgments

We thank Merry Schrumpf, Kevin Lawrence, Aimee Giessler, and Gail Sylva for excellent technical assistance; Patti Rosa, Jim Bono, Abe Elias, and Philip

Stewart for the generous donation of plasmid constructs and advice regarding genetic manipulation of *Borrelia*; Steve Porcella for assisting with *B. hermsii* sequences; and Mike Minnick for critical review of this text. The animal use protocol utilized in this study was approved by the Animal Care and Use Committee at the Rocky Mountain Laboratories. This work was supported by the Division of Intramural Research, National Institute of Allergy and Infectious Diseases, National Institutes of Health.

References

1. Falkow, S. (1988) Molecular Koch's postulates applied to microbial pathogenicity. *Rev. Infect. Dis.* **10** Suppl. **2**, S274–276.
2. Dworkin, M.S., Schwan, T.G., and Anderson, D.E. (2002) Tick-borne relapsing fever in North America. *Med. Clin. North Am.* **86**, 417–433.
3. Brouqui, P., Stein, A., Dupont, H.T., Gallian, P., Badiaga, S., Rolain, J.M., Mege, J.L., La Scola, B., Berbis, P., and Raoult, D. (2005) Ectoparasitism and vector-borne diseases in 930 homeless people from Marseilles. *Medicine (Baltimore)* **84**, 61–68.
4. Cutler, S.J. (2006) Possibilities for relapsing fever reemergence. *Emerg. Infect. Dis.* **12**, 369–374.
5. Carter, C.J., Bergström, S., Norris, S.J., and Barbour, A.G. (1994) A family of surface-exposed proteins of 20 kilodaltons in the genus *Borrelia. Infect. Immun.* **62**, 2792–2799.
6. Schwan, T.G., and Hinnebusch, B.J. (1998) Bloodstream- versus tick-associated variants of a relapsing fever bacterium. *Science* **280**, 1938–1940.
7. Schwan, T.G., Piesman, J., Golde, W.T., Dolan, M.C., and Rosa, P.A. (1995) Induction of an outer surface protein on *Borrelia burgdorferi* during tick feeding. *Proc. Natl. Acad. Sci. USA* **92**, 2909–2913.
8. Grimm, D., Tilly, K., Byram, R., Stewart, P.E., Krum, J.G., Bueschel, D.M., Schwan, T.G., Policastro, P.F., Elias, A.F., and Rosa, P.A. (2004) Outer-surface protein C of the Lyme disease spirochete: a protein induced in ticks for infection of mammals. *Proc. Natl. Acad. Sci. USA* **101**, 3142–3147.
9. Pal, U., Yang, X., Chen, M., Bockenstedt, L.K., Anderson, J.F., Flavell, R.A., Norgaard, M.V., and Fikrig, E. (2004) OspC facilitates *Borrelia burgdorferi* invasion of *Ixodes scapularis* salivary glands. *J. Clin. Invest.* **113**, 220–230.
10. Tilly, K., Krum, J.G., Bestor, A., Jewett, M.W., Grimm, D., Bueschel, D., Byram, R., Dorward, D., Vanraden, M.J., Stewart, P., et al. (2006) *Borrelia burgdorferi* OspC protein required exclusively in a crucial early stage of mammalian infection. *Infect. Immun.* **74**, 3554–3564.
11. Vieira, J., and Messing, J. (1991) New pUC-derived cloning vectors with different selectable markers and DNA replication origins. *Gene* **100**, 189–194.
12. Stoenner, H.G., Dodd, T., and Larsen, C. (1982) Antigenic variation of *Borrelia hermsii. J. Exp. Med.* **156**, 1297–1311.
13. Altschul, S.F., Gish, W., Miller, W., Myers, E.W., and Lipman, D.J. (1990) Basic local alignment search tool. *J. Mol. Biol.* **215**, 403–410.

14. Stewart, P.E., Thalken, R., Bono, J.L., and Rosa, P. (2001) Isolation of a circular plasmid region sufficient for autonomous replication and transformation of infectious *Borrelia burgdorferi. Mol. Microbiol.* **39**, 714–721.

15. Bono, J.L., Elias, A.F., Kupko, J.J., III, Stevenson, B., Tilly, K., and Rosa, P. (2000) Efficient targeted mutagenesis in *Borrelia burgdorferi. J. Bacteriol.* **182**, 2445–2452.

16. Elias, A.F., Bono, J.L., Kupko, J.J., Stewart, P.E., Krum, J.G., and Rosa, P.A. (2003) New antibiotic resistance cassettes suitable for genetic studies in *Borrelia burgdorferi. J. Mol. Microbiol. Biotechnol.* **6**, 29–40.

17. Fraser, C.M., Casjens, S., Huang, W.M., Sutton, G.G., Clayton, R., Lathigra, R., White, O., Ketchum, K.A., Dodson, R., Hickey, E.K., et al. (1997) Genomic sequence of a Lyme disease spirochaete, *Borrelia burgdorferi. Nature* **390**, 580–586.

18. Casjens, S., Palmer, N., van Vugt, R., Huang, W.M., Stevenson, B., Rosa, P., Lathigra, R., Sutton, G., Peterson, J., Dodson, R.J., et al. (2000) A bacterial genome in flux: the twelve linear and nine circular extrachromosomal DNAs in an infectious isolate of the Lyme disease spirochete *Borrelia burgdorferi. Mol. Microbiol.* **35**, 490–516.

19. Dai, Q., Restrepo, B.I., Porcella, S.F., Raffel, S.J., Schwan, T.G., and Barbour, A.G. (2006) Antigenic variation by *Borrelia hermsii* occurs through recombination between extragenic repetitive elements on linear plasmids. *Mol. Microbiol.* **60**, 1329–1343.

20. Mullis, K., Faloona, F., Scharf, S., Saiki, R., Horn, G., and Erlich, H. (1986) Specific enzymatic amplification of DNA in vitro: the polymerase chain reaction. *Cold Spring Harb. Symp. Quant. Biol.* **51**, 263–273.

21. Samuels, D.S. (1995) Electrotransformation of the spirochete *Borrelia burgdorferi*, in *Methods in Molecular Biology* (Nickoloff, J.A., ed.), Humana, Totowa, NJ, pp. 253–259.

22. Tilly, K., Elias, A.F., Bono, J.L., Stewart, P., and Rosa, P. (2000) DNA exchange and insertional inactivation in spirochetes. *J. Mol. Microbiol. Biotechnol.* **2**, 433–442.

23. Elias, A.F., Stewart, P.E., Grimm, D., Caimano, M.J., Eggers, C.H., Tilly, K., Bono, J.L., Akins, D.R., Radolf, J.D., Schwan, T.G., et al. (2002) Clonal polymorphism of *Borrelia burgdorferi* strain B31 MI: implications for mutagenesis in an infectious strain background. *Infect. Immun.* **70**, 2139–2150.

24. Hinnebusch, B.J., Barbour, A.G., Restrepo, B.I., and Schwan, T.G. (1998) Population structure of the relapsing fever spirochete *Borrelia hermsii* as indicated by polymorphism of two multigene families that encode immunogenic outer surface lipoproteins. *Infect. Immun.* **66**, 432–440.

7

Transposon Mutagenesis of the Lyme Disease Agent *Borrelia burgdorferi*

Philip E. Stewart and Patricia A. Rosa

Summary

Borrelia burgdorferi, the causative agent of Lyme disease, is an obligate parasite that cycles between vertebrate hosts and tick vectors. Attempts to understand the genetic factors that allow *B. burgdorferi* to sense, adapt to, and survive in different environments have been limited by a relatively low transformation rate. Here, we describe a *mariner*-based transposon system that achieves saturating levels of random mutagenesis in *B. burgdorferi*. In comparison with allelic exchange, which targets a single locus, transposon mutagenesis can create libraries of mutants encompassing disruptions of all genes. Suitably designed screens or selections of such a library permit the recovery of mutants exhibiting a desired phenotype. The system described here allows rapid identification of the genetic locus responsible for the mutant phenotype. With appropriate modifications, this *mariner*-based transposon can be adapted to other spirochetes and bacteria with inefficient genetic transformation methods.

Key Words: *Borrelia burgdorferi*; spirochete; DNA transposable element; *Himar1*; insertional mutagenesis.

1. Introduction

Transposons are DNA elements with the ability to move, or transpose, to new locations within a genome. Several modified transposon systems have proven useful as genetic tools for insertional inactivation of genes, identifying conditionally regulated genes, and genome sequencing (reviewed in **ref. 1**). These techniques have identified microbial genes essential for growth and virulence, as well as loci affecting physiology and morphology *(2–9)*.

From: *Methods in Molecular Biology, vol. 431: Bacterial Pathogenesis*
Edited by: F. DeLeo and M. Otto © Humana Press, Totowa, NJ

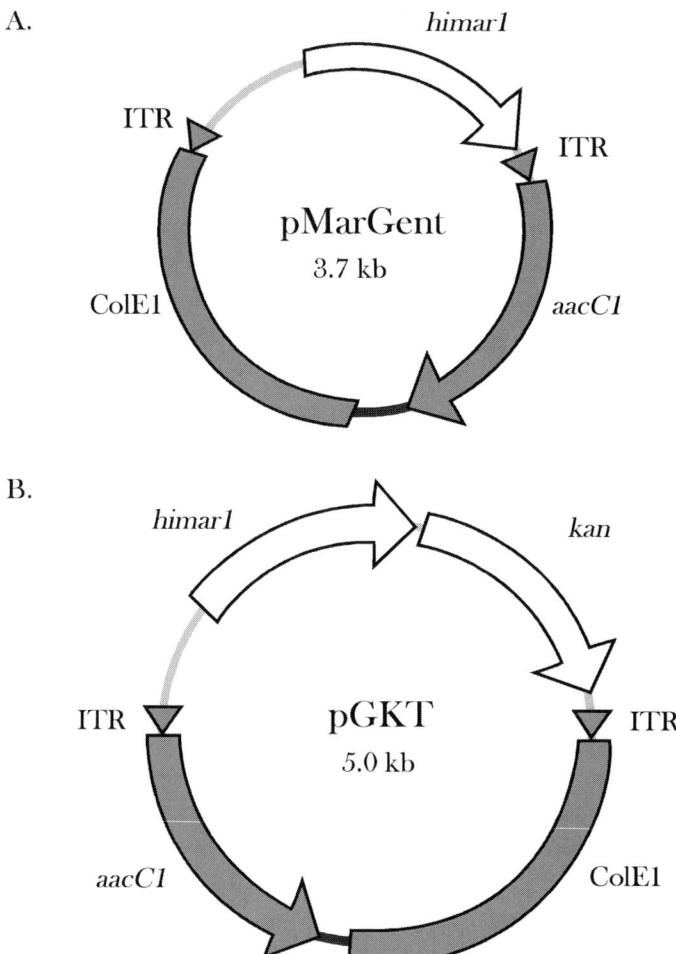

Fig. 1. Vectors for transposon mutagenesis of *Borrelia burgdorferi*. Both pMarGent **(A)** and pGKT **(B)** contain the high-copy origin of replication (ColE1) in *Escherichia coli* but are suicide vectors in *B. burgdorferi*. Therefore, gentamicin-resistance (encoded by *aacC1*) in *B. burgdorferi* can only be conferred by transposition of the transposon (region shaded gray). In *E. coli*, double selection of the antibiotics kanamycin and gentamicin can be imposed for pGKT, making it more stable than the progenitor vector pMarGent, which is only gentamicin resistant. Note that for pGKT transformation of *B. burgdorferi*, kanamycin resistance is lost once transposition has occurred. ITR, inverted terminal repeat; *kan*, kanamycin-resistance cassette; *himar1*, gene encoding transposase.

Members of the *mariner* family of transposons are particularly useful as genetic tools in heterologous hosts because they do not require host cofactors to be functional and because they integrate randomly into the genome, recognizing a T-A dinucleotide sequence to insert. Recently, hyperactive mutants of the *mariner* element *Himar1* were isolated that possessed an increased frequency of transposition over the wild-type allele *(10,11)*. These features have

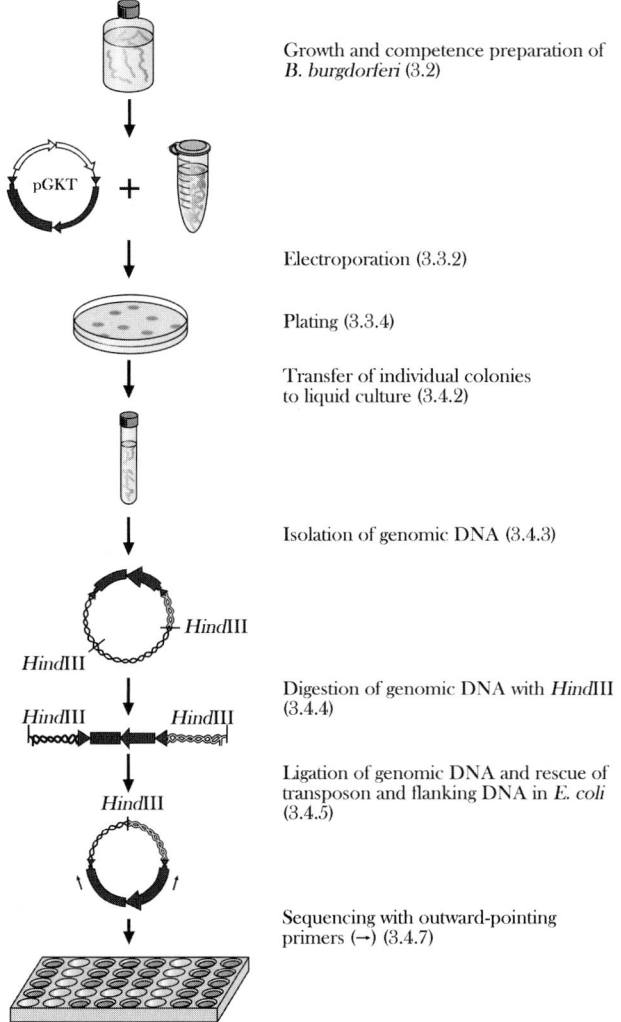

Growth and competence preparation of
B. burgdorferi (3.2)

Electroporation (3.3.2)

Plating (3.3.4)

Transfer of individual colonies
to liquid culture (3.4.2)

Isolation of genomic DNA (3.4.3)

Digestion of genomic DNA with *Hind*III
(3.4.4)

Ligation of genomic DNA and rescue of
transposon and flanking DNA in *E. coli*
(3.4.5)

Sequencing with outward-pointing
primers (→) (3.4.7)

Fig. 2. Strategy for the generation of transposon mutants in *Borrelia burgdorferi* and identification of the transposon insertion site. Numbers in parenthesis refer to the specific **Subheading** and step of the Methods section in which the procedure is described.

allowed *Himar1* to be functional in a wide range of prokaryotes *(3,11–14)*. The characteristics and promiscuity of *Himar1* make it a useful choice for rapidly generating a large number of gene knockouts.

The utility of *Himar1* in diverse organisms prompted us to adapt the *Himar1* element to generate a saturated library of random *Borrelia burgdorferi* mutants *(8)*. The transposon vectors, pMarGent (*see* **Fig. 1A**) and the more stable pGKT (*see* **Fig. 1B**) (*see* **Note 1**), have several advantageous features including random insertions in both the linear and the circular DNA molecules present in the *B. burgdorferi* genome, and a high transposition frequency that yields saturating levels of mutants from a single transformation.

The overall strategy for transposon mutagenesis (*see* **Fig. 2**) is similar to that used to obtain allelic exchange mutants. That is, *B. burgdorferi*-competent cells are prepared and transformed with pMarGent or pGKT plasmid DNA isolated from *Escherichia coli*. After an overnight recovery period, cells are plated in the presence of selective antibiotic(s). Colonies that arise are transferred to liquid growth medium and genomic DNA isolated. Digestion of the genomic DNA, followed by ligation and transformation into competent *E. coli* cells, allows the recovery and characterization of the *B. burgdorferi* DNA flanking the site in which the transposon inserted. Subsequently, *Himar1*-based systems have been used successfully in mutagenesis systems for various spirochetes and are readily adaptable to other microorganisms with limited genetic transformation systems *(5,6,15)*.

2. Materials

2.1. Preparation of Plasmid DNA

1. Gentamicin and kanamycin are available from various sources. Gentamicin selection for *E. coli*: 5 μg/mL; kanamycin selection for *E. coli*: 50 μg/mL.
2. We routinely use TOP10 chemically competent *E. coli* cells (Invitrogen, Carlsbad, CA); however, other strains, such as DH5α, should work.
3. Plasmid DNA may be isolated by various techniques or commercial kits, as long as they produce highly pure DNA. We routinely use the Hispeed Maxi Plasmid Kit (Qiagen, Valencia, CA).

2.2. Competent Cell Preparation

1. *B. burgdorferi* liquid cultures are grown in Barbour–Stoenner–Kelly (BSK)-H medium (Sigma, St. Louis, MD).
2. Petroff-Hausser chambers may be purchased from Electron Microscopy Sciences, Hatfield, PA.
3. Electroporation solution (EPS): 0.27 *M* sucrose, 15% (v/v) glycerol. Sterilize by filtration and store at room temperature.

2.3. Transformation and Plating of B. burgdorferi Cells

1. Gentamicin selection for *B. burgdorferi*: 40 µg/mL; kanamycin selection for *B. burgdorferi*: 200 µg/mL.
2. Solid (plating) BSK medium: Weigh 69.37 g bovine serum albumin, fraction V (Celliance, Kankakee, IL), 6.9 g Neopeptone, 8.3 g HEPES acid, 6.9 g glucose, 1.0 g sodium citrate, 1.1 g sodium pyruvate, 0.56 g *N*-acetylglucosamine, 6.4 g sodium bicarbonate, 3.5 g Yeastolate, and 12.7 g powdered Connaught Medical Research Laboratories (CMRL) without glutamine (US Biological, Swampscott, MA). Solubilize in a final volume of 1 L H_2O and stir 2–4 h. Adjust pH to 7.5 with 1N NaOH and sterilize by filtration. Add 12 mL rabbit serum (Pel-Freeze Biologicals, Rogers, AK) and aliquot into 310-mL portions. Store frozen at –20 °C until ready to use. Thaw at ≥55 °C and add 200 mL sterile, molten 1.7% agarose (55 °C) and antibiotic, if desired. Aliquots of transformed cells are mixed directly with 30 mL medium per plate (*see* **Note 2**).
3. *Borrelia burgdorferi* plates should be held in a CO_2 incubator (2.5%) at 35 °C.

2.4. Identification of Transposon Insertion Site

1. Gentamicin selection for *E. coli*: 5 µg/mL.
2. PCR reaction and cycling conditions: 1× PCR reaction conditions (20 µL volume)—1× Taq Polymerase buffer, 0.2 m*M* each deoxynucleotide, 10 pmoles each primer, and 0.5 U Taq polymerase. Cycling conditions are 94 °C (1 min), followed by 30 cycles of 94 °C (45 s), 55 °C (1 min), and 68 °C (2 min).
3. Primers may be obtained from any vendor and, after resuspension in H_2O, should be stored at –20 °C. Sequence of primers:
   col: 5′-CAGCAACGCGGCCTTTTTACG
   flg: 5′-GCTTAAGCTCTTAAGTTCAACC
   Primers col and flg are used during sequencing (*see* **Fig. 2**).
   JK62 (Gent F): 5′-GGCAGTCGCCCTAAAACAAAGTT
   JK61 (Gent RC): 5′-TCTCGGCTTGAACGAATTGTTAGGT
   Primers JK61 and JK62, used for amplifying a fragment of the gentamicin-resistance gene, produce a 520 base pair PCR product.

3. Methods

Borrelia burgdorferi strain B31 contains a large complement of linear and circular plasmids, two of which (lp25 and lp56) correlate with a reduced transformation efficiency with some constructs, including shuttle vectors and pMarGent *(8,16)*. The transformation barrier is presumably due to the presence of restriction/modification enzymes encoded on these *B. burgdorferi* plasmids. Therefore, generating transposon mutants requires the use of *B. burgdorferi* strains that lack these plasmids. However, lp25 and lp56 also encode factors that enhance or are required for *B. burgdorferi* survival in the mammal and tick hosts, and the loss of these plasmids reduces or eliminates the infectivity of these strains *(17–20)*. An infectious strain, 5A18 NP1, has been engineered that lacks the restriction/modification systems encoded on lp25 and lp56, and has a high transformation frequency with shuttle vector DNA *(21)*. 5A18 NP1 provides a genetic background in which virulence factors may be assessed by transposon mutagenesis *(22)*.

3.1. Preparation of Plasmid DNA

1. Transform pMarGent or pGKT (*see* **Fig. 1**) into competent *E. coli* cells and plate in the presence of the appropriate antibiotic (i.e., gentamicin for pMarGent or gentamicin and kanamycin for pGKT).
2. Inoculate *E. coli* colonies into growth medium with antibiotic selection and isolate plasmid DNA.

3.2. Competent Cell Preparation

1. Grow *B. burgdorferi* cultures to mid- to late-exponential phase (approximately $5–9 \times 10^7$ cells/mL as determined by darkfield microscopy using a Petroff-Hausser chamber) (*see* **Note 3**). Usually a 100 mL culture provides sufficient cell numbers for about two transformations.
2. Pellet cells by centrifugation ($6000 \times g$, 10 min) and wash by resuspension in 0.25 volumes EPS, i.e., 25 mL EPS for a 100 mL culture (*see* **Note 4**).
3. Pellet cells again (as in **step 2**) and resuspend in 0.1 volume EPS (i.e., 10 mL for an original culture volume of 100 mL).
4. Pellet cells for a final resuspension in ~0.001 volumes EPS, i.e., 100 µL EPS. Cells may be used immediately or stored at –80 °C.

3.3. Transformation and Plating of B. burgdorferi Cells

1. Add transposon plasmid DNA (10–20 µg in a 5 µL volume of sterilized H_2O) to 50–100 µL of competent cells (*see* **Note 5**).
2. Electroporate cell/DNA mix in a prechilled 0.2-cm gapped cuvette. Settings for the electroporator are: 2.5 kV, 25 µF, and 200 Ω.
3. Immediately after transformation, quickly resuspend cells in a final volume of 5 mL BSK medium and incubate overnight at 35 °C (*see* **Note 6**).
4. After overnight recovery, enumerate *B. burgdorferi* cells using a Petroff-Hausser chamber and plate in solid BSK medium with appropriate antibiotic selection. Subsurface colonies develop in ~6–10 days, depending on the strain.

3.4. Identification of Transposon Insertion Site

1. Confirm that colonies arising under selective pressure are transposon mutants (and not spontaneously-arising antibiotic-resistant variants) by PCR for the presence of the gentamicin-resistance marker. Stab the *B. burgdorferi* colony with a sterile toothpick, then swirl the toothpick in a tube containing the PCR reaction mix.

2. Once confirmed as transposon mutants by PCR, aspirate colonies from the plate using a sterile Pasteur pipet and transfer to liquid BSK medium (5–15 mL) containing appropriate antibiotic(s). Incubate cultures at 35 °C until cells reach high densities (between 5×10^7 and 1×10^8 cells/mL) (*see* **Note 7**).
3. Isolate genomic DNA from 5 mL cultures by pelleting the cells, removing the supernatant, and resuspending in 500 μL sterile H_2O. Lyse cells by the addition of 25 μL of 10% SDS and 5 μL of a 10 mg/mL RNase solution, and incubate at room temperature for 5 min. Phenol:chloroform extractions are repeated (usually two to three times) to remove protein and purify genomic DNA. Precipitate DNA with ethanol or isopropanol, wash with 70% ethanol, and resuspend in 50 μL sterile H_2O or TE. Alternatively, genomic DNA can be isolated from a larger 15 mL volume using the Wizard Genomic DNA Purification Kit (Promega, Madison, WI), following manufacturer's recommendations for gram-negative bacteria.
4. To isolate the transposon and DNA flanking the insertion site (*see* **Fig. 2**), digest approximately 500 ng of genomic DNA overnight with the restriction enzyme *Hind*III in a 15 μL volume (*see* **Note 8**).
5. Remove about 8 μL of digested DNA and ligate in ~10 μL volume at 14 °C for 6 h (or overnight). Transform entire ligation into chemically competent *E. coli* cells and plate about half of the transformed cells on LB plates containing gentamicin. Incubate plates at 37 °C overnight.
6. Isolate plasmid DNA from *E. coli* colonies.
7. Sequence the *B. burgdorferi* DNA flanking the transposon insertion from the plasmid DNA (*see* **Fig. 2**) using primers flg and col (*see* **step 3, Subheading 2.4.**). Identify the transposon insertion site by submitting the sequence results to the TIGR-CMR BLAST and search against the *B. burgdorferi* genome (http://tigrblast.tigr.org/cmr-blast/) (*see* **Note 9**).

4. Notes

1. The *Borrelia* promoters used to drive expression of the *Himar* transposase, the gentamicin-resistance marker, and the kanamycin-resistance marker on pMarGent and pGKT (*see* **Fig. 1**) are recognized and functional in *E. coli*. The transposase is therefore active in *E. coli* and allows transposition of the gentamicin marker during growth of the culture for plasmid DNA isolation, maintaining the gentamicin resistance of the cells but resulting in an inactive variant of pMarGent (*see* **Fig. 1A**). The plasmid preparations derived from different *E. coli* colonies, therefore, yield a range of plasmid forms and transposase activities. We generally isolate plasmid DNA from about six different *E. coli* clones and test each plasmid preparation in *B. burgdorferi* to determine which have a high transposition frequency. To reduce the chances of isolating inactive plasmid preparations from *E. coli*, we constructed the vector pGKT (*see* **Fig. 1B**) by cloning the kanamycin-resistance marker adjacent to the transposase gene of pMarGent. Growth of *E. coli* colonies harboring pGKT in the presence of both antibiotics requires that both the transposase (linked to the kanamycin marker) and the transposon portion of the plasmid

(inverted terminal repeats, origin of replication, and the gentamicin marker) be retained. We have found pGKT appears to be more stable than pMarGent and produces more plasmid preparations with a higher transposition frequency.

2. The bovine serum albumin and rabbit serum are critical components of this medium, and the ability to support *B. burgdorferi* growth to high cell densities varies by source and from lot to lot. Typically, we reserve large batches from the sources cited and test small samples.

3. Elias et al. tested a variety of growth stages and electroporation conditions and concluded that mid- to late-exponential growth phase was optimal for *B. burgdorferi* transformation and yielded the maximum number of competent cells *(23)*. Transposition frequencies for strains lacking lp25 and lp56 and prepared in this manner are $\sim 5 \times 10^{-5}$ *(8)*.

4. The objective of the washes is to remove medium components from the cells that might cause arcing during the electroporation, while concentrating the bacterium in electroporation buffer. The number of washes and volumes of EPS used were empirically determined and are a balance between efficient removal of the medium, yet limiting mechanical damage to *B. burgdorferi* cells caused by centrifugation and resuspension. The final volume that competent cells are resuspended in varies, but typically ranges between a 500-fold and a 1000-fold concentration of cells. Under darkfield microscopy, an ideal competence preparation should look like a continuous flowing sheet of spirochetes. Problems arise when the cells clump (fewer spirochetes accessible to the DNA during transformation) or if the cells are too concentrated (which might result in arcing). If clumping occurs, try moderate vortexing and repeated pipetting. If the cells are too concentrated, then add more EPS until the cells begin to flow when viewed under darkfield microscopy.

5. Generally, 10–20 μg of transposon plasmid DNA is alcohol-precipitated, washed with 70% ethanol to remove salts that may cause arcing during electroporation, and resuspended in a small volume (e.g., 5 μL) of sterile H_2O. The final volume of DNA must be small relative to the volume of competent cells, again to avoid arcing.

6. Transformed cells are generally allowed to recover overnight before selection is imposed. The length of the recovery period can be shortened to 8 h, which allows transformation and plating to occur in the same day.

7. Isolating genomic DNA from small culture volumes (5–15 mL) produces relatively low yields (a few micrograms). Harvest cells in late-exponential phase to obtain the highest cell numbers and maximize DNA yields. However, allowing cells to enter stationary phase usually results in lower yields and poor quality DNA.

8. This procedure was empirically determined and may be optimized. The restriction enzyme *Hind*III was chosen because the recognition sequence is absent from the transposon but is relatively frequent in *B. burgdorferi* DNA.

9. Sequence data using these primers will contain a small amount of vector DNA (including the inverted terminal repeat sequence that serves as a recognition site for the transposase: 5′-ACAGGTTGGCTGATAAGTCCCCGGTCT). Also, because of the large number of highly related gene families in the *B. burgdorferi* genome, some BLAST results will return multiple genes, requiring more detailed analysis to identify the exact location of the transposon insertion.

Acknowledgments

We gratefully acknowledge the technical expertise of Virginia Tobiason for construction of pGKT, and Jon Krum for design of gentamicin-specific oligonucleotides. We thank Gary Hettrick for graphical expertise and Kit Tilly for critical review of the manuscript. This research was supported by the Intramural Research Program of the National Institute of Allergy and Infectious Diseases, National Institutes of Health.

References

1. Hayes, F. (2003) Transposon-based strategies for microbial functional genomics and proteomics. *Annu. Rev. Genet.* **37**, 3–29.
2. Akerley, B. J., Rubin, E. J., Camilli, A., Lampe, D. J., Robertson, H. M., and Mekalanos, J. J. (1998) Systematic identification of essential genes by *in vitro mariner* mutagenesis. *Proc. Natl. Acad. Sci. USA* **95**, 8927–8932.
3. Ashour, J., and Hondalus, M. K. (2003) Phenotypic mutants of the intracellular actinomycete *Rhodococcus equi* created by in vivo *Himar1* transposon mutagenesis. *J. Bacteriol.* **185**, 2644–2652.
4. Lamichhane, G., Zignol, M., Blades, N. J., Geiman, D. E., Dougherty, A., Grosset, J., Broman, K. W., and Bishai, W. R. (2003) A postgenomic method for predicting essential genes at subsaturation levels of mutagenesis: application to *Mycobacterium tuberculosis*. *Proc. Natl. Acad. Sci. USA* **100**, 7213–7218.
5. Louvel, H., Saint Girons, I., and Picardeau, M. (2005) Isolation and characterization of FecA- and FeoB-mediated iron acquisition systems of the spirochete *Leptospira biflexa* by random insertional mutagenesis. *J. Bacteriol.* **187**, 3249–3254.
6. Morozova, O. V., Dubytska, L. P., Ivanova, L. B., Moreno, C. X., Bryksin, A. V., Sartakova, M. L., Dobrikova, E. Y., Godfrey, H. P., and Cabello, F. C. (2005) Genetic and physiological characterization of 23S rRNA and *ftsJ* mutants of *Borrelia burgdorferi* isolated by mariner transposition. *Gene* **357**, 63–72.
7. Sassetti, C. M., Boyd, D. H., and Rubin, E. J. (2001) Comprehensive identification of conditionally essential genes in mycobacteria. *Proc. Natl. Acad. Sci. USA* **98**, 12712–12717.
8. Stewart, P. E., Hoff, J., Fischer, E., Krum, J. G., and Rosa, P. A. (2004) Genomewide transposon mutagenesis of *Borrelia burgdorferi* for identification of phenotypic mutants. *Appl. Environ. Microbiol.* **70**, 5973–5979.

9. Wong, S. M., and Mekalanos, J. J. (2000) Genetic footprinting with *mariner*-based transposition in *Pseudomonas aeruginosa*. *Proc. Natl. Acad. Sci. USA* **97**, 10191–10196.

10. Akerley, B. J., and Lampe, D. J. (2002) Analysis of gene function in bacterial pathogens by GAMBIT, in *Bacterial Pathogenesis, Part C: Identification, Regulation and Function of Virulence Factors* (Clark, V. L., and Bavoil, P. M., eds.), Academic Press, San Diego, CA, pp. 100–108.

11. Lampe, D. J., Akerley, B. J., Rubin, E. J., Mekalanos, J. J., and Robertson, H. M. (1999) Hyperactive transposase mutants of the *Himar1 mariner* transposon. *Proc. Natl. Acad. Sci. USA* **96**, 11428–11433.

12. Maier, T. M., Pechous, R., Casey, M., Zahrt, T. C., and Frank, D. W. (2006) In vivo *Himar1*-based transposon mutagenesis of *Francisella tularensis*. *Appl. Environ. Microbiol.* **72**, 1878–1885.

13. Rubin, E. J., Akerley, B. J., Novik, V. N., Lampe, D. J., Husson, R. N., and Mekalanos, J. J. (1999) *In vivo* transposition of *mariner*-based elements in enteric bacteria and mycobacteria. *Proc. Natl. Acad. Sci. USA* **96**, 1645–1650.

14. Zhang, J. K., Pritchett, M. A., Lampe, D. J., Robertson, H. M., and Metcalf, W. W. (2000) *In vivo* transposon mutagenesis of the methanogenic archaeon *Methanosarcina acetivorans* C2A using a modified version of the insect *mariner*-family transposable element *Himar1*. *Proc. Natl. Acad. Sci. USA* **97**, 9665–9670.

15. Bourhy, P., Louvel, H., Saint Girons, I., and Picardeau, M. (2005) Random insertional mutagenesis of *Leptospira interrogans*, the agent of leptospirosis, using a *mariner* transposon. *J. Bacteriol.* **187**, 3255–3258.

16. Lawrenz, M. B., Kawabata, H., Purser, J. E., and Norris, S. J. (2002) Decreased electroporation efficiency in *Borrelia burgdorferi* containing linear plasmids lp25 and lp56: impact on transformation of infectious *Borrelia*. *Infect. Immun.* **70**, 4851–4858.

17. Jacobs, M. B., Norris, S. J., Phillippi-Falkenstein, K. M., and Philipp, M. T. (2006) Infectivity of the highly transformable BBE02⁻ lp56⁻ mutant of *Borrelia burgdorferi*, the Lyme disease spirochete, via ticks. *Infect. Immun.* **74**, 3678–3681.

18. Purser, J. E., Lawrenz, M. B., Caimano, M. J., Radolf, J. D., and Norris, S. J. (2003) A plasmid-encoded nicotinamidase (PncA) is essential for infectivity of *Borrelia burgdorferi* in a mammalian host. *Mol. Microbiol.* **48**, 753–764.

19. Purser, J. E., and Norris, S. J. (2000) Correlation between plasmid content and infectivity in *Borrelia burgdorferi*. *Proc. Natl. Acad. Sci. USA* **97**, 13865–13870.

20. Revel, A. T., Blevins, J. S., Almazan, C., Neil, L., Kocan, K. M., de la Fuente, J., Hagman, K. E., and Norgard, M. V. (2005) *bptA* (bbe16) is essential for the persistence of the Lyme disease spirochete, *Borrelia burgdorferi*, in its natural tick vector. *Proc. Natl. Acad. Sci. USA* **102**, 6972–6977.

21. Kawabata, H., Norris, S. J., and Watanabe, H. (2004) BBE02 disruption mutants of *Borrelia burgdorferi* B31 have a highly transformable, infectious phenotype. *Infect. Immun.* **72**, 7147–7154.

22. Botkin, D. J., Abbott, A., Stewart, P. E., Rosa, P. A., Kawabata, H., Watanabe, H., and Norris, S. J. (2006) Identification of potential virulence determinants by *HimarI* transposition of infectious *Borrelia burgdorferi* B31. *Infect. Immun.* **74,** 6690–6699.

23. Elias, A. F., Stewart, P. E., Grimm, D., Caimano, M. J., Eggers, C. H., Tilly, K., Bono, J. L., Akins, D. R., Radolf, J. D., Schwan, T. G., and Rosa, P. (2002) Clonal polymorphism of *Borrelia burgdorferi* strain B31 MI: implications for mutagenesis in an infectious strain background. *Infect. Immun.* **70,** 2139–2150.

8

The Biofilm Exopolysaccharide Polysaccharide Intercellular Adhesin—A Molecular and Biochemical Approach

Cuong Vuong and Michael Otto

Summary

Exopolysaccharides play a crucial role in the formation of biofilms and biofilm resistance to antimicrobials and innate host defense. Here we describe methods to analyze and quantify polysaccharide intercellular adhesin (PIA), a biofilm exopolysaccharide made of *N*-acetylglucosamine that is found in staphylococci and many other bacterial biofilm-forming pathogens.

Key Words: *Staphylococcus*; polysaccharide intercellular adhesin; deacetylation; biofilm.

1. Introduction

Many pathogenic bacteria are capable of synthesizing various extracellular, cell surface-associated polysaccharides. Such exopolysaccharides have been recognized to impact bacterial virulence significantly, for example, by determining biofilm architecture and protecting from host defenses and antimicrobial components. Polysaccharide intercellular adhesin (PIA) is produced by several different Gram-positive and Gram-negative bacteria, e.g., *Staphylococcus* sp., *Escherichia coli*, *Yersinia pestis*, *Actinobacillus actinomycetemcomitans (1)*. In staphylococci, the production of PIA is encoded by the *ica* gene locus, which comprises the *icaA*, *icaD*, *icaB*, and *icaC* genes *(2)*. For the most part, the

From: *Methods in Molecular Biology, vol. 431: Bacterial Pathogenesis*
Edited by: F. DeLeo and M. Otto © Humana Press, Totowa, NJ

respective genes in other bacteria are similar to those of staphylococci but have been given different names, such as *pga* in *E. coli* and *Actinobacillus*. PIA is a positively charged homopolymer of β-1,6-linked *N*-acetylglucosamine (NAG) residues *(3)*. The positive charge of PIA is enzymatically introduced by the IcaB protein through a mechanism of deacetylation *(4)*. IcaA and IcaD form a NAG-transferase responsible for the oligomerization of monomeric NAG units. IcaC is a putative transporter thought to be involved in the export of elongated PIA strands through the bacteria cell membrane *(5)*. Our current knowledge of the chemical structure of PIA, its biological function with regard to biofilm formation and immune evasion, and the processes involved in the regulation of PIA expression is based on intensive studies carried out predominantly in staphylococci. Here we describe a range of techniques to study PIA with an emphasis on molecular and biochemical techniques applied for staphylococci. These techniques are likely transferable to investigate PIA-related structures in other bacteria.

2. Materials

1. Tryptic soy broth (TSB) (Difco, Franklin Lakes, NJ), supplemented with 0.5% glucose after sterilization by autoclaving.
2. Enzymes: DNase (0.5 mg/mL final concentration), RNase (0.5 mg/mL final concentration), lysostaphin (0.5 mg/mL final concentration), lysozyme (0.5 mg/mL final concentration), and proteinase K (4 mg/mL final concentration); store at $-20\,°C$.
3. Standard (or wash) buffer: phosphate-buffered saline (PBS, 10 mM sodium phosphate, pH 7.4, 150 mM NaCl) or Tris-buffered saline (TBS, 150 mM NaCl, 10 mM Tris–HCl, pH 7.4), and 0.5 M EDTA, pH 8.0; prepare fresh as required.
4. Ultrafiltration devices: Amicon Ultrafree-MC-YM-10, and Amicon Centriprep-YM-10 (Millipore Corporation, Bedford, MA).
5. Tangential flow filtration: regenerated cellulose Prep/Scale Spiral Wound TFF-6 cartridge, Prep/Scale Holder, and peristaltic pump (Millipore Corporation).
6. Miscellaneous materials: dialysis membranes (~10 kDa cut off), skim milk, spectrophotometer, and nitrocellulose membranes.
7. Miscellaneous chemicals: 2,3,6-trinitrobenzenesulfonic acid (TNBS) and sodium bicarbonate (NaHCO$_3$).
8. Chromatography equipment: high-performance liquid chromatography (HPLC) and coupled electro-spray ionization mass spectrometer, Jordi PolarPac WAX 10,000A 300 × 7.8 mm column (Alltech, Deerfield, IL), and HiLoad 26/60 Superdex 200 gel filtration column (Amersham Pharmacia Biotech, Piscataway, NJ).
9. *Staphylococcus epidermidis* strains 1457 and 1457Δ*icaB*.

3. Methods

The methods described below outline (1) the construction of an isogenic mutant defect in the deacetylation of PIA, (2) different purification and (3) detection techniques, and (4) chemical analysis of PIA. Bacterial cultures for each experiment are inoculated from pre-cultures grown overnight at a dilution of 1:100 and incubated at 37 °C with shaking at 120 rpm for 16 h, unless otherwise noted.

3.1. Generation of an icaB Mutant in Staphylococcus epidermidis

DNA manipulations were performed by standard procedures for recombinant DNA to construct a temperature-sensitive plasmid for homologous recombination *(6)* and are not described here in detail. Briefly, PCR-amplified regions flanking the *icaB* gene, which encodes a PIA-deacetylating enzyme, and an erythromycin resistance cassette, *ermB*, were cloned into the plasmid pBT2, leading to pBTΔ*icaB*. Plasmid pBTΔ*icaB* was transformed into the wild-type *S. epidermidis* strain 1457 by electroporation. Allelic replacement of the *icaB* gene by the erythromycin resistance cassette was performed as described *(4)*. Successful deletion of *icaB* was verified by sequencing of the genomic DNA of the putative *icaB* mutant and determination of *icaB* expression by TaqMan analysis *(4)*. The wild-type strain *S. epidermidis* 1457 and its isogenic mutant strain *S. epidermidis* 1457Δ*icaB* were used for further investigations (*see* **Note 1**).

3.2. Purification of PIA

PIA of staphylococci can be isolated from the cell surface or from the bacterial culture filtrate. The majority of partially deacetylated PIA is attached to the bacterial cell wall, whereas completely acetylated PIA (from the *icaB* deletion mutant strain) is released into the culture filtrate *(4)*. In addition, small amounts of deacetylated PIA released from the cell surface by mechanical shear force might be detectable in the culture filtrates *(1)*.

3.2.1. Isolation of Crude PIA

Isolation of crude PIA is suitable to screen an extended range of samples with methods that do not require highly pure PIA preparations.

3.2.1.1. Isolation of Crude Surface-Attached PIA

1. Harvest 1 mL of a 16-h staphylococcal culture by centrifugation at 10,000 × *g* for 1 min at 4 °C.
2. Wash bacterial cell pellet with 1 mL of PBS buffer and repeat centrifugation step.

3. Resuspend cell pellet in 20 µL 0.5 *M* EDTA, pH 8.0 (final volume: 1/50 of growth culture volume).
4. Extract surface-located PIA by boiling cells for 5 min at 100 °C in a 2-mL (safelock) microtube.
5. Centrifuge samples at 10,000 × *g* for 5 min at 4 °C.
6. Transfer clear supernatant to a new microtube.
7. Freeze the extract at –20 °C for months (*see* **Note 2**).

3.2.1.2. ISOLATION OF PIA FROM THE CULTURE FILTRATE
1. Harvest 1 mL of a 16-h staphylococcal culture at 10,000 × *g* for 5 min at 4 °C.
2. Transfer clear supernatant into a new microtube.
3. Concentrate supernatant with an ultrafiltration device (Amicon Ultrafree-MC, YM-10) by centrifugation at 4000 × *g* for 30 min at 4 °C.
4. Repeat **step 3** until the final concentration ratio is about 50-fold (*see* **Note 3**).
5. Freeze the extract can be frozen at –20 °C.

3.2.2. Highly Pure PIA for Immunization and Chemical Analysis

Highly pure PIA is required to raise specific antibodies against PIA and for chemical analysis of PIA deacetylation. For standard calibration, high-quality PIA is preferable.

3.2.2.1. ISOLATION OF CELL-SURFACE-ATTACHED HIGHLY PURE PIA
1. Harvest 1 L of a 16-h staphylococcal culture by centrifugation at 3000 × *g* for 15 min at 4 °C.
2. Wash bacterial cell pellet with 500 mL PBS buffer and repeat centrifugation step.
3. Resuspend bacterial cell pellet in 20 mL 0.5 *M* EDTA, pH 8.0 (final volume: 1/50 of growth culture volume).
4. Extract surface-associated PIA by boiling cells for 5 min at 100 °C.
5. Centrifuge sample at 3000 × *g* for 30 min at 4 °C and transfer clear supernatant to a dialysis membrane (~10 kDa cut off).
6. Dialyze PIA-containing extracts against distilled water for 2 × 12 h at 4 °C and subsequently digest with DNase (0.5 mg/mL final concentration), RNase (0.5 mg/mL final concentration), lysostaphin (0.5 mg/mL final concentration), and lysozyme (0.5 mg/mL final concentration) at 37 °C for 16 h, followed by incubation with proteinase K (4 mg/mL final concentration) at 37 °C for 16 h.
7. Centrifuge sample at 28,000 × *g* at 4 °C for 30 min.
8. Concentrate clarified supernatant about fivefold with Amicon Centriprep YM-10 centrifugal concentrators.
9. Perform size exclusion chromatography (SEC) (Akta Chromatography system, GE Healthcare Life Sciences, Piscataway, NJ) of PIA in maximally 10-mL aliquots on a HiLoad 26/60 Superdex 200 gel filtration column. Use 20 m*M* sodium phosphate (pH 7.0) and 150 m*M* sodium chloride as a buffer at a flow rate of 3 mL/min.

10. Perform immuno-dot blot analysis (described in **Subheading 3.3.2.**) of chromatography fractions to determine PIA-containing samples. Most PIA elutes shortly after the exclusion volume.
11. Dialyze PIA-containing fractions against water for 2 × 12 h (10 kDa cut off) at 4 °C.
12. Lyophilize, then dissolve PIA in concentrated hydrochloric acid, neutralize with sodium hydroxide, and buffer the samples with 100 mM sodium phosphate buffer (pH 7.0).
13. Use highly pure PIA for immunization or for further experiments (*see* **Note 4**).

3.2.2.2. ISOLATION OF HIGHLY PURE PIA FROM THE CULTURE FILTRATE

1. Centrifuge 5 L of a 24-h staphylococcal culture by centrifugation at 3000 × g at 4 °C for 15 min.
2. Sterilize bacterial supernatant by filtration using filtration units (0.22 μm).
3. Concentrate bacterial supernatant to a final volume of 1/20 of the initial volume by tangential flow filtration using a regenerated cellulose Prep/Scale Spiral Wound TFF-6 cartridge, a Prep/Scale Holder, and a peristaltic pump at a flow rate of 8 mL/min.
4. Follow **steps 6–13** in **Subheading 3.2.2.1** (*see* **Note 5**).

3.3. PIA Detection by Immunological Assays

3.3.1. Generation of αPIA Antisera

1. Isolate PIA as described in **Subheading 3.2.2.**
2. To produce αPIA antiserum, immunize rabbits with 2 mg of PIA using a standard protocol.
3. Dilute αPIA antiserum in TBS buffer (1:100).
4. Absorb antiserum with the following extracts of a PIA-negative staphylococcal mutant strain (e.g., *S. epidermidis* 1457 M10) for 16 h to reduce non-specific binding. Prepare all extracts from cells of 50 mL of *S. epidermidis* 1457 M10 cultures or from 200 mL of bacterial supernatant, respectively:

   a. Isolate extract by boiling cells with 1 mL 0.5 M EDTA for 5 min at 100 °C.
   b. Isolate extract by boiling cells in 1 mL 1% sodium dodecyl sulfate (SDS) for 5 min at 100 °C.
   c. Isolate extract isolated from cells treated with 1 mL lysostaphin.
   d. Isolate 1 mL crude cell extract prepared by breaking cells with glass beads.
   e. Isolate extract obtained from 200 mL culture medium precipitated by trichloroacetic acid and resolve dry pellets in 1 mL PBS.

5. Sediment precipitated material by centrifugation (30 min, 28,000 × g, 4 °C) and add 1 mM of sodium azide to the clear supernatant that will be used for further investigation.

3.3.2. Immuno-Dot Blot Analysis

To quantify PIA production, use equal amounts of bacterial cells and culture supernatants. The protocol is modified from Gerke et al. *(5)*.

1. Spot 2-μL sample aliquots to a nitrocellulose membrane, let the membrane dry, and block with 0.5% skim milk in TBS buffer overnight.
2. Incubate the membrane for at least 2 h with blocked αPIA-antiserum (1:5000) and wash membrane for 3 × 5 min with TBS buffer.
3. Incubate membrane for 2 h with αIgG–alkaline–phosphatase conjugate (1:5000).
4. Detect spots by chromogenic reaction with 5 mL 5-bromo-4-chloro-3-indolyl phosphate/nitroblue tetrazolium premix solution.
5. Quantify immunological reaction of PIA (spot), e.g., by using a scanner and Total Lab Version 2003 software (Nonlinear USA, Durham, NC) (*see* **Note 6**).

3.3.3. Immunofluorescence Microscopy

1. Heat-fix 15 μL of aliquots of bacterial cell suspensions onto glass microscope slides.
2. Incubate fixed cells with 30 μL of αPIA antiserum (diluted 1:50 in TBS) for 30 min at 37 °C in a wet chamber.
3. Wash slides for 3 × 5 min with TBS buffer and let them air-dry.
4. Incubate bound antibodies with 30 μL of fluorescein isothiocyanate-conjugated anti-rabbit immunoglobulin G antibody, diluted at 1:80 in TBS, for 30 min at 37 °C in a wet chamber.
5. Wash glass slide twice with TBS and double-distilled water for 5 min, air-dry glass slide.
6. Detect PIA under a fluorescence microscope at a magnification of ×1000.

3.4. Analysis and Detection of PIA by Chromatography

3.4.1. PIA Analysis by Size Exclusion Chromatography/Electro-Spray Ionization Mass Spectrometry

This method employs SEC coupled to electro-spray ionization mass spectrometry (ESI/MS) and can be applied to culture filtrate samples and crude PIA extracts isolated from cell surfaces. The acquisition of ESI mass spectra of PIA in positive ion mode is based on the ionization of groups such as hydroxyl groups on sugar residues, or in the case of deacetylated PIA on already ionized amino groups. On the size exclusion columns used under the specific buffer conditions, PIA elutes earlier and separate from most other molecules, soon after the exclusion volume.

1. Purify PIA as described in **Subheading 3.2**.
2. Perform SEC/ESI-MS at a flow rate of 0.5 or 1 mL/min using 0.2% acetic acid. An Agilent 1100 system coupled to a Trap VL or SL mass spectrometer was used.

Separate completely acetylated PIA from other molecules on a Superdex 200 10/300 GL column, partially or completely deacetylated PIA on a Jordi PolarPac WAX 10,000A 300 × 7.8 mm column (*see* **Note 7**).

3.4.2. Colorimetric Detection of the Degree of PIA Deacetylation by Analysis of Free Amino Groups

First, create a standard curve by using completely acetylated or deacetylated PIA as calibration standard (0–1 mg/mL) (*see* **Note 8**).

1. Add 100 µL of 4% NaHCO$_3$ (pH 8.5) and 100 µL of 0.1% TNBS to 100 µL of sample.
2. Allow mixture to react at 40 °C for 2 h.
3. Add 50 µL HCl to stop the reaction and measure the absorbance at 335 nm with a UV spectrophotometer.

4. Notes

1. Recent achievements in molecular biology have created multiple methods that can be utilized to construct mutants by allelic replacements in different bacteria. A variety of temperature-sensitive plasmids, antibiotics resistance markers, and transformation protocols are now available.
2. Boiling cells with 0.5 *M* EDTA, pH 8.0, *(5)* is the best method known to date for the isolation of crude PIA from the bacterial cell surface and might be applicable for staphylococcal cells only. PIA-related polysaccharide polymers have been purified from *E. coli* by incubating cells in 50 m*M* Tris buffer, pH 8.0, 100 mg lysozyme, and 0.1 *M* EDTA at room temperature for 2 h *(7)*.
3. PIA is a β-1,6 linked *N*-acetyl glucosamine homopolymer. Deacetylated PIA is cationic because of the free amino groups with a theoretical pK = 6.9 that become protonated at neutral or acidic pH. Acetylated PIA is insoluble at neutral pH, especially after precipitation from the culture filtrate with ethanol, or with increasing concentration of PIA after reducing the volume by ultrafiltration devices.
4. The initial PIA purification method was developed by Mack et al. *(3)*. These authors used a different, two-step chromatography protocol involving size-exclusion and ion exchange chromatography on Sephadex G-200, Q-Sepharose, and S-Sepharose. A similar purification method has been described recently to isolate a PIA-related polysaccharide polymer in *E. coli (7)*. Briefly, *E. coli* cells were incubated in 50 m*M* Tris–HCL buffer (pH 8.0), 100 mg lysozyme, and 0.1 *M* EDTA at room temperature for 2 h. Phenol/chloroform extraction steps were performed to separate protein and debris contamination from the polysaccharide. Samples were concentrated by ultrafiltration devices (10,000 MW cut off) and fractionated on a fast protein liquid chromatography (FPLC) system with a Sephacryl S-2000 column (equilibration and elution buffer: 0.1 *M* PBS, pH 7.4).

5. A similar method has been employed to purify poly-N-acetylglucosamine (PNAG from *Staphylococcus aureus* strain MN8 *(8)*. Production of PNAG and PIA is encoded by the same gene locus in staphylococci, the *icaADBC* operon. Both polysaccharides are considered to be the same. PNAG-containing samples were fractionated by SEC with a Sephacryl S-300 column and 0.1N HCl/0.15 M NaCl buffer *(8,9)*.

6. Incubation with the primary αPIA antiserum up to 16 h increases the strength of the immunological reaction.

7. These methods do not require further purification of PIA before chromatography. However, purging the columns extensively after several runs is necessary, as particularly the PolarPac Wax column tends to bind negatively charged polymers (such as teichoic acids) that need to be removed because they interfere with the repulsion mechanism of positively charged PIA on this column, which is needed for optimal purification of deacetylated PIA *(4)*.

8. This assay is a modified ninhydrin test to determine the degree of deacetylation in PIA. The reagent TNBS has been shown to react specifically with free amino groups to give trinitrophenyl (TNP) derivatives that can be measured with a UV spectrophotometer at 335 nm *(10)*. Chemically, complete deacetylation of PIA can be achieved by treatment with strong acid or base for several hours.

Acknowledgments

This work was supported by the Intramural Research Program of the National Institutes of Allergy and Infectious Diseases, National Institutes of Health.

References

1. Vuong, C., Voyich, J. M., Fischer, E. R., Braughton, K. R., Whitney, A. R., DeLeo, F. R., and Otto, M. (2004) Polysaccharide intercellular adhesin (PIA) protects *Staphylococcus epidermidis* against major components of the human innate immune system. *Cell. Microbiol.* **6,** 269–275.

2. Heilmann, C., Schweitzer, O., Gerke, C., Vanittanakom, N., Mack, D., and Gotz, F. (1996) Molecular basis of intercellular adhesion in the biofilm-forming *Staphylococcus epidermidis*. *Mol. Microbiol.* **20,** 1083–1091.

3. Mack, D., Fischer, W., Krokotsch, A., Leopold, K., Hartmann, R., Egge, H., and Laufs, R. (1996) The intercellular adhesin involved in biofilm accumulation of Staphylococcus epidermidis is a linear beta-1,6-linked glucosaminoglycan: purification and structural analysis. *J. Bacteriol.* **178,** 175–183.

4. Vuong, C., Kocianova, S., Voyich, J. M., Yao, Y., Fischer, E. R., DeLeo, F. R., and Otto, M. (2004) A crucial role for exopolysaccharide modification in bacterial biofilm formation, immune evasion, and virulence. *J. Biol. Chem.* **279,** 54881–54886.

5. Gerke, C., Kraft, A., Sussmuth, R., Schweitzer, O., and Gotz, F. (1998) Characterization of the N-acetylglucosaminyltransferase activity involved in the biosynthesis

of the *Staphylococcus epidermidis* polysaccharide intercellular adhesin. *J. Biol. Chem.* **273**, 18586–18593.

6. Bruckner, R. (1997) Gene replacement in Staphylococcus carnosus and Staphylococcus xylosus. *FEMS Microbiol. Lett.* **151**, 1–8.
7. Wang, X., Preston, J. F. I., and Romeo, T. (2004) The pgaABCD locus of Escherichia coli promotes the synthesis of a polysaccharide adhesin required for biofilm formation. *J. Bacteriol.* **186**, 2724–2734.
8. Maira-Litran, T., Kropec, A., Abeygunawardana, C., Joyce, J., Mark, G., 3rd, Goldmann, D. A., and Pier, G. B. (2002) Immunochemical properties of the staphylococcal poly-N-acetylglucosamine surface polysaccharide. *Infect. Immun.* **70**, 4433–4440.
9. Maira-Litran, T., Kropec, A., Goldmann, D. A., and Pier, G. B. (2005) Comparative opsonic and protective activities of *Staphylococcus aureus* conjugate vaccines containing native or deacetylated Staphylococcal Poly-N-acetyl-beta-(1–6)-glucosamine. *Infect. Immun.* **73**, 6752–6762.
10. Habeeb, A. F. S. A. (1966) Determination of free amino groups in proteins by trinitrobenzenesulfonic acid. *Anal. Biochem.* **14**, 328–336.

II

HOST-PATHOGEN INTERACTION

9

Analysis of *Staphylococcus aureus* Gene Expression During PMN Phagocytosis

Jovanka M. Voyich, Dan E. Sturdevant, and Frank R. DeLeo

Summary

Staphylococcus aureus is a leading cause of human infections worldwide and causes a variety of diseases ranging in severity from mild to life-threatening. The ability of *S. aureus* to cause disease is based in part on its ability to subvert the innate immune system. Advances in genome-wide analysis of host–pathogen interactions have provided the necessary tools to investigate molecular factors that directly contribute to *S. aureus* pathogenesis. This chapter describes methods to analyze gene expression in *S. aureus* during interaction with human neutrophils.

Key Words: *Staphylococcus aureus*; neutrophil; polymorphonuclear leukocytes; microarray; gene expression; Affymetrix; transcript.

1. Introduction

Polymorphonuclear leukocytes (PMNs or neutrophils) are key effectors of the human innate immune response against bacteria and fungi. At sites of infection, PMNs bind and ingest bacteria by a process known as phagocytosis *(1)*. Phagocytosis of bacteria elicits production of reactive oxygen species and enriches phagocytic vacuoles with peptide microbicides *(2)*. Although PMNs kill most bacteria, pathogens such as group A *Streptococcus* and *Staphylococcus aureus* have evolved means to avoid destruction by neutrophils and thereby cause human disease *(3–6)*. These pathogens have multiple virulence determinants that work in concert to promote infection. Until recently, the importance of bacterial genes in virulence and host interaction was determined

From: *Methods in Molecular Biology, vol. 431: Bacterial Pathogenesis*
Edited by: F. DeLeo and M. Otto © Humana Press, Totowa, NJ

either on an individual gene basis or for a limited number of genes in any single study. However, this "single-target" approach is being supplanted by microarray-based technologies, which have dramatically enhanced our ability to interrogate host–pathogen interactions on a genome-wide scale *(3,4,6–12)*.

Methods for measuring gene expression in Gram-positive bacteria have existed for several decades *(13,14)*. Because of the presence of thick cell walls not easily susceptible to lysis, early methods to isolate RNA from Gram-positive organisms were very tedious and time consuming. Procedures included incubating bacteria with detergent, using cesium chloride step-gradients in combination with ultracentrifugation, and use of enzymes such as lysostaphin to disrupt bacteria *(15,16)*. Because 90% of the *S. aureus* transcripts produced during log-phase growth have a half-life less than 5 min *(17)*, enzymatic treatment of bacteria is not ideal for isolation of RNA due to the time it takes for the enzyme to lyse the bacteria. Moreover, designing experiments to measure changes in gene expression in *S. aureus* during interaction with host factors should not be complicated by additional treatments that may independently cause changes in gene expression. Therefore, complex studies investigating host–pathogen interactions require non-enzymatic means of lysing *S. aureus*. Procedures using rapid physical disruption of the lysis-resilient *S. aureus* cell walls *(18)* are more appropriate for investigating pathogen–host interactions *(3,4,6–8,10–12)*. This is accomplished via high-speed (6 m/s) reciprocating shaking of the sample with 0.1-m silica spheres in a suitable RNA lysis buffer.

In this chapter, we describe a method for analyzing *S. aureus* gene expression during interaction with human PMNs using Affymetrix oligonucleotide microarrays. Except for human PMNs, the methods described herein utilize commercially available reagents and can be reproduced by most standard laboratories.

2. Materials

2.1. Bacterial Culture

1. Tryptic soy broth (BD Biosciences, San Jose, CA).
2. Dulbecco's phosphate-buffered saline (DPBS, Gibco/Invitrogen Corporation, Carlsbad, CA).
3. Normal human serum.
4. RPMI 1640 medium (Gibco/Invitrogen) containing 10 m*M* HEPES (RPMI/H): made by adding 5 mL of sterile 1 *M* HEPES (Sigma-Aldrich, St. Louis, MO) to 500 mL of sterile RPMI 1640 medium.

2.2. Isolation of Human PMNs and Serum, and Phagocytosis Assays

1. Dextran 500 (GE Healthcare Biosciences, Piscataway, NJ).
2. Ficoll–Paque^{PLUS} (GE Healthcare Bioscience).
3. Injection- or irrigation-grade 0.9% NaCl solution (w/v) (Hospira, Inc., Lake Forest, IL).
4. 3% Dextran–0.9% NaCl solution (w/v). To make this solution, dissolve 30 g of Dextran 500 into 1 L injection- or irrigation-grade 0.9% NaCl (final volume).
5. Injection- or irrigation-grade water (Abbot Laboratories, Inc., North Chicago, IL).
6. DPBS.
7. 1.7% NaCl solution (w/v): Dissolve 17 g of RNase-free NaCl into 1 L (final volume) of injection- or irrigation-grade water. Filter-sterilize solution using a 0.2-μm bottle top filter.
8. 50- and 250-mL conical polypropylene tubes (Corning Costar, Acton, MA).
9. 12-well flat-bottom non-pyrogenic polystyrene cell culture plates (Corning Costar).
10. RPMI 1640 medium (Invitrogen Life Technologies, Carlsbad, CA): Prepare as described under **Subheading 2.1**.
11. 15-mL glass tube (Corning Costar) for isolating serum.
12. *Limulus* Amebocyte Lysate assay (Thermo Fisher, Waltham, MA). This assay is used to periodically check reagents for the presence of endotoxin.

2.3. Purification of RNA from S. aureus

1. Lysing matrix B (0.1-mm silica spheres) (MP Biomedicals, Irvine, CA).
2. FastPrep FP120 A Instrument (MP Biomedicals).
3. RNeasy Mini Kit (Qiagen, Valencia, CA).
4. β-Mercaptoethanol (β-ME) (Sigma-Aldrich, St. Louis, MO).
5. Ethanol, 200-proof (Sigma-Aldrich, St. Louis, MO).
6. 1.7-mL microcentrifuge tubes, RNase-free and non-pyrogenic (Molecular BioProducts, San Diego, CA).
7. Extra collection tubes (Qiagen).
8. Aerosol-resistant micropipette tips, RNase-free, non-stick (Ambion, Austin, TX).

2.4. Removal of Contaminating DNA from S. aureus RNA Preparations

1. RNase-Free DNase Set, on-column digestion of DNA to be used during RNA purification (Qiagen). Set includes DNase I (1500 Kunitz Units), DNase buffer RDD, and nuclease-free water.

2.5. cDNA Synthesis of S. aureus RNA

1. dNTP mix: Combine 100 mM each of dATP, dCTP, dGTP, dTTP (GE Healthcare Biosciences). For a working 10 mM dNTP mix, add 100 μL each of 100 mM

dATP, dCTP, dGTP, and dTTP to 600 μL nuclease-free water. Store this solution at −20 °C in a non-frost-free freezer.

2. Random primers, 3 μg/μL (Invitrogen, Life Technologies). For 75 ng/μL of working stock, mix 25 μL of 3 μg/μL random primers with 975 μL of nuclease-free water. Store this solution at −20 °C in a non-frost-free freezer.

3. Superscript II or Superscript III Reverse Transcriptase (Invitrogen Life Technologies).

4. 5× 1st Strand Buffer (Invitrogen Life Technologies).

5. 100 m*M* DTT (Invitrogen Life Technologies).

6. SUPERase·In, (Ambion).

7. Nuclease-free water (Ambion).

2.6. Removal of RNA

1. 1*N* NaOH solution (VWR Scientific Products, West Chester, PA).

2. 1*N* HCl solution (VWR Scientific Products).

2.7. Purification of cDNA

1. QIAquick PCR Purification Kit (Qiagen).

2.8. cDNA Fragmentation

1. 10× One-Phor-All Buffer (GE Healthcare Biosciences, Piscataway, NJ).

2. DNase I (GE Healthcare Biosciences, Piscataway, NJ).

2.9. Terminal Labeling of Fragmented cDNA

1. Enzo BioArray Terminal Labeling Kit (Enzo Biochem, Inc. New York, NY): 5× reaction buffer, 10× $CoCl_2$, biotin-ddUTP, and terminal deoxynucleotide transferase.

2.10. Hybridization of S. aureus cDNA to Affymetrix GeneChips

1. Nuclease-free water (Ambion).

2. Sterile, RNase-free microcentrifuge vials (1.7-mL, Ambion).

3. Aerosol-resistant micropipette tips, RNase-free, non-stick (Ambion).

4. Bovine serum albumin (BSA) solution, 50 mg/mL (Invitrogen Life Technologies).

5. Herring sperm DNA (Promega Corporation, Madison, WI).

6. Control Oligo B2, 3 n*M*, including 20× hybridization controls (Affymetrix).

7. NaCl, 5 *M*, Rnase/Dnase free (Ambion).

8. Tough-Spots label dots (Diversified Biotech, Boston, MA).

9. Hybridization oven 640 (Affymetrix).
10. Surfact-Amps 20 (Tween-20), 10% (Pierce Chemical, Rockford, IL).
11. 12× 2-(N-morpholino) ethanesulfonic acid (MES) stock buffer (1.22 *M* MES, 0.89 *M* Na$^+$): 64.61 g of MES hydrate, 193.3 g of MES sodium salt, and 800 mL of nuclease-free water. Adjust the pH to 6.5–7.0, mix and adjust volume to 1000 mL. Filter-sterilize (0.2 µm) and store at 2–8 °C protected from light. Discard if the solution turns yellow.
12. 2× hybridization buffer (50 mL): 8.3 mL of 12× MES stock buffer, 17.7 mL of 5 *M* NaCl, 4.0 mL of 0.5 *M* EDTA, 0.1 mL of 10%-Tween 20, and 19.9 mL nuclease-free water. Store hybridization buffer at 2–8 °C protected from light.

2.11. Washing and Staining of S. aureus Affymetrix GeneChips

1. Nuclease-free water (Ambion).
2. BSA solution, 50 mg/mL (Invitrogen Life Technologies).
3. NaCl, 5 *M*, RNase/DNase free (Ambion).
4. R-Phycoerythrin Streptavidin (Molecular Probes/Invitrogen Life Technologies).
5. Fluidics Station 450 (Affymetrix).
6. 20× Saline-Sodium Phosphate EDTA buffer (SSPE) (BioWhittaker Molecular Applications, Walkersville, MD/Cambrex): 3 *M* NaCl, 0.2 *M* NaH$_2$PO$_4$, 0.02 *M* EDTA).
7. Goat IgG, Reagent Grade (Sigma-Aldrich, St. Louis, MO).
8. For 10 mg/mL stock reagent, resuspend 50 mg of goat IgG in 5 mL 150 mM NaCl. Store at 4 °C.
9. Biotinylated anti-streptavidin goat antibody (Vector Laboratories, Burlingame, CA).
10. Surfact-Amps 20 (Tween-20), 10% (Pierce Chemical, Rockford, IL).
11. Wash Buffer A, non-stringent wash buffer (6× SSPE, 0.01% Tween-20): 300 mL of 20× SSPE, 1.0 mL of 10% Tween-20, and 699 mL of water. Filter buffer using a 0.2-µm filter and store at room temperature.
12. Wash Buffer B, stringent wash buffer (100 m*M* MES, 0.1 *M* Na$^+$, 0.01% Tween-20): 83.3 mL 12× MES stock buffer, 5.2 mL 5 *M* NaCl, 1.0 mL 10% Tween-20, and 910.5 mL water. Filter buffer using a 0.2-µm filter and store at 2–8 °C protected from light.
13. 2× Stain Buffer (100 m*M* MES, 1 *M* Na+, 0.05% Tween-20): 41.7 mL 12× MES stock buffer, 92.5 mL 5 *M* NaCl, 2.5 mL 10% Tween-20, and 113.3 mL water. Filter buffer using a 0.2-µm filter and store at 2–8 °C protected from light.

2.12. GeneChip Scanning and Conversion of Image Files

1. Affymetrix GeneChip Scanner 3000 7G Autoloader capable of high-resolution scanning (Affymetrix).
2. Affymetrix GeneChip Operating Software (GCOS) and the GeneChip Scanner 3000 high-resolution scanning patch (Affymetrix).

3. Methods

3.1. Isolation of Human Neutrophils and Serum

It is best to make certain all reagents are free of endotoxin contamination before attempting to isolate neutrophils. This can be done using the *Limulus* Amebocyte Lysate assay (as described by the manufacturer).

1. Add whole heparinized blood at a 1:1 ratio to a sterile solution of 3% dextran–0.9% NaCl (use 50- or 250-mL tubes depending on the blood volume). Mix gently and let stand at room temperature for 20–25 min.
2. During the dextran sedimentation, prepare the following in 50-mL conical tubes: (1) 35 mL of 0.09% NaCl, (2) 20 mL of 1.7% NaCl, (3) 12 mL of Ficoll–Paque^{PLUS}, and (4) 20 mL of injection- or irrigation-grade water. These solutions are normally stored at 4 °C, but to prevent priming and aggregation of PMNs, it is optimal to warm the solutions to room temperature.
3. After the 20–25 min incubation, transfer the top layer from the dextran–blood mixture to fresh tube(s) and centrifuge at 500–700 × g (with low or no break) for 10 min at room temperature. Aspirate supernatant and discard.
4. Resuspend pellet in 10 mL of NaCl from the 35 mL prepared in **step 2**. Add remaining NaCl to 35-mL total volume.
5. Underlay 10 mL of Ficoll–Paque^{PLUS} beneath the NaCl cell suspension. Spin at 500–700 × g (with low or no break) for 25 min at room temperature.
6. Carefully aspirate the supernatant. Using a sterile swab, wipe the inside of the tube to remove any residual peripheral blood mononuclear cells. This step greatly increases the purity of the PMN prep.
7. Lyse the red blood cells by resuspending the pellet in 20 mL of water. Mix gently by pipetting for 20–30 s.
8. Immediately add 20 mL of 1.7% NaCl to prevent lysis of the PMNs and centrifuge sample at 380 × g for 10 min at room temperature.
9. Aspirate supernatant and reususpend pellet in RPMI/H.
10. Count cells using a hemacytometer.
11. To isolate human serum, collect blood without coagulant and incubate at 37 °C in a glass tube for 30 min. This incubation causes the whole blood to clot. After incubation, centrifuge sample at 2000–3000 × g for 10 min. Serum can be used immediately or stored at –20 °C for later use.

3.2. Phagocytosis of S. aureus by Human PMNs

1. Inoculate flasks of fresh tryptic soy broth (TSB) containing 5% glucose with a 1:200 dilution of an overnight culture of *S. aureus*. Incubate cultures at 37 °C with shaking (250 rpm) until the optical density at 600 nm (OD_{600}) reaches 0.75. Typically, this is mid-exponential growth phase for *S. aureus*. For example, an OD_{600} of 0.75 corresponds to 2.5 × 10^8 colony-forming units (CFUs)/mL for *S. aureus* strain MW2 (USA400) (**6**). However, the growth kinetics of *S. aureus* varies between strains and depends on the growth medium used. Therefore, it

is necessary to perform traditional growth curves with the particular strain of *S. aureus* of interest to your laboratory to calculate number of CFUs/mL.

2. Opsonize *S. aureus* in 50% normal human serum for 30 min at 37 °C. Following opsonization, centrifuge bacteria at $\sim 5000 \times g$ for 5 min and resuspend at 10^9/mL in RPMI/H.

3. Coat the bottom of 12-well cell culture plates with 20% normal human serum. Incubate at 37 °C for at least 30 min. Wash plates (three times) with sterile DPBS. Keep plates on ice or at 4 °C until use. Coat enough wells for all samples during all time points. This includes samples of PMNs only, PMNs + *S. aureus*, and *S. aureus* only.

4. Place the serum-coated 12-well culture plates on ice. To one plate, add 1 mL of RPMI/H containing 10^7 purified PMNs to each well. To another plate, add 900 µL of RPMI/H.

5. Add 100 µL of 10^9/mL opsonized *S. aureus* to all wells except for PMN only control wells.

6. Centrifuge plates at 4 °C at $500 \times g$ for 8 min to synchronize phagocytosis. Spin all plates even if wells contain only *S. aureus* or only PMNs.

7. After centrifugation, incubate plates at 37 °C for desired amounts of time. For example, a time course may consist of 30, 60, and 180 min.

3.3. Purification of RNA from S. aureus During/After Phagocytosis by Human PMNs

1. At desired time points, harvest samples by removing all liquid from the wells ($\sim$1.1 mL) and centrifuging at $5000 \times g$ for 5 min (at room temperature). Immediately after removing all supernatant, add 700 µL of RLT lysis buffer (containing 143 m*M* β-ME) to the wells (*see* **Note 1**). Swirl lysis buffer to completely coat the bottom of the wells. Pipet lysate to ensure complete lysis of the adherent portion of the sample. When centrifugation is finished, decant supernatant and lyse pellet with the lysis buffer used to lyse the adherent cells in the corresponding well. Vigorously pipet and vortex the sample.

2. Add each sample to a 2-mL tube of Lysing Matrix B. Physically disrupt sample using a FastPrep instrument at 6 m/s for 20 s.

3. Centrifuge sample (in the Lysing Matrix B tube) at $600 \times g$ for 4 min (this centrifugation is simply to "settle" the silica spheres).

4. Pipet sample into new 1.7-mL microcentrifuge tubes. Approximately 500 µL of lysate will be recovered.

5. Add 350 µL of ethanol to the sample. Mix sample by pipetting.

6. Apply 700 µL of the sample (including any precipitate) to an RNeasy mini column placed in a 2-mL collection tube. Centrifuge sample for 60 s at $\geq$8000 $\times$ *g* and discard the flow-through. Run any excess sample through the same column as described.

7. Wash the column with 700 µL of buffer RW1 (from RNeasy kit) and centrifuge sample for 60 s at $\geq$8000 $\times$ *g*. Discard the flow-through.

8. Transfer RNeasy column into a new 2-mL collection tube. Apply 500 μL of RPE buffer (from RNeasy kit). Wash the column by centrifugation at ≥8000 × g for 60 s. Discard flow-through.

9. To remove residual ethanol, place the RNeasy column in a new 2-mL collection tube and centrifuge at maximum speed for 2 min.

10. To elute purified RNA, transfer the RNeasy column to a new 2-mL collection tube, pipet 40 μL of RNase-free water directly onto the silica-gel membrane, and centrifuge at ≥8000 × g for 1 min. Run eluted RNA over the same column again to increase RNA yield.

3.4. DNase Digestion and Clean-Up of S. aureus RNA

DNA contamination is difficult to remove from *S. aureus* RNA preparations. The protocol below uses a single on column DNase digestion followed by two consecutive RNA clean-up assays. The RNeasy silica-membrane removes DNA; therefore, additional clean-up steps are used to aid in removal of contaminating *S. aureus* DNA.

1. To purified *S. aureus* RNA, add 60 μL of RNase-free water (total sample volume should be 100 μL), add 350 μL RLT to the sample, and mix thoroughly by pipetting.

2. Add 250 μL of ethanol and mix thoroughly by pipetting.

3. Apply the sample to an RNeasy mini column placed in a collection tube (2-mL) and centrifuge for 60 s at ≥8000 × g. Discard flow-through.

4. Wash the column by adding 350 μL Buffer RW1 and centrifuging for 60 s at ≥8000 × g.

5. Add 80 μL of DNase solution. DNase solution contains ~27 Kunitz U resuspended in RDD buffer. It is made by adding 10 μL of DNase stock solution (750 Kunitz U/mL) to 70 μL RDD buffer.

6. Incubate sample with DNase for 15 min at room temperature.

7. Wash column with 350 μL Buffer RW1 and centrifuge for 60 s at ≥8000 × g and discard flow-through.

8. Transfer the RNeasy column into a new collection tube (2-mL). Wash column with 500 μL Buffer RPE by centrifugation for 60 s at ≥8000 × g, discard flow-through, and repeat RPE wash.

9. Place the RNeasy column into a new collection tube (2-mL) and centrifuge for 2 min at maximum speed to remove residual ethanol.

10. Elute purified RNA by transferring the RNeasy column to a new 2-mL collection tube, pipet 40 μL of RNase-free water directly onto the silica-gel membrane, and centrifuge at ≥8000 × g for 1 min. Run eluted RNA over the same column again to increase RNA yield.

11. Repeat RNA clean-up procedure. Follow **steps 1–3** and **8–10** (skip DNase **steps 4–7**). To ensure removal of DNA contamination, perform one additional clean-up step. Although this procedure is time consuming, it results in very high quality

RNA. Purity of RNA sample is vital for microarray analysis when working with heterogeneous samples (*see* **Note 2** for RNA quantification).

3.5. cDNA Synthesis from S. aureus–PMN RNA

1. Add 10 µg of purified *S. aureus*–PMN RNA (can use up to 20 µL, final concentration will be ˜0.33 µg/µL) to a 0.2-mL thin-walled PCR tube.
2. To the sample, add 10 µL of 75 ng/µL random primers (final concentration 25 ng/µL).
3. Add enough nuclease-free water to make the final volume of the sample 30 µL.
4. Incubate the RNA–primer mix at 70 °C for 10 min and 25 °C for 10 min, and chill to 4 °C. Use a thermocycler for all cDNA and target labeling reactions.
5. Make a master mix of cDNA synthesis components as follows (final concentrations are listed in parenthesis): 12 µL of 5× 1st Strand Buffer (1×), 6 µL of 100 mM DTT (10 mM), 3 µL of 10 mM dNTPs (0.5 mM), 1.5 µL SUPERase·In at 20 U/µL (0.5 U/µL), and 7.5 µL of SuperScript II (or III) at 200 U/µL (25 U/µL).
6. To the RNA–primer mix (30 µL), add 30 µL of master mix from **step 5**.
7. Synthesize cDNA by PCR as follows: 25 °C for 10 min, 37 °C for 60 min, 42 °C for 60 min. Stop reaction by inactivating SuperScript II (or III) at 70 °C for 10 min. Cool sample to 4 °C.
8. Remove RNA by adding 20 µL of 1N NaOH to the newly synthesized cDNA sample. Incubate at 65 °C for 30 min.
9. Neutralize the reaction by adding 20 µL of 1N HCl.
10. Clean-up the synthesis product using a QIAquick Column (as described by the manufacturer). Elute cDNA with 40 µL of elution buffer (Buffer EB included in the kit) (*see* **Note 2** about cDNA quantification).
11. Store purified cDNA at –20 °C.

3.6. cDNA Fragmentation

1. To the 40 µL of purified cDNA (from the previous step) containing 2–10 µg, add 5 µL of 10× One-Phor-All Buffer (1× final concentration).
2. Add 0.6 U of DNase I for each µg of cDNA. Dilute DNase in 1× One-Phor-All Buffer (*see* **Note 3** for information about checking the activity of the DNase I enzyme).
3. Add nuclease-free water until the final volume reaches 50 µL.
4. To fragment the cDNA, incubate the reaction mixture at 37 °C for 10 min.
5. Stop the reaction by inactivating the DNase at 98 °C for 10 min.
6. Store the fragmented cDNA at –20 °C until use (*see* **Note 3** about checking fragmented cDNA).

3.7. Labeling Fragmented cDNA with Biotin–ddUTP

1. To label fragmented cDNA, add 20 µL 5× reaction buffer, 10 µL 10× CoCl$_2$, and 1 µL biotin–ddUTP, and 2 µL terminal deoxynucleotide transferase to 39 µL

fragmented cDNA. Bring sample up to 100 μL with nuclease-free water (*see* **Note 4**).

2. Incubate the terminal labeling reaction at 37 °C for 60 min.

3. Stop the reaction by storing the sample at –80 °C until hybridization. The labeling efficiency can be checked by performing a gel-shift assay. As a general rule, over 90% of the fragments should be labeled and thus shifted.

3.8. Hybridization of cDNA to Affymetrix GeneChips

The following protocol is designed for analysis of *S. aureus* transcripts on Affymetrix 49 (standard) arrays.

1. Remove GeneChips from 4 °C storage. Allow chips to warm to room temperature prior to hybridization (allow at least 60 min for GeneChips to equilibrate). Likewise, thaw reagents from cold storage at room temperature. Set heat block at 65 °C and set the hybridization oven to 45 °C.

2. Place the 20× hybridization controls tube into the 65 °C heat block and incubate 5 min.

3. Mix the hybridization cocktail for each GeneChip. Use the following reaction mix: 3–7 μg (~50 μL) of cDNA, 3.3 μL of control oligonucleotide B2 (3 n*M*), 10 μL of 20× hybridization controls, 2 μL of herring sperm (10 mg/mL), 2 μL of BSA (50 mg/mL), 100 μL of 2× hybridization buffer, and 32.7 μL of H_2O for a final volume of 200 μL.

4. Place the GeneChips face down on the bench. Place a 200-μL pipette tip in one of the septa. Using the other septum, add 200 μL of the appropriate hybridization cocktail.

5. Place the GeneChips into the hybridization oven (45 °C) for 16 h at 60 rpm.

3.9. Processing Affymetrix GeneChips

Begin processing the Affymetrix GeneChips immediately following hybridization.

1. Turn on the fluidics station, while also assuring the tubing is placed into wash bottles A and B, water, and waste.

2. Turn on the workstation and open GCOS.

3. Click on the "fluidics" icon and prime the workstation(s) intended for use.

4. For each chip, using an amber tube, mix 600 μL of 2× stain buffer, 48 μL of BSA (50 mg/mL), 12 μL of streptavidin phycoerythrin (SAPE 1 mg/mL), and 540 μL of distilled H_2O. This is the SAPE solution mix.

5. Remove 600 μL of the SAPE solution and place into another amber tube.

6. For each GeneChip, using a clear tube, mix 300 µL of 2× stain buffer, 24 µL of BSA (50 mg/mL), 6 µL Goat IgG stock (10 mg/mL), 6 µL of biotiny-lated antibody (0.5 mg/mL), and 264 µL of distilled H_2O. This is the antibody solution mix.

7. Remove the GeneChips from the hybridization oven and place a 200-µL pipette tip into one of the septa. Remove the hybridization cocktail and place into the appropriate sample tube. These samples can be run on other chips at a later date. Store unused samples at –80 °C.

8. Fill the GeneChip with ~250 µL of wash buffer A without air bubbles (*see* **Note 5**). If processing more GeneChips than allowed in the fluidics station, the remaining filled chips can be stored up to 4 h at 4 °C.

9. In GCOS, click on the "expts" icon and create an experiment (*.EXP) for each chip.

10. In GCOS, click on "tools," and then "filters." Set the filters so your experiments can be viewed.

11. In GCOS fluidics window, select the appropriate protocol for each module and run.

12. Place the appropriate GeneChip into the fluidics modules and place the SAPE solution mix tubes into sample holder positions 1 and 3 for each module.

13. Place the antibody solution mix tube into sample holder position 2 for each module.

14. After about 90 min, the protocol should be finished. Once the fluidics station displays "remove cartridge," take the chip out and inspect it for air bubbles. If any are present, place back into fluidics module and let it refill. If there are still bubbles, then the chip must be filled manually with wash buffer A (*see* **Note 5**).

15. Place a tough spot over each septum for each GeneChip to be scanned.

16. Gently wipe the glass of the GeneChip with a tissue in one direction. Do not change direction, as the paraffin at the edges may coat the glass, making scanning difficult.

3.10. Scanning

The scanning protocol is written using the 3000 7G (plus) with autoloader. If an autoloader is not present, the protocol must be altered to accommodate single-chip scanning. Instruments with serial number starting with 501 or earlier will require the high-resolution scanning update.

1. Press the start button on the scanner.
2. Open GCOS and assure the connection between the scanner and workstation has been made.
3. Allow scanner to warm up for 10 min.
4. Place chips into the autoloader starting with position one.

5. In GCOS, press the "start" icon, and while the scanner is performing its automated tasks, click on "view" and "scan in progress" to visualize the results once scanning begins.

6. A successful scan will create the *.DAT and *.CEL files. The *.CEL files are now ready for analysis to create the *.CHP file for subsequent data analysis.

4. Notes

1. Because microarray analysis requires the use of fluorescence, the lysis buffer must be chosen carefully. For instance, many lysis reagents contain phenol and chloroform. Residual phenol and or chloroform can interfere with fluorescent labeling. This usually equates to less signal and more background and hence difficulty in interpreting microarray results. The lysis buffer RLT (provided with the RNeasy Mini Kit) does not interfere with subsequent labeling of the samples.

2. It is difficult to determine RNA quantity from *S. aureus* during *S. aureus*–PMN interactions because the *S. aureus* RNA isolation procedure will also isolate PMN RNA. Because of the development of very specific oligonucleotide-based bacterial gene chips, it is no longer necessary to remove host RNA *(3)*. RNA isolated as described above typically yields 100–200 µg/mL of total RNA (RNA from *S. aureus* and human PMNs). RNA can be measured by absorbance at 260 nm on a spectrophotometer (1 absorbance U = 40 µg/mL RNA). The ratio of 260/280 should be approximately 2.0. Alternatively, a fluorescence-based assay such as Quant-iT RiboGreen RNA assay (Invitrogen Life Technologies) can be used. This assay requires the use of a microplate spectrofluorometer (e.g., Molecular Devices Gemini XPS, Sunnyvale, CA). Quantifying cDNA only provides an estimate of how much of the sample is *S. aureus* cDNA. Like RNA, cDNA can be measured by absorbance at 260 nm (1 absorbance U = 33 µg/mL of single-stranded DNA). Alternatively, cDNA can be quantitated using a fluorescence-based assay such as Quant-iT OliGreen ssDNA assay (Invitrogen Life Technologies).

3. Lot-to-lot variation of DNase I enzyme activity is possible. A simple titration assay can be performed to independently determine DNase I activity. Fragmented samples (200 ng) can be loaded onto a 4–20% acrylamide gel and stained with SYBR Gold. Size of fragmented cDNA will be ~50–200 base pairs.

4. Conditions for the terminal labeling reaction are determined by the format of the specific Affymetrix GeneChip being used. The reaction described above is optimized for the Standard (49) Format.

5. In order to fill the 49 format GeneChip without air bubbles, approximately 300 µL of wash A is required. Place the chip flat on the bench with the septa facing up. Place a 200 µL pipette tip in a septum and fill a 200 µL pipette with wash A. Hold the GeneChip parallel to the floor and insert the full pipette into the open septum. Dispense only about 180 µL of the 200 µL and remove the pipette assuring no air bubble will be in the septa region. Switch to a new tip, then load another 200 µL of wash A. Making sure no air bubbles are at the tip of the pipette, insert it once again into the open septum and hold the chip perpendicular to the floor with

the existing air bubble at the top and center of the viewing glass (level). Begin dispensing the remaining wash A into the chip while holding the bubble steady at the top center. Dispense until a small amount of wash A is visible in the empty tip (opposite septum), and remove the pipette. A clear glass without air should remain.

Acknowledgments

This work was supported by the NIH-NRRI grant P20RR020185 (J.M.V.) and the Intramural Research Program of the National Institute of Allergy and Infectious Diseases, National Institutes of Health (F.R.D.).

References

1. Allen, L. A. and Aderem, A. (1996) Mechanisms of phagocytosis. *Curr. Opin. Immunol.* **8**, 36–40.
2. Nauseef, W. M. and Clark, R. A. (2000) *Granulocytic Phagocytes* (Mandel, G. L., Bennett, J. E. and Dolin, R., eds.), Churchill Livingstone, pp. 89–112.
3. Voyich, J. M., Sturdevant, D. E., Braughton, K. R., et al. (2003) Genome-wide protective response used by group A Streptococcus to evade destruction by human polymorphonuclear leukocytes. *Proc. Natl. Acad. Sci. U.S.A.* **100**, 1996–2001.
4. Voyich, J. M., Braughton, K. R., Sturdevant, D. E., et al. (2004) Engagement of the pathogen survival response used by group A *Streptococcus* to avert destruction by innate host defense. *J. Immunol.* **173**, 1194–1201.
5. Voyich, J. M., Musser, J. M., and DeLeo, F. R. (2004) *Streptococcus pyogenes* and human neutrophils: a paradigm for evasion of innate host defense by bacterial pathogens. *Microbes. Infect.* **6**, 1117–1123.
6. Voyich, J. M., Braughton, K. R., Sturdevant, D. E., et al. (2005) Insights into mechanisms used by *Staphylococcus aureus* to avoid destruction by human neutrophils. *J. Immunol.* **175**, 3907–3919.
7. Graham, M. R., Smoot, L. M., Migliaccio, C. A., et al. (2002) Virulence control in group A Streptococcus by a two-component gene regulatory system: global expression profiling and in vivo infection modeling. *Proc. Natl. Acad. Sci. U.S.A.* **99**, 13855–13860.
8. Graham, M. R., Virtaneva, K., Porcella, S. F., et al. (2005) Group A Streptococcus transcriptome dynamics during growth in human blood reveals bacterial adaptive and survival strategies. *Am. J. Pathol.* **166**, 455–465.
9. Musser, J. M. and DeLeo, F. R. (2005) Toward a genome-wide systems biology analysis of host-pathogen interactions in group A *Streptococcus. Am. J. Pathol.* **167,** 1461–1472.
10. Sumby, P., Whitney, A. R., Graviss, E. A., DeLeo, F. R., and Musser, J. M. (2006) Genome-wide analysis of group A *Streptococci* reveals a mutation that modulates global phenotype and disease specificity. *PLoS Pathog.* **2,** 0041–0049.

11. Virtaneva, K., Graham, M. R., Porcella, S. F., et al. (2003) Group A *Streptococcus* gene expression in humans and cynomolgus macaques with acute pharyngitis. *Infect. Immun.* **71**, 2199–2207.
12. Virtaneva, K., Porcella, S. F., Graham, M. R., et al. (2005) Longitudinal analysis of the group A Streptococcus transcriptome in experimental pharyngitis in cynomolgus macaques. *Proc. Natl. Acad. Sci. U.S.A.* **102**, 9014– 9019.
13. Jones, A. S. and Walker, R. T. (1968) The isolation of nucleic acids from gram-positive bacteria. *Arch. Biochem. Biophys.* **128**, 579–582.
14. Marmur, J. and Doty, P. (1961) Thermal renaturation of deoxyribonucleic acids. *J. Mol. Biol.* **3**, 585–594.
15. Kornblum, J. S., Projan, S. J., Moghazeh, S. L., et al. (1988) A rapid method to quantitate non-labeled RNA species in bacterial cells. *Gene* **63**, 75–85.
16. Jordan, D. C. and Inniss, W. E. (1959) Selective inhibition of ribonucleic acid synthesis in *Staphylococcus aureus* by vancomycin. *Nature* **184 (Suppl 24)**, 1894–1895.
17. Roberts, C., Anderson, K. L., Murphy, E., et al. (2006) Characterizing the effect of the Staphylococcus aureus virulence factor regulator, SarA, on log-phase mRNA half-lives. *J. Bacteriol.* **188**, 2593–2603.
18. Cheung, A. L., Eberhardt, K. J., and Fischetti, V. A. (1994) A method to isolate RNA from Gram-positive bacteria and mycobacteria. *Anal. Biochem.* **222**, 511–514.

10

Examining the Vector–Host–Pathogen Interface With Quantitative Molecular Tools

Jason E. Comer, Ellen A. Lorange, and B. Joseph Hinnebusch

Summary

We developed PCR assays to detect and quantitate *Yersinia pestis*, the bacterial agent of plague, in flea vector and mammalian host tissues. Bacterial numbers in fleas, fleabite sites, and infected lymph nodes were determined using real-time PCR with primers and probes for a gene target on a multi-copy plasmid specific to *Y. pestis*. Tissue-matched standard curves used to determine absolute bacterial numbers in unknown samples were linear over at least five orders of magnitude. The methods were applied to studies of transmission of *Y. pestis* by the rat flea *Xenopsylla cheopis*, but should be generally useful to investigate the transmission dynamics of any arthropod-borne disease.

Key Words: Arthropod-borne disease; quantitative real-time polymerase chain reaction; vector-borne transmission; vector competence.

1. Introduction

Pathogens transmitted by blood-feeding arthropods cause serious public health problems worldwide. Despite their importance, many basic epidemiological parameters of arthropod-borne diseases remain poorly defined because transmission cycle dynamics have been difficult to study. For example, quantitative estimates of vector competence, such as susceptibility to infection, permissiveness of pathogen reproduction, extent of dissemination in the arthropod, and comparative transmission efficiencies of different vector species of a particular pathogen are lacking for many arthropod-borne diseases. Epidemiological modeling of vector-borne diseases therefore often relies on indirect estimates of these parameters *(1)*. Much also remains to be learned about how specific microbial factors

From: *Methods in Molecular Biology, vol. 431: Bacterial Pathogenesis*
Edited by: F. DeLeo and M. Otto © Humana Press, Totowa, NJ

influence the ability to produce a transmissible infection in the vector and to disseminate from a bite site after transmission.

Many arthropod-borne viruses, bacteria, and parasites are difficult to culture in the laboratory, and microscopic direct counts of microbes in infected tissues are labor-intensive and relatively insensitive. The advent of quantitative molecular tools has provided rapid and sensitive alternative means to determine microbial numbers at different stages of infection in arthropod and mammalian hosts. Prominent among these new tools is real-time quantitative PCR (qPCR). Several instrumentation and detection systems devoted to real-time PCR are commercially available and have been used widely to quantitate gene and transcript copy numbers (reviewed in **refs. 2–5**). Although these systems differ in their details, they share the same basic principles: The accumulation of a target sequence is measured in real time during the exponential phase of PCR via a fluorescent label. The number of PCR cycles needed for amplification-associated fluorescence to reach a specific threshold level of detection (the C_t value) is inversely proportional to the amount of target sequence in the sample. Therefore, the copy number of the target in an unknown sample can be determined by interpolation of its C_t value versus a standard curve of C_t values obtained from a serially diluted solution containing known amounts of target.

A variety of real-time qPCR strategies have been successfully applied to detect and quantify arthropod-borne pathogens in their vectors and hosts. From these studies has come important new information on the kinetics and progression of infection in both the arthropod and the mammal *(6–12)*, the number of microbes transmitted by an infected vector during a blood meal *(13–15)*, the number of microbes taken up by an uninfected vector in an infected blood meal *(13,16)*, and the comparative competence and transmission efficiency of different arthropod vectors for an agent *(17,18)*. In this chapter, we detail real-time qPCR protocols used to investigate flea-borne transmission dynamics and pathogenicity of *Yersinia pestis (13,16)*. A key feature is the use of tissue-matched standards containing known numbers of *Y. pestis*, minimizing the effect of PCR inhibitors in different sample tissues and allowing absolute quantification of bacterial cells. We have found that estimation of *Y. pestis* in flea, skin, and lymph node samples by qPCR compared favorably with the "gold standard" colony-forming unit quantitation method (*see* **Fig. 1**).

2. Materials
1. ABI Prism 7700 or 7900HT Sequence Detection System (Applied Biosystems, Foster City, CA) and associated data analysis software.
2. Primer Express software v2.0 (Applied Biosystems).

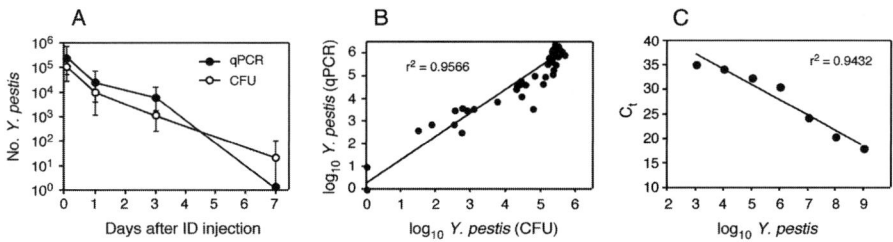

Fig. 1. Evaluation of quantitative PCR (qPCR) methods to quantify *Yersinia pestis* in host and vector tissues. (**A**) Comparison of qPCR and colony-forming unit (CFU) methods to track the decline in numbers of an attenuated *Y. pestis* strain after intradermal (ID) injection. Skin biopsies from individual mice were taken 0–7 days after injection with 10^5 *Y. pestis* KIM6$^+$. Biopsies were triturated with a tissue homogenizer and divided equally for qPCR and plate count assays. (**B**) Comparison of qPCR and CFU methods to quantitate *Y. pestis* in infected fleas. Flea triturates (*see* **step 4, Subheading 3.1.2.**) were divided equally for qPCR and plate count assays. (**C**) Standard curve prepared for qPCR of *Y. pestis* in rat lymph nodes as described in **Subheading 3.1.4**.

3. TaqMan® primers and probe specific for the plasmid-borne plasminogen activator (*pla*) gene of *Y. pestis*:

    Pla Forward Primer: 5´-CAAATATATCCCCTGACAGCTTTACA-3´
    Pla Reverse Primer: 5´-AAGCATTTCATGAGACTTTCCACTC-3´
    Pla Probe: 5´-6FAM-TGCAGCCCTCCACCGGGATGC-TAMRA-3´

4. TaqMan® Universal PCR Master Mix (Part No. 4304437, Applied Biosystems).
5. MicroAmp® Optical 96-well reaction plates and optical adhesive covers (Part No. N801-0560 and 4311971, Applied Biosystems).
6. RNA protect (Qiagen, Valencia, CA).
7. DNA purification kit [we used the Puregene Kit D-50KA (Gentra, Minneapolis, MN) for flea and skin biopsy samples and the Qiaprep Spin Miniprep Kit (Qiagen) for the lymph node samples].
8. Qiashredder spin-column homogenizers (Qiagen, used for skin samples); a slurry of 0.1-mm glass beads (BioSpec Products, Bartlesville, OK) in sterile H_2O (used for flea samples); cell strainer, 70-µm nylon mesh (BD Falcon, Bedford, MA; used for lymph node samples).
9. Plastic pestles that conform to the interior of a 1.5-mL microfuge tube.
10. Brain–heart infusion (BHI) broth (Difco, Sparks, MD) and sterile PBS.
11. 3% H_2O_2 and 70% ethanol.
12. Petroff-Hausser bacterial counting chamber.
13. Spectrophotometer.

3. Methods

3.1. Preparation of Standard Curve

The standard curve is determined from a set of samples containing known numbers of *Y. pestis* in uninfected tissue of the same type and amount as the infected (unknown) samples. Both standard and unknown samples are then processed identically for qPCR. Careful matching of the tissue background of both standard and unknown samples cancels out the effects of any PCR inhibitors inherent in a particular tissue type. Because the *pla* gene target occurs on a plasmid of unknown but reportedly high copy number per cell, the standards are based on numbers of bacteria rather than numbers of gene target copies. This maintains the increased sensitivity provided by a multi-copy target while permitting estimation of the absolute number of bacterial cells per unknown sample.

3.1.1. Preparation of Stock Bacterial Standards

1. Grow *Y. pestis* KIM6$^+$ (or other attenuated, *pla*$^+$ strain) in liquid BHI medium at 28 °C to stationary phase (overnight).
2. Determine the number of *Y. pestis* per mL by direct count, using a Petroff-Hausser bacterial counting chamber and a phase-contrast microscope (400× magnification).
3. Centrifuge a known volume of culture to pellet the bacterial cells, remove the culture supernatant, and resuspend the bacteria in sterile PBS to a concentration of 1.0×10^9 per mL.
4. Aliquot this stock suspension and store the aliquots in a −80 °C freezer.

3.1.2. Preparation of Standard Curve Samples to Quantitate Y. pestis in Fleas

1. Surface-sterilize uninfected fleas (one for each standard and one for the negative control) by sequentially washing them for 1 min in 3% H_2O_2 and 70% ethanol. After the ethanol wash, allow the fleas to air-dry and rinse them in sterile distilled H_2O.
2. Thaw a stock bacterial standard aliquot, vortex, and make 1:10 serial dilutions in sterile BHI broth. Six standards containing 1×10^2 to 1×10^7 bacteria in 100 µL BHI in a sterile 1.5-mL microfuge tube are needed, plus a negative control tube containing 100 µL sterile BHI.
3. Add 20 µL of sterile glass bead slurry and an uninfected, surface-sterilized flea to each of the six standards and the negative control.
4. Triturate each flea thoroughly by grinding it with a pestle into the glass beads against the bottom of the microfuge tube. Vortex well, then allow glass beads and flea debris to settle out before transferring the supernatant to a clean microfuge tube. Proceed to DNA extraction (*see* **Subheading 3.3.**).

3.1.3. Preparation of Standard Curve Samples to Quantitate Y. pestis in Fleabite-Site Skin Biopsies

1. Collect skin biopsies from a euthanized, uninfected mouse of the same size as the skin biopsies to be collected from flea-bitten mice. Several skin biopsies can be collected from a single animal. Freeze individual skin biopsies in sterile microfuge tubes for future use.
2. Thaw a stock bacterial standard aliquot, vortex, and make 1:10 serial dilutions in sterile PBS. Six standards containing 1×10^2 to 1×10^7 bacteria in 100 μL PBS in a sterile 1.5-mL microfuge tube are needed.
3. Add a thawed, uninfected skin biopsy to each of the six standard sample tubes and to a tube containing 100 μL sterile PBS for use as a negative control. Proceed to DNA extraction.

3.1.4. Preparation of Standard Curve Samples to Quantitate Y. pestis in Rat Lymph Nodes

1. Aseptically dissect lymph nodes from euthanized, uninfected rats. Several lymph nodes can be collected from an individual animal. For example, bilateral inguinal, axillary, and maxillary lymph nodes are easy to locate and can be dissected rapidly.
2. Immediately after dissection, disrupt each lymph node by pressing it through a 70-μm cell strainer into 10 mL of RNA protect. Transfer the suspension to a 15-mL conical tube, mix, and dispense 1-mL aliquots to 1.5-mL microfuge tubes.
3. Centrifuge tubes at $13.5 \times g$ for 5 min. Remove the RNA protect supernatant and store the pellets at –80 °C.
4. Thaw a stock bacterial standard aliquot (*see* **Subheading 3.1.1.**), vortex, and make 1:10 serial dilutions in sterile PBS. Seven standards containing 1×10^3 to 1×10^9 bacteria in 500 μL PBS are needed.
5. Thaw seven uninfected lymph node pellets from **step 3**. Add one of the 500 μL PBS bacterial suspensions from **step 5** to the seven lymph node pellets and resuspend. Proceed to DNA extraction.

3.2. Preparation of Unknown Samples

After they are prepared, samples can be stored at –80 °C prior to DNA extraction.

3.2.1. Fleas

1. Surface-sterilize fleas as described in **step 1, Subheading 3.1.2**.
2. Add individual fleas to 1.5-mL microfuge tubes containing 100 μL BHI and 20 μL glass bead slurry. Triturate fleas and collect supernatant as described in **step 4, Subheading 3.1.2**.

3.2.2. Skin Biopsies of Fleabite Site

1. After a flea has attempted to feed on a mouse, euthanize the animal, wipe the area with an alcohol pad, and aseptically collect a skin biopsy of the bite site.
2. Place the biopsy in a 1.5-mL microfuge tube.

3.2.3. Lymph Nodes from Infected Rats

1. Aseptically dissect lymph nodes from euthanized, infected rats.
2. Immediately after dissection, disrupt each lymph node by pressing it through a 70-μm cell strainer into 10 mL of RNA protect.
3. Transfer one-tenth (1 mL) of the lymph node suspension to a 1.5-mL microfuge tube and centrifuge at $13.5 \times g$ for 5 min. Remove and discard the RNA protect supernatant.

3.3. DNA Extraction

1. A number of commercially available DNA extraction kits are suitable to prepare purified total DNA from tissue samples. The method chosen can be based on individual preference, sample characteristics (e.g., ratio of microbial to host cells), qPCR target, and specific experimental needs (*see* **Notes**).
2. Assess purity and quantity of purified DNA by A_{260}/A_{280} spectrophotometry.

3.4. Real-time qPCR

1. Prepare a sufficient volume of 1× TaqMan master mix containing 500 nM of forward and reverse primers and 200 nM of probe to perform qPCR on each control, standard, and unknown sample in triplicate. The total volume of individual TaqMan reactions is 25 μL for the ABI Prism 7700 and 20 μL for 7900HT.
2. Pipet master mix for individual reactions to wells of the 96-well reaction plate.
3. Add 1–5 μL of template DNA prepared from negative control, standard, and unknown samples. Do not add any template to three "no-template" control wells.
4. For the *Y. pestis pla* gene primer and probe set and the ABI 7700 instrument, the reaction program consists of 95 °C for 10 min followed by 45 cycles of 94 °C for 20 s, 60 °C for 20 s, and 72 °C for 30 s. For the 7900HT instrument, the program is 50 °C for 2 min, 95 °C for 10 min, and then 40 cycles of 95 °C for 15 s and 60 °C for 1 min.

3.5. Data Analysis

1. Using the software package supplied with the TaqMan instrument, generate a standard curve by plotting the C_t values of the standards on the *y*-axis and the log10 number of *Y. pestis* cells they contained on the *x*-axis.
2. Determine the number of *Y. pestis* present in the unknown samples by using their C_t value in the linear equation of the standard curve to solve for *x*, the log number of *Y. pestis*.

4. Notes

1. Primer Express software v2.0 was used to select TaqMan primer and probe sequences for the *Y. pestis pla* gene.
2. The PCR reaction can be optimized by empirically predetermining the annealing temperature and $MgCl_2$ concentration that results in the minimum C_t values.
3. Skin biopsy samples can be minced or homogenized with a tissue grinder before DNA extraction.
4. Sensitivity of qPCR may also be improved by diluting the DNA template purified from skin samples, which may contain PCR inhibitors *(19)*. Prior to use as qPCR template, we routinely diluted DNA extracts from skin samples 1:10, 1:50, or 1:100 in DNA resuspension buffer, depending on the weight (<5, 5–7, or 7–10 mg, respectively) of the original skin biopsy.
5. Mechanical disruption of fleas using a bead mill is an alternative to manual titration using a pestle *(20)*.
6. The limit of detection from skin biopsy samples was 10^3 *Y. pestis*. The use of reverse transcriptase qPCR of a highly expressed gene may result in greater sensitivity *(21)*.
7. The limit of detection from lymph node samples was also 10^3 *Y. pestis*. However, only one-tenth of the sample was used for qPCR; the majority of the sample was used for other experimental purposes. Extracting DNA from the entire sample would increase the sensitivity.
8. We used the Puregene cell, tissue, body fluid, and Gram-negative bacteria DNA purification kit from Gentra for mouse skin and flea samples. The manufacturer's protocol was used with the following modification for use with skin samples: following proteinase K and RNase treatment, the samples were passed through a Qiashredder column to fragment the DNA and to remove mouse hair.
9. We used the Qiaprep spin miniprep kit from Qiagen for lymph node samples. This extraction method enriches for bacterial plasmid DNA, which may be an advantage because the *pla* gene target is on a *Y. pestis*-specific plasmid.
10. Sample collection and treatment can be tailored to fit additional experimental requirements. For example, we collected lymph node contents in RNA protect reagent, because part of the sample was used for gene expression analysis and part for qPCR.

References

1. Dye, C. (1992) The analysis of parasite transmission by bloodsucking insects. *Annu. Rev. Entomol.* **37**, 1–19.
2. Bell, A.S. and Ranford-Cartwright, L.C. (2002) Real-time quantitative PCR in parasitology. *Trends Parasitol.* **18**, 337–342.
3. Edwards, K., Logan, J. and Saunders, N. (eds.) (2004) *Real-Time PCR. An Essential Guide*. Horizon Bioscience, Norfolk, UK.
4. Kochanowski, B. and Reischl, U. (eds.) (1999) *Quantitative PCR Protocols*. Humana Press, Totowa, NJ.

5. Walker, N.J. (2002) A technique whose time has come. *Science* **296**, 557–559.
6. Bell, A.S. and Ranford-Cartwright, L.C. (2004) A real-time PCR assay for quantifying *Plasmodium falciparum* infections in the mosquito vector. *Int. J. Parasitol.* **34**, 795–802.
7. Hodzic, E., Feng, S., Freet, K.J., Borjesson, D.L. and Barthold, S.W. (2002) *Borrelia burgdorferi* population kinetics and selected gene expression at the host-vector interface. *Infect. Immun.* **70**, 3382–3388.
8. Nicolas, L., Sidjanski, S., Colle, J.H. and Milon, G. (2000) *Leishmania major* reaches distant cutaneous sites where it persists transiently while persisting durably in the primary dermal site and its draining lymph node: a study with laboratory mice. *Infect. Immun.* **68**, 6561–6566.
9. Piesman, J., Schneider, B.S. and Zeidner, N.S. (2001) Use of quantitative PCR to measure density of *Borrelia burgdorferi* in the midgut and salivary glands of feeding tick vectors. *J. Clin. Microbiol.* **39**, 4145–4148.
10. Pusterla, N., Huder, J.B., Leutenegger, C.M., Braun, U., Madigan, J.E. and Lutz, H. (1999) Quantitative real-time PCR for detection of members of the *Ehrlichia phagocytophila* genogroup in host animals and *Ixodes ricinus* ticks. *J. Clin. Microbiol.* **37**, 1329–1331.
11. Richardson, J., Molina-Cruz, A., Salazar, M.I. and Black, W., IV. (2006) Quantitative analysis of dengue-2 virus RNA during the extrinsic incubation period in individual *Aedes aegypti*. *Am. J. Trop. Med. Hyg.* **74**, 132–141.
12. Zeidner, N.S., Schneider, B.S., Dolan, M.C. and Piesman, J. (2001) An analysis of spirochete load, strain, and pathology in a model of tick-transmitted Lyme borreliosis. *Vector Borne Zoonotic Dis.* **1**, 35–44.
13. Lorange, E.A., Race, B.L., Sebbane, F. and Hinnebusch, B.J. (2005) Poor vector competence of fleas and the evolution of hypervirulence in *Yersinia pestis*. *J. Infect. Dis.* **191**, 1907–1912.
14. Medica, D.L. and Sinnis, P. (2005) Quantitative dynamics of *Plasmodium yoelii* sporozoite transmission by infected anopheline mosquitoes. *Infect. Immun.* **73**, 4363–4369.
15. Vanlandingham, D.L., Schneider, B.S., Klingler, K., Fair, J., Beasley, D., Huang, J., Hamilton, P. and Higgs, S. (2004) Real-time reverse transcriptase-polymerase chain reaction quantification of West Nile virus transmitted by Culex pipiens quinquefasciatus. *Am. J. Trop. Med. Hyg.* **71**, 120–123.
16. Engelthaler, D.M., Hinnebusch, B.J., Rittner, C.M. and Gage, K.L. (2000) Quantitative competitive PCR as a technique for exploring flea-*Yersinia pestis* dynamics. *Am. J. Trop. Med. Hyg.* **62**, 552–560.
17. Scoles, G.A., Ueti, M.W. and Palmer, G.H. (2005) Variation among geographically separated populations of *Dermacentor andersoni* (Acari: Ixodidae) in midgut susceptibility to *Anaplasma marginale* (Rickettsiales: Anaplasmataceae). *J. Med. Entomol.* **42**, 153–162.
18. Teglas, M.B. and Foley, J. (2006) Differences in the transmissibility of two *Anaplasma phagocytophilum* strains by the North American tick vector species, *Ixodes pacificus* and *Ixodes scapularis* (Acari: Ixodidae). *Exp. Appl. Acarol.* **38**, 47–58.

19. Cogswell, F.B., Bantar, C.E., Hughes, T.G., Gu, Y. and Philipp, M.T. (1996) Host DNA can interfere with detection of *Borrelia burgdorferi* in skin biopsy specimens by PCR. *J. Clin. Microbiol.* **34**, 980–982.

20. Alexander, C.J., Easterday, W.R., Van Ert, M.N., Wagner, D.M. and Keim, P. (2004) High-throughput extraction of arthropod vector and pathogen DNA using bead milling. *Biotechniques* **37**, 730–734.

21. Ornstein, K. and Barbour, A.G. (2006) A reverse transcriptase-polymerase chain reaction assay of *Borrelia burgdorferi* 16S rRNA for highly sensitive quantification of pathogen load in a vector. *Vector Borne Zoonotic Dis.* **6**, 103–112.

11

Intracellular Localization of *Brucella abortus* and *Francisella tularensis* in Primary Murine Macrophages

Jean Celli

Summary

Intracellular bacterial pathogens have evolved sophisticated strategies to survive and proliferate within cells of their hosts. Studying their intracellular life cycle is key to understanding virulence and requires methodologies that can identify the compartments in which they localize and characterize the replicative niche they generate. Here, we describe immunofluorescence-based microscopy techniques applied to the intracellular pathogens *Brucella abortus* and *Francisella tularensis* during their respective intracellular cycles inside murine bone marrow-derived macrophages. Standard immunofluorescence techniques are used to define the intracellular localization of the pathogens based on their co-localization with specifically labeled macrophage organelles. In addition, we describe an assay to assess the integrity of *Francisella*-containing phagosomes and bacterial release into the macrophage cytoplasm, which is a hallmark of *Francisella* intracellular pathogenesis.

Key Words: *Brucella*; *Francisella*; macrophages; intracellular cycle; endosomes, endoplasmic reticulum; confocal immunofluorescence microscopy; digitonin.

1. Introduction

Survival and proliferation within mammalian cells are key virulence features of bacterial pathogens with an intracellular life cycle *(1)*. Such bacteria have evolved sophisticated mechanisms to avoid intracellular bactericidal functions of host cells, many of which aim to modulate host functions to control their intracellular trafficking and reach or generate a niche permissive for replication. An essential step in our understanding of the intracellular pathogenesis

From: *Methods in Molecular Biology, vol. 431: Bacterial Pathogenesis*
Edited by: F. DeLeo and M. Otto © Humana Press, Totowa, NJ

of these bacteria is to define their intracellular trafficking, which is key to comprehending their mechanisms of survival. *Brucella abortus* and *Francisella tularensis* are the causative agents of brucellosis and tularemia, respectively. Upon infection of their host, both bacteria encounter phagocytic cells, of which macrophages constitute a major target for survival and replication *(2,3)*. *B. abortus* ensures its intracellular survival by inhibiting the fusion of its vacuole, the *Brucella*-containing vacuole (BCV), with degradative lysosomes *(4,5)*. In addition to this early survival mechanism, *Brucella* uses a type IV secretion system (VirB) *(6)* to control interactions and fusion of its vacuole with the macrophage endoplasmic reticulum (ER) and generates an ER-derived organelle that segregates *Brucella* from the degradative endocytic pathway and allows unrestricted multiplication *(4,7,8)*. A different survival strategy is used by *F. tularensis*, which is capable of rapid physical escape from its initial phagosome following phagocytic uptake *(9,10)*. Once in the macrophage cytoplasm, intracellular *Francisella* undergoes replication. As exemplified by *Brucella* and *Francisella* , such pathogens traffic through, and interact with, various intracellular compartments to complete their cycle. The ability to localize these bacteria within the host cell at various stages of the infection cycle is a major asset toward understanding pathogenesis. Here, we detail fluorescence microscopy-based methods to localize *B. abortus* or *F. tularensis* in various compartments of primary macrophages.

2. Materials

2.1. Bacterial Cultures

1. Tryptic soy broth (TSB) or agar (TSA) (Sigma, St Louis, MO).
2. Cystine heart agar blood (CHAB) made of cystine heart agar (CHA; Difco, Beckton Dickinson, Sparks, MD, USA) supplemented with 9% sheep blood. The sheep blood is chocolatized by adding it to autoclaved CHA before it cools below 75 °C.
3. Kanamycin solution, 50 µg/mL (K0254; Sigma).

2.2. Culture and Infection of Primary Macrophages

1. C57BL/6J (or BALB/c) 6- to 10-week-old female mice for bone marrow harvesting.
2. For bone marrow harvesting, sterile dissection tweezers, scissors, scalpels, and 1-mL syringe fitted with a $25G^{5/8}$-gauge needle.
3. Dulbecco's modified Eagle's medium (DMEM) containing 1 g/L D-glucose (Invitrogen, Carlsbad, CA) supplemented with 10% fetal bovine serum (FBS; Invitrogen) and 2 mM L-glutamine (Invitrogen).
4. Gentamicin solution, 10 µg/mL (G1272; Sigma).

5. Dulbecco's phosphate-buffered saline (D-PBS) with or without $MgCl_2$ and $CaCl_2$ (Invitrogen).
6. D-PBS saline without $MgCl_2$ and $CaCl_2$, supplemented with 1g/L D-glucose: dissolve 2.5 g of D-glucose in 10 mL of D-PBS and add it to 500 mL of D-PBS by filtering it through a 0.2-μm filter in a tissue culture cabinet.
7. L-929 mouse fibroblasts (ATCC No CCL-1) grown in DMEM supplemented with 10% FBS and 2 mM L-glutamine.

2.3. Immunofluorescence Microscopy

1. 12-mm round 0.1-mm thick glass coverslips: degrease coverslips in 100% acetone for 5 min, rinse them five times in 70% ethanol, dry them on Whatman paper, and sterilize them by autoclaving between Whatman paper discs in a glass Petri dish.
2. High-Precision DUMONT tweezers #5 and #7 (Ted Pella Inc., Redding, CA).
3. PBS: Prepare 10× stock with 1.37 M NaCl, 27 mM KCl, 100 mM Na_2HPO_4, and 18 mM KH_2PO_4 (adjust to pH 7.4 with HCl if necessary). Autoclave before storage at room temperature. Prepare working solution by diluting one part 10× stock with nine parts of water (*see* **Note 1**).
4. Paraformaldehyde (PFA; Sigma, P6148): Prepare a 25% stock by dissolving 10 g of PFA in 30 mL of water and heat at 70 °C on a stirring hot plate in a fume hood. Add drops of 10N NaOH until PFA completely dissolves. Aliquot in Eppendorf tubes and store at –20 °C. A 2.5% working solution is prepared by thawing a 25% stock aliquot at 70 °C and diluting it in 9 mL of water. pH is adjusted to 7.2–7.4 with 1N HCl (~15 μL) using a pH strip.
5. Quenching buffer: 50 mM NH_4Cl in PBS.
6. Permeabilization/blocking buffer: 10% horse serum (Sigma) and 0.1% (v/v) saponin in PBS.
7. Antibody dilution buffer: 10% horse serum (Sigma) and 0.1% (v/v) saponin in PBS.
8. Primary antibodies: mouse monoclonal anti-*F. tularensis* LPS (U.S. Biological, Swampscott, MA), rat monoclonal anti-mouse LAMP1 (1D4B; developed by J. T. August and obtained from the Developmental Studies Hybridoma Bank developed under the auspices of the NICHD and maintained by The University of Iowa, Department of Biological Sciences, Iowa City, IA), rabbit polyclonal anti-calnexin (SPA-860; Stressgen Bioreagents, Ann Arbor, MI), rabbit polyclonal anti-calreticulin (PA3-900; Affinity BioReagents, Golden, CO), mouse anti-PDI, (SPA-891; Stressgen Bioreagents), goat polyclonal anti-EEA1(N-19; Santa Cruz Biotechnology, Santa Cruz, CA), rabbit anti-β-COP (PA1-061; Affinity BioReagents), and mouse anti-GM130 (BD Transduction Laboratories, San Jose, CA).
9. Secondary antibodies: Alexa Fluor™488-conjugated, Alexa Fluor™ 568-conjugated (Molecular Probes, Carlsbad, CA), and cyanin-5-conjugated (Jackson ImmunoResearch Laboratories, West Grove, PA) donkey anti-mouse, anti-rat, anti-rabbit, and anti-goat antibodies.

10. Mowiol mounting medium: add 2.4 g of Mowiol 4–88 (Calbiochem, San Diego, CA) to 6 g of glycerol and stir to mix. Add 6 mL of water and leave stirring at room temperature for several hours. Add 12 mL of 0.2 *M* Tris–HCl (pH 8.5) and heat to approximately 50 °C until Mowiol dissolves. Clarify by centrifugation at 5000 × *g* for 15 min, aliquot in airtight 1.5 mL microtubes, and store at –20 °C. Once thawed, the Mowiol mounting medium is stable at room temperature for several weeks (*see* **Note 2**).

2.4. Phagosomal Integrity Assay

1. KHM buffer: 110 m*M* potassium acetate, 20 m*M* HEPES, and 2 m*M* MgCl2, pH 7.3. Adjust pH using 1 *M* KOH and sterilize by 0.2 μm filtration. KHM buffer can be stored at 4 °C for several weeks.
2. Digitonin solution: Prepare extemporaneously a 10 mg/mL solution of digitonin (Sigma, D141) in sterile bi-distilled water. Heat to 95 °C for complete dissolution and let cool to room temperature before use.
3. Mouse anti-*Francisella* LPS monoclonal antibody (U.S. Biological).
4. Rabbit polyclonal anti-calnexin CT (Stressgen Bioreagents), which recognizes the cytoplasmic tail of the ER chaperone calnexin.
5. Reagents used for immunofluorescence staining (*see* **Subheading 2.3.**).

3. Methods

Intracellular localization of *B. abortus* or *F. tularensis* is performed on infected murine bone marrow-derived macrophages at various stages of their intracellular life cycle, using immunofluorescence staining of bacteria and various intracellular organelles. Specific intracellular compartments are labeled by the immunodetection of host proteins enriched or specifically present on (or in) these compartments or by loading with specific fluorescent probes (*see* **Table 1**). The presence of bacteria in a given intracellular organelle is evaluated by co-localization of bacteria- and organelle-specific fluorescent signals, either by indirect epifluorescence or laser-scanning confocal fluorescence microscopy (*see* **Fig. 1**). Following phagocytic uptake, both *Brucella* and *Francisella* are located within an initial phagosome that interacts to various extents with the endocytic pathway *(4,9)*. Maturation of the BCV into a replicative organelle can be monitored through the subsequent acquisitions and exclusions of markers of various organelles, namely early endocytic, late endocytic, and ER markers *(4,5,8,11)* (*see* **Fig. 1**). *Francisella* phagosomal escape into the macrophage cytoplasm limits its interactions with membrane-bound compartments, yet early interactions with the endocytic compartment are detectable and phagosomal escape can be assessed using an assay based on the specific detection of

Table 1
Commercially Available Antibodies and Fluorescent Probes that Label Intracellular Organelles of Murine Macrophages

Compartment	Marker/probe	Antibodies	Comments/reference
Early endosomes	EEA-1	Goat α-EEA-1 (N-19)	Rab5 effector; Santa Cruz Biotechnologies (use at 1:50)
Late endosomes	LAMP-1	Rat monoclonal α-LAMP-1 (1D4B)	J. T. August, DHSB[a] (use at 1:400)
Lysosomes	LAMP-1	Rat monoclonal α-LAMP-1 (1D4B)	J. T. August, DHSB[a] (use at 1:400)
	Cathepsin D	Rabbit α-cathepsin D	Dako Corporation (use at 1:200)
Endoplasmic reticulum	Calnexin	Rabbit α-calnexin-CT	Transmembrane protein; Stressgen Bioreagents (SPA-860, use at 1:1000)
	Calreticulin	Rabbit α-calreticulin	Membrane-associated; Affinity Bioreagents (use at 1:1000)
	PDI	Mouse α-PDI	Lumenal; Stressgen Bioreagents (SPA-860, use at 1:100)
ERGIC	COPI	Rabbit α-β-COP	Affinity Bioreagents (use at 1:200, requires SDS extraction), also labels the Golgi apparatus
Golgi apparatus	GM130	Mouse α-GM130	BD Transduction Laboratories (use at 1:100)
Autophagosomes	Monodansylcadaverine		Autofluorescent, accumulates in autophagosomes

[a]Developmental Studies Hybridoma Bank, developed under the auspices of the NICHD and maintained by the University of Iowa, Department of Biological Sciences, Iowa City, IA.

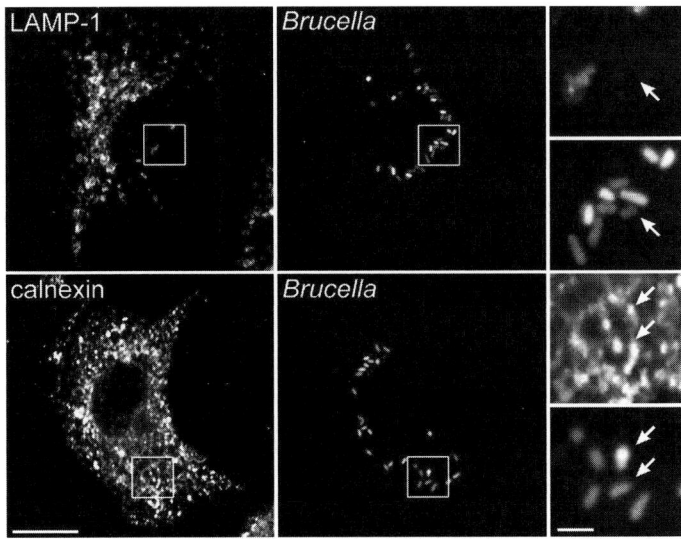

Fig. 1. Confocal laser scanning micrographs of a C57BL/6 mouse bone marrow-derived macrophage infected for 24 h with virulent *Brucella abortus* expressing GFP. Upper panels: LAMP-1 staining of the late endocytic compartment showing that replicative BCVs are LAMP-1-negative. Insets show a magnification of a LAMP-1-negative area (boxed on the whole image), where bacteria replicate (arrow). Lower panels: Calnexin staining of the endoplasmic reticulum (ER) showing recruitment of ER markers on the replicative BCVs. Insets show a magnification of calnexin-positive *Brucella*-containing vacuoles (BCVs) (arrows). Scale bars, 10 and 2 μm (insets).

cytoplasmic bacteria (phagosomal integrity assay, **Fig. 2**). The successful localization of intracellular bacteria depends on the quality of the immunostaining, which is achieved through the use of antibodies giving high specific signal-to-noise ratios. As not all antibodies are useful for immunofluorescence applications, we provide in this chapter references of commercially available antibodies that have proven to be of such quality.

3.1. Culture of Bone Marrow-Derived Macrophages

1. To prepare L-929-conditioned medium, grow L-929 mouse fibroblasts in DMEM supplemented with 10% FBS and 2 m*M* L-glutamine in tissue-culture flasks at 37 °C under 7% CO_2 with a humidified atmosphere. Expand cells into 175-cm^2 flasks seeded with 2×10^6 cells in 80 mL of medium. Grow cells for 7 days, during which they reach confluency and produce CSF-1, a growth factor required for bone marrow monocyte differentiation into macrophages and growth. At day 7, collect supernatants, filter them through 0.22 μm membranes, and freeze 10 mL aliquots at –20 °C (*see* **Note 3**).

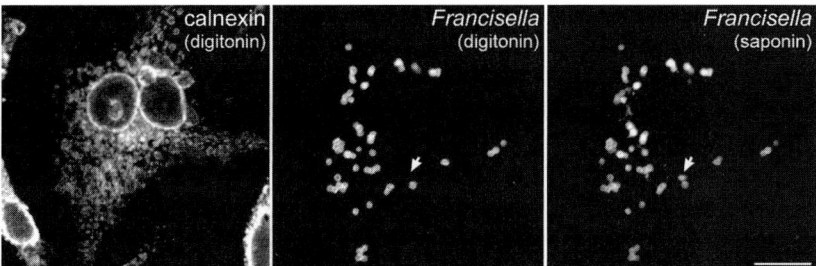

Fig. 2. Confocal laser scanning micrographs of a C57BL/6 mouse bone marrow-derived macrophage infected for 24 h with *Francisella tularensis* and subjected to the phagosomal integrity assay. Left-hand panel: Calnexin staining of the endoplasmic reticulum (ER) following digitonin treatment that demonstrates permeabilization of the cell; middle panel: *Francisella* staining following digitonin treatment that demonstrates that replicating *Francisella* are detectable using cytoplasmically-delivered antibodies; and right-hand panel: *Francisella* staining following complete permeabilization with saponin that detect all intracellular bacteria. The arrowhead indicates the only phagosomal bacterium, which is not detected following cytoplasmic delivery of anti-*Francisella* antibodies (middle panel). Scale bar, 10 μm.

2. Euthanize mice using a lethal dose of isofluorane and immediately remove femurs from hind legs using sterile dissection tools. It is important to remove and discard all the tissue around the bones to avoid further contamination of the macrophage culture with fibroblasts. Femurs can be kept in cold DMEM until further processing (not more than 1 h is preferable).

3. In a biosafety cabinet, prepare a six-well plate with one well containing 70% ethanol and four others 2 mL of DMEM. Surface sterilize each femur by rapidly passing it in 70% ethanol, rinse it with DMEM in the next well, and then transfer it in the third well. Scrape off any remaining tissue around the bone, transfer the clean bone into the next well and cut both ends with a scalpel.

4. Transfer immediately to the next well and flush cells out gently with a 1-mL syringe fitted with a $25G^{5/8}$-gauge needle, by passing DMEM 3–4 times through the bone from each side until the bone is white. Be careful not to scrape the bone with the needle, as this will release osteoclasts. Avoid bubbles as much as possible. Collect cells in a sterile tube and put on ice. You should obtain $1–2 \times 10^7$ monocytes/femur.

5. Put the cell suspension (from 1.5–2 femurs) in culture in a 150 mm non-tissue culture-treated dish in 30 mL of DMEM supplemented with 10% FBS, 2 mM L-glutamine, and 10% L-929-conditioned medium for 5 days at 37 °C under 7% CO_2 with a humidified atmosphere. Monocytes will progressively differentiate into macrophages (BMMs), which will adhere loosely to the dish.

6. On day 5, rock dishes gently back and forth to dislodge non-adherent cells (non-differentiated monocytes). Wash twice with 15 mL of D-PBS containing $MgCl_2$

and $CaCl_2$ to remove all non-adherent cells. Wash once with ice-cold D-PBS without $MgCl_2$ and $CaCl_2$ but supplemented with 1 g/L D-glucose. Add 15 mL of PBS-glucose and incubate the dish on ice with gentle rocking for 10 min. Make sure the adherent macrophages remain submerged at all time.

7. To retrieve adherent BMMs, pipette the PBS across the entire surface of the plate and monitor under the microscope that all BMMs have been detached. Avoid bubbles as much as possible. Transfer the BMM suspension into a 50-mL conical tube and centrifuge the cells at $200 \times g$ for 5 min at 15 °C. Discard the supernatant and gently resuspend the cell pellet in 5 mL of DMEM.

8. Count BMMs using a phase hemacytometer and seed 12-mm glass coverslips in wells of a 24-well plate at a density of 1×10^5/well in complete medium. You should obtain $1-2 \times 10^7$ BMMs/dish.

9. Replenish medium after 24 h in culture. The cells may then undergo one or two rounds of multiplication and then stop multiplying. On day 7 post-isolation, BMMs are ready to use (*see* **Note 4**).

3.2. Macrophage Infections

1. The following method of macrophage infection with either *B. abortus* or *F. tularensis* aims at synchronizing bacterial phagocytosis, which is important for subsequent intracellular trafficking analyses, as all bacteria have entered macrophages within a short time frame.

2. To prepare bacterial inocula, streak *B. abortus* first on TSA plates from a frozen stock vial and grow for 3 days at 37 °C. Then inoculate 2 mL of TSB in a 50-mL conical tube with two to three fresh isolated colonies and incubate for 16 h in a shaking incubator at 37 °C. Grow *F. tularensis* on CHAB plates from a frozen stock vial for 3 days at 37 °C in 7% CO_2. Right before proceeding for infections, resuspend a few colonies in TSB containing 0.1% L-cysteine. Estimate bacterial counts by measuring the optical density at 600 nm ($OD_{600\ nm}$) of the *Brucella* sub-culture or the *Francisella* suspension. An $OD_{600\ nm}$ of 1.0 corresponds to $\sim 3 \times 10^9$ bacteria/mL for both organisms, but this needs to be determined for every organism by plating serial dilutions of the bacterial inoculum on appropriate medium before hand. Determining the bacterial inoculum is essential to calculate the multiplicity of infection (MOI) to be used. Theoretical MOIs of 25 and 50 are used for *B. abortus* and *F. tularensis* , respectively (*see* **Note 5**).

3. To infect BMMs, pre-chill macrophages for 5 min on ice while preparing the bacterial suspension in ice-cold complete medium. On ice, aspirate the macrophage medium in one well at a time and immediately replace with 0.5 mL of bacterial suspension. Centrifuge bacteria onto BMMs at $400 \times g$ for 10 min at 4 °C to favor bacteria–macrophage contact.

4. Rapidly warm BMMs to 37 °C for 3 min by floating the plates in a waterbath to trigger bacterial phagocytosis. Transfer plates to a tissue-culture incubator and further incubate them for a total of 20 min at 37 °C under 7% CO_2. (*see* **Note 6**).

5. Wash BMMs five times with DMEM to remove extracellular bacteria, incubate for 40 min in complete medium and then for 60 min in complete medium

supplemented with 100 µg/mL gentamicin to kill the remaining extracellular bacteria. Thereafter, incubate infected BMMs in gentamicin-free medium.

3.3. Immunofluorescence Microscopy

1. First fix infected BMMs with PFA before proceeding for immunostaining. Alternatively, cells can be fixed using ice-cold 100% methanol for 30 s at –20 °C when the immunological detection of particular epitopes requires harsher extraction conditions. The method described herein is adapted to PFA fixation.

2. Wash infected BMMs three times with room temperature PBS and then rapidly transfer coverslips to another PBS-containing 24-well plate using high-precision tweezers. Aspirate the PBS and immediately cover the coverslip with 300–400 µL of 2.5% PFA in PBS that has been preheated to 37°C. Incubate coverslips at 37°C for 20 min (*see* **Note 7**).

3. Wash fixed samples in wells three times with PBS and leave coverslips for at least 10 min in 50 mM NH$_4$Cl-PBS to quench free aldehyde groups. This step diminishes the fluorescence background of the samples. Coverslips can be kept for up to 24 h at 4 °C in quenching buffer before immunostaining although it is preferable to perform the procedure immediately after fixation.

4. Perform all subsequent steps in a humidified, lightproof chamber such as a black shallow box containing a sheet of parafilm on top of wet 3M Whatman paper.

5. To allow antibody access to intracellular structures and simultaneously saturate antibody non-specific binding sites, retrieve each coverslip from its well, drain excess liquid onto a Kimwipe, and invert it (cells facing down) on top of a 50 µL drop of permeabilization/blocking buffer. Incubate for 30 min to 1 h at room temperature. Saponin is the mildest detergent for sample permeabilization and is reversible. Hence, it is kept present in antibody dilutions subsequently used. Alternatively, some antigens require extraction with 0.1% Triton X-100 in PBS for 5 min at room temperature, which irreversibly permeabilizes the cells (*see* **Note 8**).

6. Prepare primary antibodies dilutions in antibody dilution buffer and distribute them as 50-µL drops. For antibodies available in limited amounts, the volume of the drops can be decreased to 15 µL. Drain permeabilization/blocking buffer from each coverslip onto a Kimwipe and invert it onto the corresponding drop of primary antibody. Gently swirl the incubation box to homogenize the antibody drop with the remaining liquid on the coverslip. Incubate for 20 to 40 min at room temperature depending on the antibody affinity and antigen accessibility (*see* **Note 9** and **Table 1**).

7. Prepare secondary antibodies dilutions (1:500 dilutions for all Alexa Fluor™ and Cyanin 5-conjugated antibodies) in antibody dilution buffer and distribute them as 50-µL drops. Prepare four beakers, two containing 50–100 mL of 0.1% saponin-PBS, one containing PBS, and one containing bi-distilled water.

8. Wash each coverslip twice in 0.1% saponin-PBS and once in PBS. Each wash consists of holding the coverslip with tweezers and stirring for few seconds in each

beaker. Drain all liquid onto a Kimwipe and immediately invert it on a drop of the appropriate secondary antibodies. Incubate for 40 min at room temperature.

9. Prepare glass microscopy slides to mount stained coverslips by cleaning them with ethanol to remove any residue.

10. Wash each coverslip twice in 0.1% saponin-PBS, once in PBS, and finally once in water to remove salts. Drain all liquid onto a Kimwipe and mount coverslips on microscopy slides by inverting them (cells facing down) on 10-µL drops of Mowiol mounting medium. Allow at least 2 h of polymerization before viewing samples on an epifluorescence microscope.

3.4. Phagosomal Integrity Assay (12)

1. This assay is based on the differential labeling of intracellular bacteria depending on whether they are contained within an intact phagosome or are accessible to cytoplasmically delivered, specific antibodies due to the disruption of their phagosome. It relies on a two-step permeabilization of infected cells consisting of (1) permeabilization of the plasma membrane of live cells using digitonin, which allows cytoplasmic delivery of antibodies and detection of accessible bacteria and (2) a complete permeabilization of all eukaryotic membranes using saponin after PFA fixation, which allows antibody access to all intracellular bacteria, regardless of their location. The use of different fluorophore-conjugated secondary antibodies generates a differential labeling of cytoplasmic (double labeling) and phagosomal bacteria (single labeling), which allows direct analysis of their intracellular location.

2. Prepare dilutions of mouse anti-*F. tularensis* LPS (1:2000) and rabbit anti-calnexin CT (1:250) antibodies in KHM buffer to, respectively, detect cytoplasmically accessible bacteria and digitonin-permeabilized cells. Distribute as 30-µL drops in a humidified immunostaining chamber and preheat to 37 °C.

3. Wash infected BMMs three times with room temperature KHM buffer. Immediately transfer coverslip(s) in another 24-well plate containing KHM buffer if necessary.

4. Aspirate KHM buffer and add 500 µL of 50 µg/mL digitonin in KHM buffer for 1 min at room temperature. Incubation time and digitonin concentration may vary depending on the cells used, so it is recommended to run preliminary tests.

5. Rinse coverslips in wells with KHM buffer three to four times, carefully retrieve each coverslip using high-precision tweezers and quickly drain it onto a Kimwipe.

6. Invert coverslip(s) on pre-heated 30-µL antibody drops and incubate for 10 min at 37°C (*see* **Note 10**).

7. Put the coverslip(s) back in the 24-well plate and wash three to four times with PBS.

8. Fix with 2.5% PFA in PBS as described in **Subheading 3.3**.

9. Permeabilize cells with permeabilization/blocking buffer for 30 min at room temperature.

10. Incubate coverslips on 50 μL-drops of Alexa Fluor™ 488-conjugated anti-mouse secondary and Cyanin 5-conjugated anti-rabbit secondary antibody dilutions (1:500) for 40 min at room temperature. This step allows the detection of cytoplasmic bacteria and calnexin at the cytoplasmic face of the ER that have been recognized following digitonin permeabilization.
11. Wash each coverslip twice in 0.1% saponin-PBS and once in PBS.
12. Incubate coverslips a second time with a mouse anti-*Francisella* LPS antibody dilution for 40 min at room temperature in permeabilization/blocking buffer. Because cells are fully permeabilized with saponin, this step allows the antibody to reach all intracellular bacteria.
13. Wash each coverslip twice in 0.1% saponin-PBS and once in PBS.
14. Incubate coverslips with an Alexa Fluor™ 568-conjugated anti-mouse secondary antibody dilutions or for 40 min at room temperature in permeabilization/blocking buffer to detect all intracellular bacteria.
15. Wash each coverslip twice in 0.1% saponin-PBS, once in PBS, and finally once in water to remove salts. Drain all liquid onto a Kimwipe and mount coverslips on microscopy slides by inverting them (cells facing down) on 10-μL drops of Mowiol mounting medium. Allow at least 2 h of polymerization before viewing samples on a confocal fluorescence microscope equipped with 488-, 568-, and 643-nm laser lines.
16. Under such staining conditions, permeabilized cells are detectable with the 643-nm (or equivalent) laser line. Cytoplasmic bacteria are detectable using both 488- and 568-nm (or equivalent) laser lines while phagosomal bacteria are only detected using the 568 nm (or equivalent) laser line. Alternate combinations of secondary antibodies are possible.

4. Notes

1. All solutions should be prepared in bi-distilled sterile water, which is referred to as "water" in the text.
2. Mowiol is extremely slow to dissolve and may require overnight stirring before heating up to approximately 50 °C to complete its dissolution.
3. Once thawed, aliquots can be kept at 4 °C for about 15 days. It is best to prepare large batches from which many experiments can be performed using the same conditioned medium. The preparation of L-929-conditioned medium is crucial for the differentiation of bone marrow monocytes into macrophages. Batches should be tested for their ability to differentiate monocytes into macrophages and maintain macrophages with proper spindle-like morphology for about 10 days following differentiation.
4. The cultures must be homogeneous (i.e., only macrophages) and cells must be more or less spindle shaped. The presence of cells with heterogenous shapes indicates that the original bone marrow isolate contained too many already mature macrophages that produced M-CSF or other such molecules involved in differentiation of precursor cells into dendritic cells, granulocytes, and so on. If cells look

flattened and contained many translucent vacuoles, they are probably activated by residual endotoxins or other activating molecules in the FBS or the L-929-conditioned medium. In this case, the percentage of L-929-conditoned medium can be decreased and/or reagents with lower endotoxin levels should be used.

5. Theoretical MOIs refer to the number of bacteria added to the wells containing the macrophages. Because the synchronized infection procedure (*see* **point 3** of **Subheading 3.2**) allows bacterial uptake for a limited time, only a fraction of the bacteria added to the well is phagocytosed. Consequently, the actual MOI (number of bacteria internalized by macrophages) is lower and has to be estimated by microscopy or counts of viable intracellular bacteria following gentamicin treatment.

6. A rapid warm up of the chilled, infected BMMs is essential to induce an efficient uptake of the bacteria in close contact to the BMMs and synchronize their subsequent intracellular trafficking.

7. Shorter fixation times with PFA, such as 10 min, are possible, as they sufficiently fix intracellular structures and can help preserve sensitive epitopes. However, they might not be sufficient to kill all bacteria in the samples, which is a concern for highly infectious (Biosafety Level 3) pathogens.

8. Triton X-100 permeabilization can help extract epitopes from intracellular membranes or cytoskeletal structures and yield to better staining. Because the Triton X-100 effect on cell membranes is irreversible, it is not required to keep it in the antibody dilutions buffers.

9. Optimal dilutions and incubation times that generate strong labeling and specific detection of the epitope have to be determined for each antibody. It is essential to the quality of the immunostaining to drain all liquid from the coverslip before incubation with primary and secondary antibodies, as remaining liquid will dilute the applied antibodies further.

10. The short incubation time (10 min) required at this stage relates to the relative fragility of digitonin-permeabilized primary macrophages. Other cells might be more resistant to digitonin treatment and allow for a longer incubation with antibodies. It is nonetheless important to pre-establish and use antibody dilutions that ensure detectable signals for both bacteria and calnexin.

Acknowledgments

We are grateful to Leigh Knodler for critical reading of the manuscript. This work was supported by the Intramural Research Program of the NIH, National Institute of Allergy and Infectious Diseases.

References

1. Knodler, L. A., Celli, J., and Finlay, B. B. (2001) Pathogenic trickery: deception of host cell processes. *Nat. Rev. Mol. Cell Biol.* **2**, 578–588.

2. Celli, J. (2005) Surviving inside a macrophage: the many ways of *Brucella. Res. Microbiol.* **157**, 93–98.

3. Oyston, P. C., Sjostedt, A., and Titball, R. W. (2004) Tularaemia: bioterrorism defence renews interest in *Francisella tularensis. Nat. Rev. Microbiol.* **2**, 967–978.
4. Celli, J., de Chastellier, C., Franchini, D. M., Pizarro-Cerda, J., Moreno, E., and Gorvel, J. P. (2003) *Brucella* evades macrophage killing via VirB-dependent sustained interactions with the endoplasmic reticulum. *J. Exp. Med.* **198**, 545–556.
5. Pizarro-Cerda, J., Meresse, S., Parton, R. G., van der Goot, G., Sola-Landa, A., Lopez-Goni, I., Moreno, E., and Gorvel, J. P. (1998) *Brucella abortus* transits through the autophagic pathway and replicates in the endoplasmic reticulum of nonprofessional phagocytes. *Infect. Immun.* **66**, 5711–5724.
6. O'Callaghan, D., Cazevieille, C., Allardet-Servent, A., Boschiroli, M. L., Bourg, G., Foulongne, V., Frutos, P., Kulakov, Y., and Ramuz, M. (1999) A homologue of the *Agrobacterium tumefaciens* VirB and *Bordetella pertussis* Ptl type IV secretion systems is essential for intracellular survival of *Brucella suis. Mol. Microbiol.* **33**, 1210–1220.
7. Celli, J. and Gorvel, J. P. (2004) Organelle robbery: *Brucella* interactions with the endoplasmic reticulum. *Curr. Opin. Microbiol.* **7**, 93–97.
8. Comerci, D. J., Martinez-Lorenzo, M. J., Sieira, R., Gorvel, J. P., and Ugalde, R. A. (2001) Essential role of the VirB machinery in the maturation of the *Brucella abortus*-containing vacuole. *Cell. Microbiol.* **3**, 159–168.
9. Clemens, D. L., Lee, B. Y., and Horwitz, M. A. (2004) Virulent and avirulent strains of *Francisella tularensis* prevent acidification and maturation of their phagosomes and escape into the cytoplasm in human macrophages. *Infect. Immun.* **72**, 3204–3217.
10. Golovliov, I., Baranov, V., Krocova, Z., Kovarova, H., and Sjostedt, A. (2003) An attenuated strain of the facultative intracellular bacterium *Francisella tularensis* can escape the phagosome of monocytic cells. *Infect. Immun.* **71**, 5940–5950.
11. Celli, J., Salcedo, S. P., and Gorvel, J. P. (2005) *Brucella* coopts the small GTPase Sar1 for intracellular replication. *Proc. Natl. Acad. Sci. U.S.A.* **102**, 1673–1678.
12. C. Checroun, Wehrly, T. D., Fisher, E. R., Hayes, S. F., and Celli., J. (in press) Autophagy-mediated reentry of *Francisella tularensis* into the endocytic compartment following cytoplasmic replication. *Proc. Natl. Acad. Sci. U.S.A.* **103**, 14578–14583.

12

Rate and Extent of *Helicobacter pylori* Phagocytosis

Lee-Ann H. Allen

Summary

Helicobacter pylori is a Gram-negative bacterium that colonizes the gastric epithelium and plays a causative role in the development of peptic ulcers and gastric cancer. Phago-cytosis is an element of innate defense used by macrophages and neutrophils to engulf microorganisms. We and others have shown that strains of *H. pylori* that contain the *cag* pathogenicity island actively retard their entry into phagocytes. Consequently, there is a lag of several minutes between bacterial binding and the onset of engulfment, and relative to other particles and microbes, the rate of internalization is slow. Herein, we describe in detail the use of synchronized phagocytosis and indirect immunofluorescence microscopy to quantify the rate and extent of *H. pylori* phagocytosis. This method is appropriate for primary phagocytes as well as transformed cell lines. More importantly, the effects of opsonins, virulence factors, and other agents on infection can be measured independent of bacterial viability or intracellular locale.

Key Words: Macrophage; neutrophil; phagocytosis; immunofluorescence micros-copy; antibody; opsonin.

1. Introduction

The ability of phagocytes to rapidly ingest and kill large numbers of microor-ganisms is an essential element of innate defense. Ligation of specific receptors on macrophages, monocytes and neutrophils activates signaling cascades that drive local actin polymerization and plasma membrane expansion during phago-cytosis *(1,2)*. The morphology of the forming phagosome is dictated by the receptor engaged and can be modulated further by virulence factors of certain pathogens *(1,2)*. Typically, ingestion is very rapid; actin polymerization is

From: *Methods in Molecular Biology, vol. 431: Bacterial Pathogenesis*
Edited by: F. DeLeo and M. Otto © Humana Press, Totowa, NJ

apparent within 30 s of microbe binding and engulfment is complete within 2–5 min (depending on particle size) *(3–5)*. Indeed, the speed and efficiency of phagocytosis distinguish this process from pathogen-driven invasion of other cell types.

   We have shown that *Helicobacter pylori* has the unusual ability to actively retard its entry into phagocytes *(4,6)*. Delayed phagocytosis requires bacterial protein synthesis, and genes within the *cag* pathogenicity island have been implicated in this process *(4,6–8)*. Herein, we describe methods used in our laboratory to measure the rate and extent of *H. pylori* phagocytosis that utilize synchronized phagocytosis, differential staining, and immunofluorescence microscopy. Advantages of this assay are the fact that very short infection times can be evaluated, large numbers of samples can be analyzed relatively quickly, and measurement of uptake is not coupled to intracellular survival or residence of bacteria in a specific compartment.

2. Materials

2.1. Macrophage Cultivation

1. The method described here is suitable for primary murine bone marrow-derived macrophages, resident or elicited peritoneal macrophages, or any other adherent macrophage or macrophage-like cell line such as J774A.1 and RAW264.7.
2. Tissue-culture media: endotoxin-free Dulbecco's modified Eagles medium (DMEM), minimal essential medium-alpha (MEMα) or Hepes-buffered RPMI-1640 supplemented with 2 mM L-glutamine, and heat-inactivated fetal bovine serum (FBS, to 10% final concentration) (*see* **Note 1**). Store at 4 °C.

2.2. Growth of Helicobacter pylori

1. Blood agar plates: Adjust trypticase soy agar base to pH 6.0 and then autoclave. Cool sterilized agar in a 50 °C water bath. Add defibrinated sheep blood to 5% final concentration immediately prior to pouring plates. Store at 4 °C.
2. Purchase *Helicobacter pylori* from the American Type Culture Collection (Manassas, VA). Methods described here were optimized using strains NCTC 11637, 60190, and Tx30a *(6)*. Clinical isolates of *H. pylori* can also be used.
3. CampyPak bags (Becton-Dickinson, Sparks, MD, USA) or an incubator that can achieve a microaerophilic (5% O_2, 10% CO_2, 85% N_2) atmosphere such as an Heraeus Trigas (Sorvall Instruments, Thermo Scientific, Wlatham, MA, USA) equipped with an oxygen electrode.

2.3. Microscopy Supplies

1. Round glass coverslips (12 mm diameter, 1.0 μm thick).
2. Straight needle-point stainless steel forceps (from a surgical supply company).

3. Lids from 24-well tissue-culture cluster plates. These do not need to be sterile and can be saved after use in other experiments.

4. Phosphate-buffered saline (PBS): 137 mM NaCl, 2.68 mM KCl, 8.1 mM Na$_2$HPO$_4$, and 1.47 mM KH$_2$PO$_4$ (pH 7.4) in tissue-culture grade deionized water. Also, PBS supplemented with 10 mM glucose. Sterile filter (0.2 μm) and store at room temperature and 4 °C.

5. "PAB" washing/blocking buffer: PBS supplemented with 0.5 mg/mL NaN$_3$ and 5 mg/mL bovine serum albumin. Sterile filter and store at 4 °C.

6. Anti-*H. pylori* antibody (Ab): Rabbit anti-*H. pylori* polyclonal Ab (#YVS6601) from Accurate Chemical and Scientific Corp. (Westbury, NY) (*see* **Note 2**).

7. Secondary Ab: Fluorescein isothiocyanate (FITC)- and rhodamine-conjugated F(ab′)$_2$ secondary Abs (Jackson ImmunoResearch Laboratories, West Grove, PA). Rehydrate with sterile deionized water according to the manufacture's directions and store at 4 °C.

8. Gelvatol mounting medium (*see* **Note 3**): Mix 2.4 g of polyvinyl alcohol with 6 g of glycerol, 6 mL of deionized water, and 12 mL of 200 mM Tris–HCl (pH 8.0) in a beaker on a stir plate for several hours until dissolved. Transfer the solution into a 50-mL polypropylene tube, heat the solution to 50 °C for 10 min, and then clarify by centrifugation at 5000 × g for 15 min. Decant the supernatant into a new 50-mL polypropylene tube and add 625 mg of 1,4-diazobicyclo-[2.2.2]-octane (DABCO). Invert the tube to dissolve DABCO. Store 1 mL aliquots of mounting medium in a frost-free freezer at –20 or –80 °C. DABCO is essential as it reduces photobleaching of fluorescence.

9. Fluorescence microscope: Upright fluorescence microscope with 40×, 63×, and 100× phase contrast objectives and filters for detection of both FITC and rhodamine fluorescence.

3. Methods

3.1. Preparation of Acid-Washed Coverslips

Acid washing removes manufacturing residue, oils, and other contaminants such as lipopolysaccharide (LPS) that may inadvertently activate macrophages or cause high non-specific background fluorescence. Transfer one package (1 ounce) of 12-mm round glass coverslips into a small glass bottle. Pour nitric acid over the coverslips making sure they are all submerged. Cap the bottle and incubate in the fume hood for at least 48 h. Carefully decant the acid and then rinse the coverslips with 15–20 changes of sterile deionized tissue-culture grade water (pour water over the coverslips, cap bottle, rotate gently, decant water, and repeat). Rinse the coverslips with two changes of 95% ethanol to remove residual water. Store coverslips in 70% ethanol at room temperature. Use after ≥16 h in ethanol.

3.2. Plating Macrophages on Coverslips

Helicobacter pylori will infect different types of macrophages including primary human and murine macrophages and murine macrophage-like cell lines and human neutrophils isolated from peripheral blood *(4,6,8–13)*. As noted above, endotoxin-free reagents are required to prevent unintended cell activation. In addition, tissue-culture medium must not contain penicillin or streptomycin, as these drugs compromise *H. pylori* viability. Thus, all phagocytes should be maintained in antibiotic-free medium or, at a minimum, transferred to antibiotic-free medium 48 h prior to infection.

1. Flaming of coverslips and dispersal in 35-mm dishes. One 35-mm tissue-culture dish is needed per sample. Each dish will hold three coverslips that provide triplicate samples for each experimental condition or time point. Using forceps, remove a few coverslips from their ethanol storage bottle and place onto a pile of Kimwipes. Fill one small microbeaker with 70% ethanol and light a Bunsen burner. Working with one coverslip at a time, grasp with forceps, dunk into ethanol, and drain off excess ethanol by touching edge of the coverslip to the pile of Kimwipes. Pass the coverslip through the flame and then place it into a 35-mm dish. Because the 35-mm dishes are opened only briefly, this procedure can be performed on the bench top. Failure to drain off excess ethanol will cause coverslips to shatter when passed through the flame.

2. Plating macrophages: Harvest BMM, J774A.1, or RAW264.7 cells using a cell scraper. Harvest peritoneal macrophages from anaesthetized mice by lavage with 4 °C PBS. Centrifuge all cells at $400 \times g$ for 15 min at 4 °C. Resuspend the cell pellets in tissue-culture medium (DMEM for J774A.1 and RAW264.7, MEMα for peritoneal macrophages, DMEM or MEMα supplemented with CSF-1 for bone marrow-derived macrophages, and RPMI-1640 for human macrophages) to achieve $\sim$5–7 $\times$ 10^5cells/mL. A moderate cell density is optimal for microscopy. Working in the tissue-culture hood, remove lids from the 35-mm dishes. Be sure that coverslips in each dish are not touching one another or the edge of the dish. If necessary, reposition coverslips using a sterile probe (a yellow Pipetman tip or forceps can be used to push coverslips into position). Plate 50 μL of the macrophage suspension directly onto each coverslip (*see* **Note 4**). Liquid should spread evenly over the acid-washed glass, and surface tension will hold the domes of liquid in place (*see* **Note 5**). Close dishes and transfer into a 37 °C tissue-culture incubator for 90–120 min to let cells attach. Dishes can be placed in a small pan or on a tray to facilitate transport. Thereafter, add 1.5–2 mL of additional medium to each dish and leave in the tissue-culture incubator overnight. For peritoneal macrophages, remove non-adherent lymphocytes using three changes of medium or PBS before addition of 1.5–2 mL of fresh medium.

3.3. Cultivation of Helicobacter pylori

Because *H. pylori* is a biosafety level 2 pathogen, all manipulations of live bacteria should be performed in a tissue-culture hood. Streak bacteria from a frozen glycerol stock onto a blood agar plate using a sterile inoculation loop. A heated and humidified microaerophilic environment (5% O_2, 10% CO_2, 85% N_2; 37 °C; 98% humidity) is required for optimal *H. pylori* growth. These conditions can be achieved using a special incubator. Alternatively, a microaerophilic atmosphere can be generated using CampyPak pouches that can then be placed in a 37 °C tissue-culture incubator. Depending on the *H. pylori* strain employed, growth will be apparent after 1 or 2 days.

Harvest bacteria from agar plates by scraping into 1 mL of sterile PBS supplemented with 10 mM glucose. Pellet bacteria in a microfuge (7000 × *g*, 2 min, 4 °C). Discard the supernatant and resuspend the pellet in 1 mL of PBS/glucose (pipette gently, do not vortex). Pellet the bacteria once again and resuspend organisms in PBS/glucose. Repeat for a total of three or four washes. If desired, bacteria can be opsonized with specific antibody or complement (*see* **Note 6**). Determine bacterial concentration by measuring the absorbance of a 1:10 dilution of the stock at 600 nm. Store washed *H. pylori* on ice until use.

We routinely check all *H. pylori* preparations for motility and viability before infection of macrophages. To assess motility, place a drop of washed bacteria onto a microscope slide. Overlay bacteria with a coverslip and view using a phase-contrast microscope. Molecular Probes BacLight Live/Dead reagents are used to quantify viability of washed bacteria. Dilute 0.1 mL of *H. pylori* in 1 mL of PBS containing 3 μL of a 1:1 mixture of the live-dead stains (3.34 m*M* Styo9 and 20 m*M* propidium iodide). After 15 min at room temperature, place an aliquot of stained bacteria onto a microscope slide, overlay with a large coverslip, and examine the sample using an immunofluorescence microscope equipped with a ×40 dry objective and filters for FITC and rhodamine fluorescence. By this assay, live bacteria fluoresce bright green, dead bacteria are bright red, and damaged organisms may appear orange. Typically, more than 95% of *H. pylori* are viable, motile, and spiral shaped. Overgrown or spent cultures will accumulate clumps of non-viable cocci. Discard samples if fewer than 90% of bacteria are viable.

3.4. Synchronized Phagocytosis

Centrifugation of *H. pylori* onto macrophages at low temperature allows particle binding but not phagocytosis, and rapid transfer of samples to 37 °C supports synchronized ingestion. Aspirate medium from dishes of macrophages. Place 1 mL of fresh 37 °C medium into each dish and return dishes to the incubator. Dilute *H. pylori* in cold tissue-culture medium to achieve a moderate multiplicity of infection (MOI). From an MOI of 15:1, we obtain ∼8 bacteria per cell. Add 1 mL of cold, diluted *H. pylori* to each dish of macrophages (already containing 1 mL of warm medium). Mixing warm and cold medium reduces the overall temperature below the threshold for phagocytosis (16 °C).

Rapidly and carefully transfer the dishes onto microplate carriers in a refrigerated tissue-culture centrifuge pre-cooled to 10–12 °C. Centrifuge for 3–4 min at $600 \times g$ with maximum braking. Carefully transfer dishes to 37 °C incubator to allow phagocytosis or transfer immediately to ice ($t = 0$ min sample). If dishes remain cold during centrifugation, bacteria in the 0-min samples should be cell associated but not internalized.

3.5. Differential Staining of Bound and Ingested Bacteria

The method described here stains sequentially extracellular and intracellular bacteria and as such allows the rate and extent of phagocytosis to be quantified.

1. At the desired time points (e.g., 0, 1, 3, 5, 7, 10, 20, 30 min; determine empirically for each strain and cell type) move samples from 37 °C into a metal pan placed over a bed of ice. Working quickly, aspirate medium and wash samples with three changes of 4 °C PBS to remove unbound organisms and stop further uptake of any uningested, surface-associated bacteria.
2. Dilute anti-*H. pylori* Ab in PBS/glucose: We use pAb YVS6601 at 1:2000 for strains 11637 and 60190 and 1:750 for strain Tx30a. Aspirate cold PBS from each dish and replace with 1 mL of antibody solution. Leave dishes on a bed of ice and incubate in a refrigerator or cold room for 30–60 min (*see* **Note 7**). Because the macrophages are intact, the Ab has access only to extracellular organisms.
3. Aspirate the primary Ab solution from each dish and wash samples gently with three changes of ice-cold PBS/glucose. Fix and permeabilize samples using methanol: Pour ∼3 mL of methanol into each dish and let sit for 10 min. From this point on, samples can be manipulated at room temperature. Aspirate off the methanol, rinse samples once with PBS to rehydrate and then block in PAB for 15–30 min.
4. While samples are blocking, dilute the rhodamine-conjugated secondary Ab 1:200 in PAB. Thirty microliters is needed per coverslip. Also, set up a pile of Kimwipes and label a 24-well dish lid that will be used to hold coverslips during remaining incubations (*see* **Fig. 1**).
5. Working with one dish at a time, pick up a coverslip using forceps and drain off PAB by touching the edge of the coverslip to the pile of Kimwipes. Dry the back of each coverslip and place cell-side up on the 24-well dish lid (*see* **Fig. 1**). Overlay with 30 µL of secondary Ab. Repeat for the other samples. Place coverslips inside a Tupperware type box lined with damp paper towels to prevent evaporation of liquid. Incubate for 30–60 min at room temperature. At the end of this incubation, extracellular *H. pylori* will be stained red. Meanwhile, prepare a

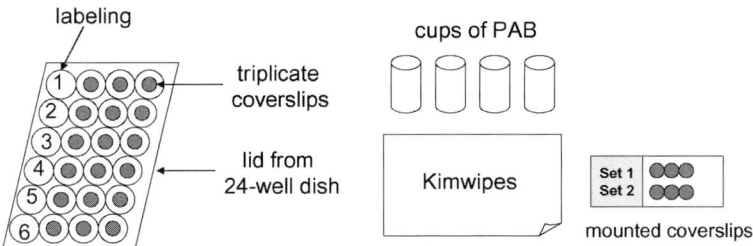

Fig. 1. Set up for sample staining and washing. Materials for cell staining, washing, and mounting coverslips are shown.

second stock of anti-*H. pylori* Ab. You will need 30 µL of Ab diluted in PAB for each coverslip.

6. Using forceps, lift a coverslip off of the dish lid and drain Ab by touching the edge of the coverslip to the pile of Kimwipes. Wash coverslips sequentially in four microbeakers of PAB, draining thoroughly between each wash. Dry the back of the coverslip and return it to the 24-well dish lid. Overlay with 30 µL of anti-*H. pylori* pAb and incubate for 30–60 min. Because macrophages are permeabilized, the anti-*H. pylori* Ab now has access to both intracellular and extracellular bacteria.

7. Prepare FITC-conjugated secondary Ab diluted 1:200 in PAB. Wash coverslips in microbeakers of PAB as described above and return to the dish lid. Overlay with 30 µL of secondary Ab and incubate for 30–60 min.

3.6. Mounting Stained Coverslips onto Slides

Up to six coverslips will fit onto one slide. It is important to place the coverslips as close as possible to the marking area of the slide (*see* **Fig. 1**). Samples mounted near the far right edge of the slide cannot be viewed because this part of the slide rests on the microscope stage.

Thaw one tube of mounting medium. Working with coverslips for one slide at a time, place six small drops (∼15 µL) of gelvatol mounting medium onto the glass slide. Wash each coverslip sequentially in four beakers of PAB as described above followed by a final rinse in deionized water. The water rinse prevents a dried salt crust from forming on mounted coverslips. Blot excess water off the back of the coverslip and then invert it (cell side down) onto a bead of mounting medium (start from one edge of the coverslip and lower it slowly to avoid trapping air bubbles). Repeat for the other coverslips on this slide. Gently press down on the top of each coverslip to ensure that their edges are not overlapping (this will also push out small air bubbles). Remove any excess mounting medium using an aspirator. Place slides in a cardboard slide folder; this keeps the slides flat and protected from light.

Let the mounting medium set overnight at 4 °C. Leftover mounting medium can be refrozen for future use.

3.7. *Microscopy and Data Analysis*

Combining synchronized phagocytosis with differential staining allows precise quantification of total macrophage-associated bacteria as well as the fraction of intracellular and extracellular *H. pylori* at each time point. For each sample, analyze random fields of macrophages using immunofluorescence and phase-contrast microscopy. At least 50 infected macrophages should be examined per sample and the number of green (FITC stained) and red (rhodamine stained) *H. pylori* recorded. As all bacteria are stained green and only uningested (extracellular) *H. pylori* are also stained red, the rate and extent of phagocytosis at each time point can be calculated:

$$\% \text{ intracellular } H.\ pylori = \frac{\text{no. green bacteria} - \text{no. red bacteria}}{\text{no. green bacteria}} \times 100$$

Other infection parameters can be measured, including infection efficiency (fraction of infected cells), association index (total number of *H. pylori* per 100 macrophages), and phagocytic index (number of intracellular *H. pylori* per 100

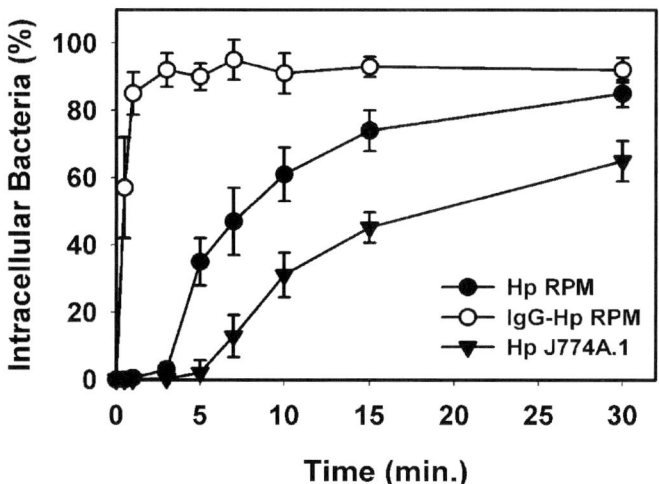

Fig. 2. Effect of cell type and opsonins on *Helicobacter pylori* phagocytosis. *H. pylori* strain 11637 (Hp) is ingested by resident peritoneal macrophages (RPM) after a lag of several min. By contrast, uptake of IgG-opsonized Hp is rapid. Compared with primary macrophages, infection of J774A.1 cells with Hp is less efficient. Data are the mean ± SEM of three independent experiments performed in triplicate.

macrophages). It is important to note that "0-min" samples should contain <3% ingested bacteria (*see* **Fig. 2**), and the number of total cell-associated bacteria should remain constant over the course of the experiment as the fraction of intracellular *H. pylori* increases. If large numbers of "green-only" bacteria are apparent at 0 min, either samples did not remain cool (<16 °C) during the centrifugation or an artifact occurred because the anti-*H. pylori* antibody used in the first round of staining did not react efficiently with surface exposed epitopes. These caveats aside, this assay is rapid and efficient and can be adapted easily to assess the effects of opsonins or signaling pathway inhibitors on bacterial binding and phagocytosis (*see* **Note 8**) to compare phagocytosis of different strains of bacteria or to compare *H. pylori* entry into different types of phagocytes (*see* **Fig. 2**) *(6,14)*. Finally, the fact that this assay is independent of phagosome composition is significant as some, but not all, strains of *H. pylori* inhibit phagosome maturation and because phagosome composition varies somewhat in primary macrophages and macrophage-like cell lines *(10,11)*.

4. Notes

1. To prevent macrophage priming or activation, it is essential to ensure that all tissue-culture media, serum, and other reagents are endotoxin free.
2. The anti-*H. pylori* Ab we prefer works well with many *H. pylori* strains and is cost effective. However, any Ab that reacts with both surface and cytoplamic (internal) bacterial epitopes would be suitable. Alternatively, two different Ab may be used to detect intracellular and extracellular organisms. In all cases, each Ab should be tested for reactivity with intact and methanol-permeabilized bacteria and the appropriate concentrations determined prior to use with infected macrophages.
3. We prefer the polyvinyl alcohol base of gelvatol mounting medium because it hardens rapidly and completely. The mounting medium we prepare gives consistent results, but, in our hands, this is not true for similar products obtained from commercial vendors. A common alternative to gelvatol is buffered glycerol (with or without an anti-fading agent). Because glycerol mounting solutions do not harden, coverslips must be attached to microscope slides using several layers of nail polish (which is time consuming to apply and apt to leak).
4. To use this assay with phagocytes that do not adhere strongly to uncoated glass, coverslips should be precoated with extracellular matrix proteins. For example, plate human monocyte-derived macrophages on coverslips coated with 0.1 mg/mL collagen and plate human neutrophils on coverslips coated with 0.1 mg/mL fibrinogen.
5. This plating method conserves macrophages, ensures that each sample contains a similar number of cells and (most importantly) is compatible with synchronized phagocytosis.
6. *Helicobacter pylori* are not opsonized in vivo *(15,16)*. Nevertheless, bacteria can be opsonized with specific IgG or complement factors in vitro *(6,13,17)*. To

opsonize with IgG, incubate washed organisms with a sub-agglutinating concentration Ab directed against *H. pylori* LPS (or another surface antigen) for 30 min at 37 °C. To opsonize bacteria with complement factors, incubate washed bacteria in 50% fresh human serum for 20–30 min at 37 °C. Relatively high concentrations of serum are required to fix complement on this organism *(17)*. Wash all opsonized bacteria prior to measurement of optical density and macrophage infection.

7. Human monocyte-derived macrophages tend to lift off coverslips when incubated at 4 °C. Thus, human macrophages (or other cells that do not adhere strongly such as neutrophils) should be fixed in 2% paraformaldehyde for 15 min at room temperature and then washed with PBS. Incubation with primary Ab can then be performed at room temperature. Subsequent steps of the protocol are unchanged.

8. When using pharmacological inhibitors of macrophage signaling pathways, it is important to note that many of these agents are inactivated by components in tissue-culture medium and/or FBS. Thus, for these types of studies, it is necessary to use serum-free medium or Hank's-buffered saline solution supplemented with 1 mM Hepes and 10 mM glucose. Washed macrophages should be serum-starved or incubated in Hank's buffer for at least 2 h prior to inhibitor treatment and infection. In general, macrophages are pretreated with inhibitor (or vehicle) for 30–60 min at 37 °C and then infected in inhibitor-containing medium. Incubate samples at 37 °C for 30 min prior to differential staining.

Acknowledgements

This work was supported by a Merit Review Grant from the Department of Veteran's Affairs.

References

1. Aderem, A., Underhill, D.M. (1999) Mechanisms of phagocytosis in macrophages. *Annu. Rev. Inmmunol.* **17**, 593–623.
2. Underhill, D.M., Ozinsky, A. (2002) Phagocytosis of microbes: complexity in action. *Annu. Rev. Immunol.* **20**, 825–852.
3. Allen, L.-A.H., Aderem, A. (1996) Molecular definition of distinct cytoskeletal structures involved in complement- and Fc receptor-mediated phagocytosis in macrophages. *J. Exp. Med.* **184**, 627–637.
4. Allen, L.-A.H., Schlesinger, L.S., Kang, B. (2000) Virulent strains of Helicobacter pylori demonstrate delayed phagocytosis and stimulate homotypic phagosome fusion in macrophages. *J. Exp. Med.* **191**, 115–127.
5. Allen, L.-A.H., Aderem, A. (1995) A role for MARCKS, the αisozyme of protein kinase C and myosin I in zymosan phagocytosis by macrophages. *J. Exp. Med.* **182**, 829–840.
6. Allen, L.-A.H., Allgood, J.A. (2002) Atypical protein kinase C-ζ is essential for delayed phagocytosis of *Helicobacter pylori*. *Curr. Biol.* **13**, 1762–1766.

7. Ramarao, N., Meyer, T.F. (2001) Helicobacter pylori resists phagocytosis by macrophages: quantitative assessment by confocal microscopy and fluorescence-activated cell sorting. *Infect. Immun.* **69**, 2604–2611.

8. Ramarao, N., Gray-Owen, S.D., Backert, S., Meyer, T.F. (2000) Helicobacter pylori inhibits phagocytosis by professional phagocytes involving type IV secretion components. *Mol. Microbiol.* **37**, 1389–1404.

9. Allen, L.-A.H. (2000) Modulating phagocyte activation: the pros and cons of *Helicobacter pylori* virulence factors. *J. Exp. Med.* **191**, 1451–1454.

10. Zheng, P.Y., Jones, N.L. (2003) *Helicobacter pylori* strains expressing the vacuolating cytotoxin interrupt phagosome maturation in macrophages by recruiting and retaining TACO (coronin 1) protein. *Cell. Microbiol.* **5**, 25–40.

11. Schwartz, J.T., Allen, L.-A.H. (2006) Role of urease in megasome formation and *Helicobacter pylori* survival in macrophages. *J. Leukoc. Biol.* **79**, 1214–1225.

12. Andersen, L.P., Blom, J., Nielsen, H. (1993) Survival and ultrastructural changes of *Helicobacter pylori* after phagocytosis by human polymorphonuclear phagocytes and monocytes. *APMIS* **101**, 61–72.

13. Allen, L.-A.H., Beecher, B.R., Lynch, J.T., Rohner, O.V., Wittine, L.M. (2005) *Helicobacter pylori* disrupts NADPH oxidase targeting in human neutrophils to induce extracellular superoxide release. *J. Immunol.* **174**, 3658–3667.

14. Allen, L.-A.H., Allgood, J.A., Han, X., Wittine, L.M. (2005) Phosphoinositide 3-kinase regulates actin polymerization during delayed phagocytosis of Helicobacter pylori. *J. Leukoc. Biol.* **278**, 220–230.

15. Darwin, P.E., Sztein, M.B., Zheng, Q.X., James, S.P., Fantry, G.T. (1996) Immune evasion by *Helicobacter pylori*: gastric spiral bacteria lack surface immunoglobulin deposition and reactivity with homologous antibodies. *Helicobacter* **1**, 20–27.

16. Berstad, A.E., Brandtzaeg, P., Stave, R., Halstensen, T.S. (1997) Epithelium related deposition of activated complement in *Helicobacter pylori* associated gastritis. *Gut* **40**, 196–203.

17. Rokita, E., Makristathis, A., Presterl, E., Rotter, M.L., Hirschl, A.M. (1998) *Helicobacter pylori* urease significantly reduces opsonization by human complement. *J. Infect. Dis.* **178**, 1521–1525.

13

Culture, Isolation, and Labeling of *Anaplasma phagocytophilum* for Subsequent Infection of Human Neutrophils

Dori L. Borjesson

Summary

Anaplasma phagocytophilum is the etiologic agent of granulocytic anaplasmosis, a tick-borne, zoonotic, emerging infectious disease. *A. phagocytophilum* is an obligate intracellular pathogen that primarily resides within membrane-bound, cytoplasmic vacuoles of host neutrophils. Closely related to *Ehrlichial* and *Rickettsial* organisms, *A. phagocytophilum* is a small, fragile, Gram-negative bacterium that presents unique challenges for culture, isolation, enumeration, and labeling. This chapter delineates pathogen-specific considerations for culture and labeling of this organism for subsequent use in assays to examine mechanisms of host cell–pathogen interactions.

Key Words: *Anaplasma phagocytophilum*; granulocytic anaplasmosis; cell culture; fluorescent labeling; neutrophils.

1. Introduction

Anaplasma phagocytophilum is the etiologic agent of granulocytic anaplasmosis, an emerging, zoonotic, tick-borne, infectious disease *(1,2)*. *A. phagocytophilum* is an obligate intracellular bacterium that principally inhabits the cytoplasm of host neutrophils. This intracellular niche is particularly noteworthy, as neutrophils are powerful cells that function as a first line of defense against bacterial pathogens. Understanding the mechanisms by which *A. phagocytophilum* circumvents neutrophil defense systems is facilitated by optimum pathogen culture, isolation, and labeling techniques.

From: *Methods in Molecular Biology, vol. 431: Bacterial Pathogenesis*
Edited by: F. DeLeo and M. Otto © Humana Press, Totowa, NJ

Successful culture of *A. phagocytophilum* was first described in 1996 *(3)*. Since that time, pathogen culture, isolation, and labeling techniques have rapidly been developed to facilitate the study of pathogen–host cell interactions. The techniques described in this chapter highlight standard and pathogen-specific culture and isolation techniques. Options for labeling *A. phagocytophilum* for use in flow cytometric and immunofluorescence protocols are described with an emphasis on a labeling technique that facilitates the distinction between intra- and extracellular organisms. This technique permits ready dissection of pathogen binding, entry, and infection of human neutrophils. In addition, the techniques described could be readily modified for study of pathogen interaction with a broad array of host cells.

2. Materials

2.1. Anaplasma phagocytophilum Culture and Propagation in HL60 Cells

1. Human myeloid leukemia cell line (HL60 cells, ATCC 240-CCL; American Type Culture Collection, Rockville, MD). Upon receipt, HL60 cells should be stored in liquid nitrogen (*see* **Note 1**).
2. RPMI-1640 media supplemented with 10% heat-inactivated fetal bovine serum (FBS) and 2 mM L-glutamine (*see* **Note 2**).
3. Nunc cryovials.
4. Standard cell-culture flasks (25 cm^2 and 75 cm^2 flasks, canted-neck polystyrene, non-pyrogenic; BD Biosciences, San Jose, CA, USA).
5. 0.2 µm vented blue plug seal cap (*see* **Note 3**).
6. Sterile 1–25-mL pipettes.
7. 15- and 50-mL polystyrene conical tubes.
8. 500-mL top filter with 45-mm neck.
9. 0.22-µm cellulose acetate, sterilizing, low protein binding membrane non-pyrogenic (Corning, NY, USA).
10. Freezing media: 90% cell-culture media (RPMI-1640 media containing 10% FBS) and 10% dimethysulfoxide (DMSO).
11. Infected, EDTA-anticoagulated patient blood. Sources of infected blood may include infected human beings, mice, horses, dogs, or ruminants. Blood must be sterile.
12. All-*trans*-retinoic acid (1 µM retinoic acid made from a 1 mM stock solution prepared in 95% (v/v) ethanol and stored at –80 degrees; Sigma) or 1.25% DMSO (Sigma-Aldrich, St.Louis, MO, USA).
13. 1,25-dihydroxyvitamin D$_3$ (2 µM Vitamin D$_3$; Calbiochem, San Diego, CA) prepared as a 10^4 M stock solution by dissolving in 95% ethanol and stored at –20 °C.

14. 12-*O*-tetradecanoylphorbol-13-acetate (TPA; Sigma) prepared as a 1.6×10^{-4} *M* stock in DMSO and stored at $-80\,^{\circ}$C for ≤ 6 months.

2.2. Isolation of Cell-Free Anaplasma phagocytophilum

1. Hematocytometer, 1.5-mL microfuge tubes, 1–500-µL pipettor, trypan blue, microscope slides, modified Romanosky stain set (i.e., Hemaquick), and glass microfiber filters (25-nm GD/X filter).
2. 3–9-mL sterile syringes and 25–27.5 gauge needles, or sonic dismembranator.
3. Cytofuge (Shandon Cytospin; ThermoShandon, Pittsburgh, PA, USA).
4. SPGN buffer: 7.5% sucrose, 3.7 m*M* KH$_2$PO$_4$, 7 m*M* K$_2$HPO$_4$, and 5 m*M* glutamine, pH 7.4) (*see* **Note 4**).
5. Diatrizoate meglumine (66% diatrizoate meglumine and 10% diatrizoate sodium in 0.32% sodium citrate and 0.04% edentate disodium) should be obtained as a 76% solution. Diatrizoate meglumine is an ionic radiopaque contrast agent. Brand names include Renografin 76, Hypaque 76, and Angiovist 370. Dilutions of diatrizoate meglumine can be made in PBS buffer.

2.3. Determining Bacterial Infectious Dose

1. Hematocytometer, microscope slides, and modified Romanosky stain set (i.e., Hemaquick).

2.4. Bacteria Labeling and Modifications

1. Specific purified monoclonal antibodies that label outer membrane proteins of *A. phagocytophilum* may be obtained from non-commercial sources (e.g., R1B10, R5E4, and R5A9) *(4)*.
2. Zenon Mouse IgG Labeling Kit (Invitrogen Corp., Carlsbad, CA, USA) to label unconjugated antibodies.
3. Serum-free protein block (DAKO, Carpinteira, CA, USA).
4. Methanol, PBS, coverslips, and bovine serum albumin (BSA).
5. Fluorescent dyes including the lipophilic fluorescent compound PKH-67 (Sigma) and Cell Tracker Green (CMFDA; Invitrogen) can be used to label *A. phagocytophilum* for immunofluorescent assays or flow cytometry *(5)*.
6. Doxycycline (10 µg/mL; Sigma) or oxytetracycline hydrochloride (10 mg/mL; Sigma).
7. Paraformaldehyde (2% solution dissolved in PBS).

2.5. Incubation of Bacteria with Neutrophils

1. RPMI-1640 media buffered with 10 m*M* HEPES, pH 7.4.
2. 24-well tissue-culture plates.
3. AlexaFluor 488, anti-AlexaFlour 488 conjugated to AlexaFlour 594 (Invitrogen, Carlsbad, CA, USA).

   4. No. 1, 12-mm round glass coverslips.
   5. Microscope slides, mounting media.

3. Methods

3.1. *Anaplasma phagocytophilum Culture and Propagation in HL60 Cells*

1. Prepare 1 L of RPMI-1640 containing 10% FBS cell-culture media (RPMI/10% FBS) by placing 100 mL of heat-inactivated FBS into 900 mL RPMI-1640 media. Filter-sterilize using a 0.22-μm pore size bottle top filter. Store at 4 °C for up to 6 months.
2. To initiate HL60 culture, pipette 9 mL of RPMI/10% FBS media into a 15-mL conical tube. Thaw one vial of HL60 cells in a 37 °C water bath for 2–3 min. Pipette thawed cells into the 9 mL of media. Spin at 450 × g for 5 min to pellet cells. Pour off freezing media supernatant. Resuspend cells in 5 mL of fresh media and place in 25-cm²flask. Cells should be maintained at 37 °C and 5% CO_2/ 95% air.
3. The initiation of infected HL60 cell culture is identical to that described above (#2); however, 5 mL of uninfected HL60 cells in fresh media should be added to the 1 mL of 90% thawed/infected cells to a final concentration of ~5 × 10^5 cells/mL in a 25-cm²flask.
4. To maintain uninfected cells in culture, keep cell density from 5 × 10^5 to 2 × 10^6 cells/mL by feeding the cells with fresh media twice a week. Uninfected cells grow slower than infected cells, so uninfected cells should be thawed and propagation initiated before infected cells. In addition, maintain at least one additional flask of uninfected HL60 cells to feed infected cell cultures. Uninfected cells can be expanded to a 75-cm² flask by taking 10 mL of uninfected cells and adding 10 mL of fresh media.
5. To maintain infected cells in culture, check infection level every 2–3 days (see Section 3.2.1). Initial infection is often noted by day 3 with peak infection by day 5–10. Culture will be > 90% infected by day 5 (*see* **Note 5; Fig. 1A** and Color Plate 5, following p. 46). Cell lysis will be noted by day 12–14. Add 1:1 fresh uninfected to infected HL60 cells twice weekly. *A. phagocytophilum* has a cytotoxic effect on HL60 cells and will induce cell lysis and apoptosis if they become too infected (*6*). Infected cells can also be expanded to a 75-cm² flask by adding 2 mL of highly infected HL60 cells diluted with 8 mL uninfected HL60 cells and 10–15 mL fresh media. Harvest pathogen from infected cell cultures at peak infection (>90% infected HL60 cells (*see* **Note 6**).
6. To freeze uninfected and infected HL60 cells, transfer media and cells from tissue-culture flask into 50 mL conical tubes. Centrifuge at 450 × g for 5 min to pellet cells. Decant or aspirate supernatant and discard. Resuspend 1 mL of cells at a final concentration of 5 × 10^6cells/mL in freezing medium. Transfer contents (~0.5 mL per tube) into NUNC cryovials. Freeze at –80 °C overnight, then transfer vials to liquid nitrogen the next day. Freeze infected HL60 cells at ~90% infection level.

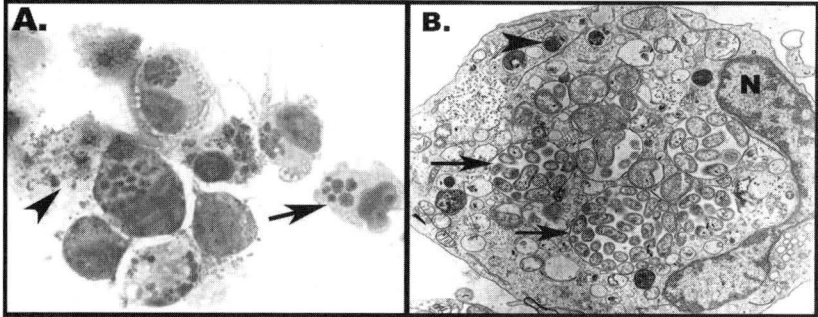

Fig. 1. Images of *Anaplasma phagocytophilum*-infected HL60 cells. (**A**) Cytofuge smear of highly infected HL60 cells. Arrow indicates intact, multiple pathogen morulae within the HL60 cell cytoplasm *(6)*. Arrowhead indicates HL60 cell lysis with abundant extracellular bacteria (1000× magnification; Hema-Quik stain). (**B**) Electron micrograph of a highly infected HL60 cell with abundant intracellular bacteria found within cytoplasmic vacuoles (morula). The N labels the cell nucleus. Arrows indicate the "reticulate," possibly replicative, stage of organism development. The arrowhead indicates the "dense cored," possibly infective, stage of organism development. (*See* Color Plate 5, following p. 46.)

3.1.1. Initiation of Infected HL60 Cells Through Cultivation of Patient Blood Samples in HL60 Cells

1. Sterile-inoculate 100 μL of patient EDTA-anticoagulated whole blood into 5 mL of 5×10^5 cells/mL HL60 cells in a 25-cm^2 cell-culture flask. Infection will typically be noted by day 5 post-inoculation; however, up to 15 days can be needed for cultures to be 90% infected. Variation in infectivity is noted with some patient samples resulting in <5% infection even after 2 weeks. Blood samples may be stored at 4 °C for up to 48 h before inoculation into culture. Diagnostic culture of organisms from patient blood may be enhanced by addition of retinoic acid to HL60 cells (*see* **Subheading 2.1.2.**). Cultures may be negative after only 24 h of appropriate antibiotic therapy.

3.1.2. Induction of Granulocytic or Monocytic Differentiation

1. Incubate 5×10^5–1×10^6 cells/mL HL60 cells for 3–6 days with 1 μ*M* all-*trans*-retinoic acid *(7)* or 1.25% DMSO *(8)*. After 6 days of exposure, approximately half of the cells will appear granulocytic and show myeloperoxidase staining (*see* **Note 7**).
2. Incubate 5×10^5–1×10^6 cells/mL HL60 for 3–6 days with 2 μ*M* 25-OH vitamin D$_3$ or with the protein kinase C activator, TPA, at 1.6×10^{-7}–10^{-10} *M* final concentration to induce macrophage differentiation *(8)* (*see* **Note 7**).

3.2. Isolation of Cell-Free Anaplasma phagocytophilum

3.2.1. Routine Infection Studies

1. Count infected HL60 cells by removing ~150 μL of cells from culture flask. Dilute 10 μL of well-mixed cells into 90 μL of trypan blue and count cells using a hematocytometer. The use of trypan blue permits concurrent assessment of cell viability.
2. Estimate percent infection. Take remaining undiluted cells and make a cytofuge smear (use a Shandon Cytospin at 70 × *g* for 5 min). Air-dry the smear, fix with methanol and stain with any modified Romanosky stain (e.g., Hema-Quik, Dif-Quick, and Wright-Giemsa). With a light microscope, determine the percent of infected HL60 cells by counting 100 cells (*see* **Fig. 1A**). Alternatively, if a cytospin is unavailable, centrifuge ~300 μL of infected HL60 cells at 450 × *g* for 5 min to sediment cells. Remove supernatant and resuspend cells in ~50 μL of saline or PBS. Place a large drop of sedimented cells unto a microscope slide and smear the cells in a fashion similar to that of a blood smear. Stain and count as described.
3. Disrupt HL60 cells for pathogen isolation by cell sonication or needle lysis. To sonicate, resuspend desired number of cells in ~ 5 mL of media and place in a 50-mL conical tube. Place tubes on ice. Sonicate cells gently with 3–5 quick pulses. Alternatively, cells can be lysed by 3–6 passages through a 25–27.5 gauge needle (*see* **Note 8**).
4. Centrifuge sonicate or lysate to pellet unwanted cellular debris (450 × *g* for 5 min; *A. phagocytophilum* is in supernatant). Filter supernatant with glass microfiber filter to remove remnant HL60 debris and further clarify the lysate. Pellet the pathogen by centrifugation of supernatant (5000 × *g* for 10–15 min). Discard supernatant and resuspend pelleted *A. phagocytophilum* in working media/buffer. Wash one time by pelleting pathogen as above (5000 × *g* for 10–15 min).

3.2.2. Purification Process

To obtain a more pure population of *A. phagocytophilum*, organisms can be isolated from host cell debris using a number of techniques (*see* **Note 9**). The most common protocol uses diatrizoate meglumine. This technique was first described for the Rickettsiae in 1981 and has been modified over time by a number of investigators *(9–12)* (*see* **Note 10**).

1. Follow steps 1–3 in **Subheading 3.2.1**. High numbers of infected HL60 cells (i.e., 2–6 × 10^8 cells) are needed for this process.
2. Centrifuge sonicate or lysate to pellet unwanted cellular debris (450 × *g* for 5–10 min; *A. phagocytophilum* is in supernatant). Pellet pathogen by centrifugation of supernatant (5000 × *g* for 10–15 min). Discard supernatant and resuspend pelleted *A. phagocytophilum* in SPGN buffer.
3. Layer resuspend *A. phagocytophilum* onto a discontinuous gradient of 6–8 mL of 42% and 10–12 mL of 30% diatrizoate meglumine (*see* **Note 11**). Ultracentrifuge gradients to equilibrium at 50,000 × *g* for 60–75 min at 4 °C. Bacterial fractions

can be collected with a sterile pipette or through punctures in the bottom or middle of the tube in the band at the 30 and 42% interface. Resuspend collected bands in 20 mL of SPGN and pellet at 20,200 × g for 30 min. Resuspend pellet in SPGN buffer to desired concentration.

3.3. Determining Bacterial Infectious Inoculum

Because of their small size and intracellular location, routine methods of bacterial enumeration (i.e., Petroff-Hauser chambers) are not possible or accurate. Two acceptable methods of approximating multiplicity of infection are as follows:

1. The most common method is to enumerate the number of infected HL60 cells used to obtain the infectious inoculum and state it as a ratio of number of infected HL60 cells: number of neutrophils (e.g., to inoculate A. *phagocytophilum* into neutrophils for a study, state "bacteria liberated from 1×10^6 infected HL60 cells were incubated with 2.5×10^5 target cells (*5*)." Immunofluorescence has been used to confirm that a ratio of 1 infected HL60 cell : 2 neutrophils is ~5–20 A. *phagocytophilum* per neutrophil (*13*).
2. Alternatively, a formula to estimate multiplicity of infection has been developed: Estimated number of A. *phagocytophilum* = total infected cell number × average number of morula in an infected cell (typically 5) × average number of A. *phagocytophilum* in a morula (typically 19) × percentage of A. *phagocytophilum* recovered as host-cell free [typically 50% as determined by using metabolically [^{35}S] methionine-labeled A. *phagocytophilum* (*14*)].

3.4. Bacteria Labeling and Modifications

3.4.1. Labeling with Monoclonal Antibody to Anaplasma phagocytophilum Conjugated to a Fluorochrome

1. Make a direct or cytofuge smear of the infected cells of interest (e.g., infected neutrophils or HL60 cells). Air dry.
2. Permeabilize cells with 2 drops of 100% methanol. Air dry.
3. Place several drops of protein block onto slide preparation. Let sit for 10 min. Remove (dab) excess solution.
4. To conjugate monoclonal antibody: Take 25 µL of R5E4 antibody and add 5 µL of fluorochrome (reagent A from the Zenon conjugation system). Incubate at room temperature in the dark for 5 min. Add 5 µL of the blocking antibody (reagent B from the Zenon conjugation system). Incubate for 5 min. Dilute conjugated antibody 1:200 in 5% BSA in PBS.
5. Add several drops of conjugated antibody to slide. Incubate the slide for 1 h in the dark.
6. Wash with PBS three times for 5 min each time.
7. Fix with 4% PFA in PBS for 20 min.
8. Wash with PBS three times for 5 min each time.

3.4.2. Labeling with Fluorescent Dyes

1. Dilute stock CMFDA (CellTracker Green at 10 m*M*) to 15 µ*M* final concentration in pre-warmed media with no FBS.
2. Add 500 µL of 15 µ*M* CMFDA to a 500 µL aliquot of bacteria (isolated as described in **Subheading 3.2.1.**).
3. Mix well. Incubate at 37 °C for 45 min.
4. Centrifuge at 5000 × *g* for 10 min to pellet bacteria.
5. Wash by resuspending in 500 µL of prewarmed media and centrifuging at 5000 × *g* for 10 min to pellet bacteria.
6. Resupend in 500 µL of working media (*see* **Note 12**).

3.4.3. Treatments Before Incubation with Neutrophils

Anaplasma phagocytophilum can be modified to assess the role of viability in measured outcomes.

1. To heat kill *A. phagocytophilum*, take isolated pathogen suspended in working buffer and heat in a water bath or heating block at 60–100 °C for 10 min *(13)*.
2. Exposure to doxycycline or oxytetracycline hydrochloride for 30 min at room temperature also inactivates *A. phagocytophilum* and renders the organism non-infective for HL60 cell culture.
3. To fix *A. phagocytophilum*, suspend isolated pathogen in 2% paraformaldehyde in working buffer for 30 min at room temperature.

3.5. Incubation of Bacteria with Neutrophils

1. Isolate human neutrophils according to standard protocols *(13)*. Resuspend neutrophils in RPMI/HEPES to 2×10^6 cells/mL. Keep cells at room temperature until they are added to the tissue-culture plate (*see* **Note 13**).
2. Resuspend isolated *A. phagocytophilum* in 1 mL PBS containing 7.5 µg/mL Alexa Fluor 488 (Invitrogen-Molecular Probes). Incubate for 15 min at room temperature. Wash twice in 1 mL PBS to remove unbound AlexaFluor 488. Resuspend bacteria in RPMI/HEPES at desired concentration to add a volume of 250 µL of suspended bacteria to each well.
3. Immerse #1, 12-mm round glass coverslips in ethanol and flame. Place one clean coverslip in each well of a 24-well tissue-culture plate and let sit under UV light in a tissue-culture hood.
4. Coat each coverslip with 30 µL of 100% autologous human serum. Incubate for at least 1 h at 37 °C. Wash twice with PBS.
5. Add 250 µL of human neutrophils to each well and let cells settle and adhere for 15 min at room temperature. Chill culture plates with neutrophils on ice until addition of bacteria.
6. Add 250 µL of AlexaFluor 488-labeled *A. phagocytophilum* to each well and centrifuge at 900 × *g* for 8 min at 4 °C to synchronize phagocytosis. Transfer all plates except 0 min time point to 37 °C.

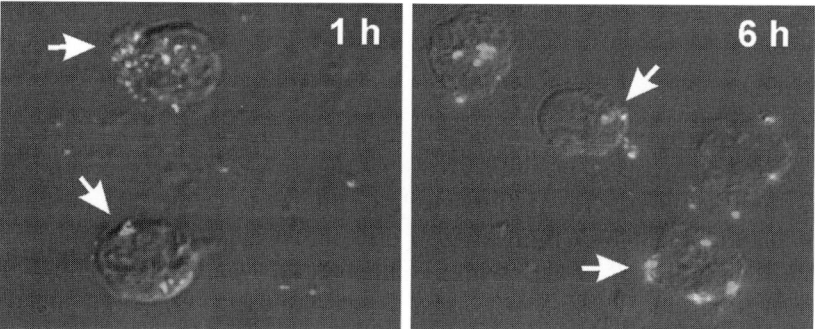

Fig. 2. Interaction of *Anaplasma phagocytophilum* with human neutrophils. (**A**) Ingestion of *A. phagocytophilum* by human neutrophils. Neutrophils were incubated with ∼5–20 *A. phagocytophilum*/neutrophil for the indicated times, and percent ingestion (number of neutrophils containing internalized *A. phagocytophilum*) was determined by fluorescence microscopy. (**B**) Micrographs illustrating ingestion of *A. phagocytophilum* by human neutrophils. Green, intracellular bacteria. Red or red-green (yellow), extracellular bacteria. Reprinted with permission from reference (*13*). Copyright 2005 The American Association of Immunologists, Inc. (*See* Color Plate 6, following p. 46.)

7. At desired time points, gently aspirate/remove supernatant and wash cells once with 500 μL of cold PBS, aspirate wash. Add 300 μL of 4% paraformaldehyde and fix cells on ice for 30 min.

8. To counterstain uningested *A. phagocytophilum,* aspirate fixative, wash once with 500 μL of PBS, and add 500 μL of PBS containing 7.5 μg/mL anti-AlexaFluor 488 (conjugated to AlexaFlour 594). Incubate for 15 min at room temperature. Aspirate stain and wash twice in 500 μL of PBS.

9. Mount coverslips onto microscope slides. Drain off excess PBS by wicking the edge on a paper towel and then dry the back with a tissue. Apply coverslip facedown onto a slide with one drop of mounting media. Incubate in the dark at 4 °C overnight to permit mounting media to fully harden.

10. Score phagocytosis on a fluorescence microscope. Count all bacteria associated with 100 cells per time point per treatment from random fields of view. Bacteria associated with neutrophils that are both red and green are uningested; bacteria staining green only are ingested (*see* **Note 14; Fig. 2A and B** and Color Plate 6, following p. 46).

4. Notes

1. *Anaplasma phagocytophilum* has also been successfully propagated in a tick cell line (ISE6 developed from *Ixodes scapularis*) grown at 34 °C (*15*), microvascular endothelial cell lines developed from human and rhesus origins (*16*), and a human immortalized megakaryocytic cell line (*17*).

2. *Anaplasma phagocytophilum* grown in HL60 cells may also be cultured in Iscove's modified Dulbecco's medium (IMDM) or Dulbecco's modified Eagle's medium (DMEM) supplemented with 20% FBS. Differences in pathogen kinetics and growth in these different media have not been appreciated or reported. Different investigators have varied the percent of FBS in the media from 1–20%.

3. Top vented cell-culture flasks are highly recommended, as antibiotics are not used in cell-culture systems for *A. phagocytophilum*. The use of these flasks decreases the risk of secondary bacterial contamination.

4. Monocytic *Ehrlichia* metabolize best in alkaline pH (7.2–8.0; adjust with 2 *M* NaOH), the use of slightly alkaline pH has been maintained for the related *A. phagocytophilum*. Sucrose phosphate buffer provides osmotic protection for these fragile organisms. Membrane active agents (phospholipase, detergent) and low-osmolarity buffers or osmotic shock are all deleterious to *Ehrlichial* viability *(10)*. Glutamine is the preferred metabolic substrate for organisms *(18)*.

5. Keep track of pathogen passage number. Organisms passaged a high number of times may have different virulence characteristics than low-passaged pathogens. In addition, highly passaged HL60 cells shed P-selectin glyco-protein ligand-1 (PSGL-1). PSGL-1 serves as a pathogen receptor *(5)*. If poor growth or propagation of organisms is noted (suspect this when cells fail to increase infectivity >50–60% in spite of high numbers of intact, uninfected HL60 cells), assess PSGL-1 expression on the surface of feeder uninfected HL60 cells.

6. Harvest pathogen when HL60 cells are at peak infection, when both intracellular and extracellular organisms are noted (*see* **Fig. 1A**). Ultrastructural evaluation of *A. phagocytophilum* reveals both reticulate and dense morphologic forms, the biologic significance of which is unknown (*see* **Fig. 1B**) *(19)*. Recent work with the related *E. chaffeensis* suggests that the dense-cored form may be the infective stage and the reticulate form may be the replicative stage (*see* **Fig. 1B**) *(20)*. Empirically, pathogen isolated from HL60 cell cultures that are only ≤50% infected (early in infection) may function differently in assay systems.

7. Pathogen growth can be sustained in both granulocyte-induced and uninduced HL60 cell cultures; however, growth of *A. phagocytophilum* will not be sustained in monocyte-differentiated HL60 cells *(8,21)*. Preferential growth of organism occurs in HL60 cells differentiated toward granulocytes, especially when differentiated with DMSO. DMSO differentiation results in heterogeneity of CD15 surface expression. However, cells with high levels of expression show greatest binding and infection of organism *(21)*.

8. Sonication conditions will vary depending on sonicator used. Cell lysis is readily evident when foaming of media occurs. The organism is fragile, and thus minimal and gentle sonication is recommended. Empirically, we have lysed cells for

up to 30–45 s beyond foaming, at high power, and recultured the sonicate successfully. If pathogen is to be used in an experiment where sterility is mandatory, the sonicator can be moved into the hood and all steps treated aseptically.

9. *Anaplasma phagocytophilum* has also been purified using a packed sterile Sephacryl S-1000 (Pharmacia, Uppsala, Sweden) chromatography column *(22)* or Percoll *(18,23)*.

10. The use of Renografin gradients permits the isolation of an enriched suspension of organisms. These purification techniques are principally applied when organisms are needed for molecular analysis (e.g., cloning, preparation of genomic DNA, monoclonal antibody production, and immunodiagnostics). For assays where viable, fully infectious organisms are needed, extensive purification may alter viability and virulence. Extracellular organisms rapidly become inactivated. Metabolism of the related monocytic *Ehrlichial* organisms is only maintained for approximately 3 h post-isolation *(10,18)*. Electron microscopy has revealed that gradient purification does not completely remove host cell material from *Ehrlichia*, and purified organisms may still be enclosed in an host-derived phagosomal membrane. Attempts to completely remove host cell membranes by DNAase or trypsin have been minimally successful *(10)*.

11. Initially, this technique was described for *Rickettsiae* and monocytic *Ehrlichia* using discontinuous Renografin gradients of 42, 36, and 30%. Heavy-banding *Rickettsiae* sedimented on the 42% cushion and light-banding *Rickettsiae* were collected at the 36–30% interface. Light-banding organisms were much more infective. Heavy bands included defective or damaged *Rickettsiae* that had no functional integrity of cytoplasmic membranes and "took on" gradient sucrose (and were thus heavier) *(9)*. Modifications to collect the entire bacterial band between 30 and 42% are now most commonly used.

12. *Anaplasma phagocytophilum* has been labeled with an array of fluorescent dyes. To the author's knowledge, no alterations in bacterial function, including binding or entry into host cells, has been noted. In addition, bacteria can be fixed with paraformaldehyde after labeling.

13. *Anaplasma phagocytophilum* delays neutrophil apoptosis *(13,14)*. As such, neutrophils can be successfully cultured for up to 96 h in RPMI-1640/HEPES media supplemented with 10% FBS and 2 mM L-glutamine.

14. Pathogen uptake can be visualized and quantified by fluorescence microscopy. Neutrophil uptake of *A. phagocytophilum* is relatively slow (*see* **Fig. 2A**). However, approximately 75% of neutrophils will ingest bacteria within 6 h after initiation of neutrophil-*A. phagocytophilum* interaction (*see* **Fig. 2A and B**) *(13,24)*.

Acknowledgments

This work was supported in part by a grant from the National Institutes of Health, AI051529.

References

1. Chen, S. M., Dumler, J. S., Bakken, J. S., and Walker, D. H. (1994) Identification of a granulocytotropic *Ehrlichia* species as the etiologic agent of human disease. *J. Clin. Microbiol.* **32,** 589–95.
2. Dumler, J. S., Barbet, A. F., Bekker, C. P., Dasch, G. A., Palmer, G. H., Ray, S. C., Rikihisa, Y., and Rurangirwa, F. R. (2001) Reorganization of genera in the families Rickettsiaceae and Anaplasmataceae in the order Rickettsiales: unification of some species of *Ehrlichia* with *Anaplasma, Cowdria* with Ehrlichia and *Ehrlichia* with Neorickettsia, descriptions of six new species combinations and designation of Ehrlichia equi and 'HGE agent' as subjective synonyms of Ehrlichia phagocytophila. *Int. J. Syst. Evol. Microbiol.* **51,** 2145–2165.
3. Goodman, J. L., Nelson, C., Vitale, B., Madigan, J. E., Dumler, J. S., Kurtti, T. J., and Munderloh, U. G. (1996) Direct cultivation of the causative agent of human granulocytic ehrlichiosis. *N. Engl. J. Med.* **334,** 209–215.
4. Ravyn, M. D., Lamb, L. J., Jemmerson, R., Goodman, J. L., and Johnson, R. C. (1999) Characterization of monoclonal antibodies to an immunodominant protein of the etiologic agent of human granulocytic ehrlichiosis. *Am. J. Trop. Med. Hyg.* **61,** 171–176.
5. Herron, M. J., Nelson, C. M., Larson, J., Snapp, K. R., Kansas, G. S., and Goodman, J. L. (2000) Intracellular parasitism by the human granulocytic ehrlichiosis bacterium through the P-selectin ligand, PSGL-1. *Science* **288,** 1653–1656.
6. Hsieh, T. C., Aguero-Rosenfeld, M. E., Wu, J. M., Ng, C., Papanikolaou, N. A., Varde, S. A., Schwartz, I., Pizzolo, J. G., Melamed, M., Horowitz, H. W., Nadelman, R. B., and Wormser, G. P. (1997) Cellular changes and induction of apoptosis in human promyelocytic HL-60 cells infected with the agent of human granulocytic ehrlichiosis (HGE). *Biochem. Biophys. Res. Commun.* **232,** 298–303.
7. Breitman, T. R., Selonick, S. E., and Collins, S. J. (1980) Induction of differentiation of the human promyelocytic leukemia cell line (HL-60) by retinoic acid. *Proc. Natl. Acad. Sci. U.S.A.* **77,** 2936–2940.
8. Klein, M. B., Hayes, S. F., and Goodman, J. L. (1998) Monocytic differentiation inhibits infection and granulocytic differentiation potentiates infection by the agent of human granulocytic ehrlichiosis. *Infect. Immun.* **66,** 3410–3415.
9. Hanson, B. A., Wisseman, C. L., Jr., Waddell, A., and Silverman, D. J. (1981) Some characteristics of heavy and light bands of *Rickettsia prowazekii* on Renografin gradients. *Infect. Immun.* **34,** 596–604.
10. Weiss, E., Dasch, G. A., Kang, Y. H., and Westfall, H. N. (1988) Substrate utilization by *Ehrlichia sennetsu* and *Ehrlichia risticii* separated from host constituents by renografin gradient centrifugation. *J. Bacteriol.* **170,** 5012–5017.
11. Asanovich, K. M., Bakken, J. S., Madigan, J. E., Aguero-Rosenfeld, M., Wormser, G. P., and Dumler, J. S. (1997) Antigenic diversity of granulocytic *Ehrlichia* isolates from humans in Wisconsin and New York and a horse in California. *J. Infect. Dis.* **176,** 1029–1034.

12. Ijdo, J. W., Sun, W., Zhang, Y., Magnarelli, L. A., and Fikrig, E. (1998) Cloning of the gene encoding the 44-kilodalton antigen of the agent of human granulocytic ehrlichiosis and characterization of the humoral response. *Infect. Immun.* **66,** 3264–3269.

13. Borjesson, D. L., Kobayashi, S. D., Whitney, A. R., Voyich, J. M., Argue, C. M., and DeLeo, F. R. (2005) Insights into pathogen immune evasion mechanisms: *Anaplasma phagocytophilum* fails to induce an apoptosis differentiation program in human neutrophils. *J. Immunol.* **174,** 6364–6372.

14. Yoshiie, K., Kim, H. Y., Mott, J., and Rikihisa, Y. (2000) Intracellular infection by the human granulocytic ehrlichiosis agent inhibits human neutrophil apoptosis. *Infect. Immun.* **68,** 1125–1133.

15. Munderloh, U. G., Jauron, S. D., Fingerle, V., Leitritz, L., Hayes, S. F., Hautman, J. M., Nelson, C. M., Huberty, B. W., Kurtti, T. J., Ahlstrand, G. G., Greig, B., Mellencamp, M. A., and Goodman, J. L. (1999) Invasion and intracellular development of the human granulocytic ehrlichiosis agent in tick cell culture. *J. Clin. Microbiol.* **37,** 2518–2524.

16. Munderloh, U. G., Lynch, M. J., Herron, M. J., Palmer, A. T., Kurtti, T. J., Nelson, R. D., and Goodman, J. L. (2004) Infection of endothelial cells with *Anaplasma marginale* and *A. phagocytophilum. Vet. Microbiol.* **101,** 53–64.

17. Borjesson, D. L. and Feferman, R. (2006) *Anaplasma phagocytophilum Infects Cells of the Megakaryocytic Lineage and Alters Murine Hematopoiesis* (abstract) from American Society of Microbiology, 106th General Meeting, Orlando, FL.

18. Weiss, E., Williams, J. C., Dasch, G. A., and Kang, Y. H. (1989) Energy metabolism of monocytic Ehrlichia. *Proc. Natl. Acad. Sci. U.S.A.* **86,** 1674–1678.

19. Popov, V. L., Han, V. C., Chen, S. M., Dumler, J. S., Feng, H. M., Andreadis, T. G., Tesh, R. B., and Walker, D. H. (1998) Ultrastructural differentiation of the genogroups in the genus *Ehrlichia. J. Med. Microbiol.* **47,** 235–251.

20. Zhang, J.-Z., Popov, V. L., Walker, D. H., and Yu, X. -J. (2006) *Differential Expression of Outer Membrane Proteins in Dense-Cored Cells and Reticulate Cells of Ehrlichia chaffensis* (abstract) from American Society of Microbiology, 106th General Meeting, Orlando, FL.

21. Heimer, R., Van Andel, A., Wormser, G. P., and Wilson, M. L. (1997) Propagation of granulocytic *Ehrlichia* spp. from human and equine sources in HL-60 cells induced to differentiate into functional granulocytes. *J. Clin. Microbiol.* **35,** 923–927.

22. Rikihisa, Y., Ewing, S. A., Fox, J. C., Siregar, A. G., Pasaribu, F. H., and Malole, M. B. (1992) Analyses of *Ehrlichia canis* and a canine granulocytic *Ehrlichia* infection. *J. Clin. Microbiol.* **30,** 143–148.

23. Rikihisa, Y., Zhang, Y., and Park, J. (1994) Inhibition of infection of macrophages with *Ehrlichia risticii* by cytochalasins, monodansylcadaverine, and taxol. *Infect. Immun.* **62,** 5126–5132.

24. Carlyon, J. A., Abdel-Latif, D., Pypaert, M., Lacy, P., and Fikrig, E. (2004) *Anaplasma phagocytophilum* utilizes multiple host evasion mechanisms to thwart NADPH oxidase-mediated killing during neutrophil infection. *Infect. Immun.* **72,** 4772–4783.

14

Ultrastructural Analysis of Bacteria–Host Cell Interactions

David W. Dorward

Summary

Electron microscopy of bacterial pathogens and interactions between bacteria and host cells and tissues provides valuable insights into structural and molecular properties and processes involved in pathogenesis. Applications for electron microscopy in bacterial pathogenesis range from discovering etiologic agents and following chronological events during infections by conventional examination of clinical samples to assessing molecular host–cell responses to infection and in situ interactions between receptors and ligands using specific immune-labeling techniques. This chapter focuses on techniques for preparing samples of bacteria and host cells for conventional transmission (TEM) and scanning electron microscopy (SEM) and use of luminescent nanocrystals or "quantum dots" as specific probes for correlative light and electron microscopy. Conventional TEM and SEM are well established tools for high resolution examination of structural effects and chronological events associated with bacterial infections. The recent development of quantum dots as physiological and immunological probes in biology has provided a powerful technique for bridging fluorescent analyses of fixed and live material with preparation and examination by TEM and SEM.

Key Words: Transmission electron microscopy; scanning electron microscopy; Gram-positive; Gram-negative; immune labeling; histochemistry; colloidal gold; quantum dots; correlative microscopy; fluorescence.

1. Introduction

Ultrastructural analyses by TEM and SEM are versatile and integral tools for examining and characterizing infectious cycles and host responses to bacterial pathogens at the tissue, cellular, sub-cellular, and molecular levels. Utilizing focused electron beams with wavelengths less than 100 pm,

From: *Methods in Molecular Biology, vol. 431: Bacterial Pathogenesis*
Edited by: F. DeLeo and M. Otto © Humana Press, Totowa, NJ

electron probe microimaging augments biochemical analyses by providing structural and spatial information at resolution 2–3 orders of magnitude greater than that achievable by standard far-field optical microscopy. Together with optical imaging of fixed and live preparations, TEM and SEM provide numerous techniques for investigating and analyzing chronological and structural relationships of interactions between bacteria and their hosts and environments.

Protocols for conducting such investigations vary considerably according to particular aspects of experimental design and desired outcome *(1–7)*. Examples include imaging techniques, such as negative staining for structural or immunological analyses of bacteria, rapid freezing and freeze-substitution for preservation and immobilization of fine structures and aqueous solutes, microwave processing for rapid diffusion and infiltration of fixatives and embedding resins, variable pressure SEM and containment chambers for examining hydrated samples, and enzymatic or immunological labeling for locating and identifying specific target molecules. Such techniques are commonly used to assess bacterial morphology and phenotypic expression of surface and extracellular components such as flagella, pili, fimbrae, and capsules. Furthermore, TEM and SEM can reveal the chronology of microbial cell cycles, phase changes, sporulation, and cytopathic effects on host cells and tissues, such as cytoplasmic and cytoskeletal rearrangements, plasma membrane ruffling and pedestal formation, organelle trafficking and vesicle fusion, and cell lysis, necrosis, apoptosis, and autophagy *(1–4)*. Despite the diversity of TEM and SEM applications and approaches to microbial investigations, most protocols share common processes to address properties and requirements inherent with electron probe analyses of biological material, such as a high vacuum environment and both the damaging effects and exploitable attributes of focused cathode-ray irradiation. This chapter provides descriptions of conventional preparative and immunological-labeling techniques with which many questions can be addressed and modifications adapted. For in-depth review of theoretical and practical applications of electron microscopy, readers are directed to any of several excellent volumes *(1–4)*.

Because of the broad nature of microbial samples prepared for EM, this chapter cannot describe methods to address all types of ultrastructural studies. Rather, this chapter provides methods for preparing samples that are either cultured on or subsequently attached to a rigid or semi-rigid substrate. Examples of such substrates are silicon chips, Thermanox® cover slips, Aclar® film, filter membranes, or agar-based culture medium. Adherence to a firm substrate is advantageous for positioning TEM samples for sectioning along planes of interest, for maintaining the integrity of biofilms and host cell layers, and for interpreting structures and events that may be polarized with respect to basal, lateral, or apical cell surfaces. In many instances, such layers are appropriate and useful in vitro models for the native associations of bacteria and host cells. Microbial and cell adherence to such substrates can be promoted, if necessary, by pre-coating the substrate with a biologically relevant substance, such as fibronectin, vitronectin, collagen, or poly-L-lysine. It should be noted, however, that many situations exist in which adherence to a surface is not advantageous and use of centrifugal pellets, for example, may be preferable.

During all steps of sample preparation, care should be taken to assure that the samples, reagents, containers, and tools are kept clean and free of particulates and volatile contaminants. Dust, fingerprints, trace salts, and other materials in solution degrade image quality and can damage or destroy samples and instrumentation. To minimize problems with contamination, several preventative steps are advisable, including (1) acquiring high quality water and reagents, (2) using pre-cleaned or sterile plasticware and thoroughly cleaning and rinsing all glassware, (3) filtering buffers, stains, and sample rinse water with 20–200 nm filters prior to use, (4) handling samples, tools, and sample holders with protective gloves, and (5) reducing dust by filtering incoming ventilation and controlling foot traffic into preparative and instrument areas. Furthermore, several reagents and resins used in sample preparation are highly toxic, carcinogenic, and/or radioactive. Therefore, material safety data and recommendations from the suppliers should be obtained, understood, and followed for safe handling and disposal of all potentially hazardous reagents.

1.1. Conventional TEM

This section provides preparative methods for examining thin sections of microbial material. The examples shown in **Fig. 1** depict mixtures of bacteria with human host cells prepared by these methods. These techniques are suitable for samples that may or may not include specific staining or labeling steps for identifying or highlighting structures or molecules of interest (see **Subheading 1.3**). It is presumed that certain equipment including a fume hood, ultramicrotome and diamond or glass knife, and transmission electron microscope are available for use.

Preparative steps involve fixing with cross-linking agents, contrasting with electron-dense compounds, dehydrating, embedding, and sectioning. Typically, primary fixation involves immersion in buffered glutaraldehyde or a mixture of glutaraldehyde and paraformaldehyde. These reagents react with and can cross-link free amino groups. If needed, the primary fixative can also be used for perfusion of tissues. As the rate of diffusion and thus fixation is dependent

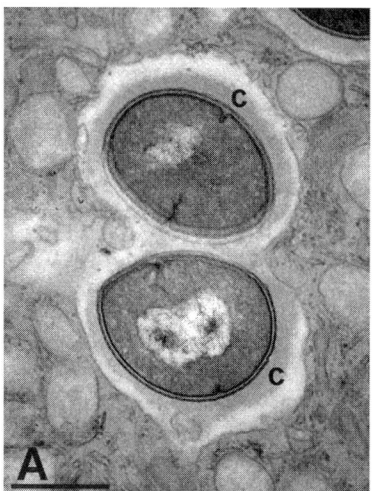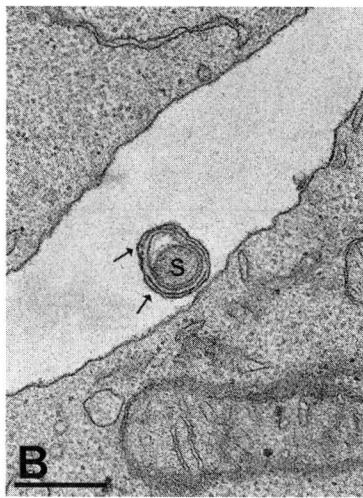

Fig. 1. Transmission electron micrographs showing sections of bacteria and host cells. (**A**) Digital image of methicillin-resistant *Staphylococcus aureus* within a primary human neutrophil. Bacterial capsular material (c) is preserved within the endosome. Cytoplasmic vesicles are evident in association with the endosomal membrane. (**B**) Photographic image of *Borrelia burgdorferi* and lymphocytes. In cross-section, two layers of extra membrane (arrows) are evident surrounding the spirochete (**s**) (unpublished data). Bars, 500 nm.

on the biochemical nature of the material and physical properties such as temperature and sample thickness, the length of time required is variable and largely unpredictable. Organisms with thickened, resistant, and/or hydrophobic cell walls such as Gram-positive bacteria, spores, or mycobacteria require extended fixation or enhanced processing, such as microwave irradiation *(4)* or both. Experimental assessment of viability during fixation can be useful for determining the necessary length of treatment. Membranes and storage lipids are stabilized by cross-linking with oxidative osmium species. Osmium treatment both reduces the quantity of lipids extracted during preparation and provides contrast for lipophillic structures. Contrast is often enhanced by subsequent staining with metallic salt solutions, such as uranyl acetate and lead citrate. To enable thin sectioning, the samples are dehydrated in organic solvents and infiltrated with epoxy or acrylic resins, which are then polymerized to provide support and minimize compression artifacts during sectioning. As with fixation steps, the duration period and processing techniques for dehydration and embedment may require adjustment depending on types of samples and viscosity of the resin.

1.2. Conventional SEM

For conventional SEM, samples must also be cross-linked and dehydrated during preparation. Treatments with aldehydes and osmium oxides and solvent dehydration can often be performed identically for both TEM and SEM. Once fully dehydrated in 100% organic solvent, the SEM samples are dried and mounted. To minimize shrinkage during drying, intermediate drying agents are used. This can be an organic compound, such as hexamethyldisilazane *(4,5)*, or more typically the agent is liquid carbon dioxide exchanged within a critical point dryer. Once dried, the samples are coated with a conductive metal or carbon before examination or examined without coating. Coating, usually performed in a magnetron or ion beam sputterer, can provide enhanced conductivity, some degree of protection against beam damage, and perhaps most importantly, increased production of secondary electrons for imaging. Although SEM offers several modes of imaging for biological samples, typically samples are examined using secondary electrons for cell and microbial surface topography and backscattered electrons for compositional contrast of immune probes and labels.

1.3. Immune Labeling with Quantum Dots

Immune labeling of biological material for electron microscopy is a well-established and powerful tool for localizing and visualizing macromolecules and molecular complexes with nanometer scale resolution. In recent years, considerable efforts have been undertaken to develop and utilize techniques for correlative microscopy involving assessment of equivalent samples by fluorescent (including single and multi-photon confocal microscopy) and electron microscopy *(1,2,6,7)*. Because of the inherent differences between light and electron beam imaging, relatively few techniques have been developed that can utilize the same specific labeling reagents. These include (1) use of fluorescein-conjugated probes for fluorescence microscopy, which can be subsequently labeled with anti-fluorescein antibodies and electron-dense metals *(8)*, (2) use of the green fluorescent protein family of expressed markers that can also be labeled with antibodies and electron-dense metals *(9)*, the ReASH® system, which utilizes a genetically engineered tetracysteine motif and a bi-arsenical fluorescent dye that can photoactively deposit diaminobenzidine for TEM detection of the expressed protein tag *(10)*, and (4) the use of fluorescent CdSe nanocrystals known as "quantum dots," which are photoluminescent, phosphorescent, and electron dense *(11–13)*. Among these options, the quantum dots are unique in that they enable fluorescent examination of labeled live or fixed samples and subsequent preparation of the same samples for electron microscopy without further manipulation of the label. This was recently described for simultaneous detection of several quantum dot probes by fluorescence and TEM of sections by the wavelength of emission and size class of particles, respectively *(7)*. To date, there are no reports of SEM studies using quantum dots as immune labels. However, phosphorescent emissions similar in wavelength to photoluminescent emissions have been detected by SEM with a cathodoluminescence spectrometer *(13)*. Studies are underway to determine whether cathodoluminescent properties of quantum dots will be adaptable for immune labeling of biological material for SEM. In the meantime, the CdSe core and ZnS coating of quantum dots provide for compositional contrast using backscattered electron detection. Labeling and imaging methods described below enable direct detection of the quantum dots by TEM and low-voltage SEM.

2. Materials

2.1. Conventional TEM

1. Hank's-buffered salt solution (HBSS) (Cambrex BioSciences, Inc., Walkersville, MD).
2. 25% aqueous glutaraldehyde, EM grade, in nitrogen-filled ampules (Ted Pella, Inc., Redding, CA).
3. 16% aqueous paraformaldehyde, EM grade, in nitrogen-filled ampules (Electron Microscopy Sciences, Hatfield, PA).
4. Sodium cacodylate (dimethyl arsenate) (Ted Pella, Inc.).
5. 4% aqueous osmium tetroxide, EM grade (Ted Pella, Inc.).
6. Potassium ferrocyanide.
7. 70 and 100% (dehydrated) ethanol.
8. Spurr's embedding medium (Ted Pella, Inc).
9. Uranyl acetate (Ted Pella, Inc.).
10. Lead citrate (Ted Pella, Inc.).
11. 10 N sodium hydroxide.
12. Disposable serological pipets.
13. Disposable transfer pipets.
14. Powder-free nitrile or other appropriate gloves.
15. Thermanox® cover slips, 13 mm (Nalge Nunc International, Rochester, NY).
16. Embedding capsules or molds (Ted Pella, Inc.).
17. Copper EM grids, 200–300 mesh (Ted Pella, Inc.).
18. Grid box(es) (Ted Pella, Inc.).
19. Laboratory film such as Parafilm®.
20. Syringe filters, 20 nm Anotop® (Whatman International, Ltd., Maidstone, UK).
21. Disposable Leur-Lock syringes, 10 mL.
22. Fine tweezers, such as anti-capillary tweezers.

2.2. Conventional SEM

1. Same as items **1–7** and **12–14** in **Subheading 2.1** (TEM).
2. Silicon chip specimen mounts (Ted Pella, Inc).
3. 24-Well tissue culture plates.
4. SEM sample stubs compatible with the microscope (Ted Pella, Inc).
5. Chromium, iridium, or other appropriate sputtering target (Refining Systems, Inc., Las Vegas, NV).
6. Carbon sputtering target (South Bay Technology, San Clemente, CA).

2.3. Immune Labeling with Quantum Dots

1. Phosphate-buffered saline (PBS).
2. Immunoglobulin-free bovine serum albumin (BSA).
3. Primary antibodies or anti-serum, for example, rabbit anti-serum.
4. Secondary antibody conjugate, for example, goat anti-rabbit 655 nm quantum dots (Invitrogen, Inc., Carlsbad, CA).
5. Parafilm® or similar laboratory film.
6. Anti-capillary tweezers (Ted Pella, Inc).

3. Methods

3.1. Conventional TEM

This section describes steps for fixating and contrasting, dehydrating, embedding, sectioning, and mounting samples for TEM. Many variations on these steps have been described depending on types of samples used, experimental goals, and personal preference *(1–4)*. This set of methods was used successfully to demonstrate stages of infection of primary human neutrophils with clinical isolates of group A *Streptococcus*, and methicillin-resistant *Staphylococcus aureus* (MRSA), and lymphocytes by *Borrelia burgdorferi*. The samples of neutrophils were allowed to adhere to 13-mm Thermanox® cover slips in the wells of a 24-well tissue-culture plate prior to exposure to bacteria. In contrast, 10^5 B cells and 10^6 spirochetes were co-incubated within liquid growth medium suspensions and then processed for TEM as pellets. In these examples, the time periods for each step are relatively long and reflect allowance for slow diffusion of reagents through Gram-positive cell walls, and pellets of cells and spirochetes, respectively. In all processing steps, care must be taken to avoid unintentional drying of samples (*see* **Note 1**). When necessary, centrifugation and resuspension should be conducted as gently as possible to minimize mechanical damage (*see* **Note 2**). Furthermore, TEM grids must be kept clean and handled very carefully, grasping only at the edges with very fine tweezers.

3.1.1. Fixation

1. Immerse samples in at least 10 volumes of Karnovsky's fixative containing 2.5% glutaraldehyde and 4% paraformaldehyde in 0.1 M sodium cacodylate buffer, pH 7.2 for 2 h (*see* **Note 3**).
2. Wash the samples twice for 30 min each time by replacing the fixative with cacodylate buffer.
3. Post-fix the samples by replacing the wash buffer with a mixture containing 1% osmium tetroxide and 0.8% potassium ferrocyanide in 0.1 M cacodylate for 2 h.
4. Wash the samples once for 30 min in cacodylate and twice in deionized or distilled water.

3.1.2. In-Block Staining (Optional)

1. Stain the samples by immersion in 1% uranyl acetate in water for 2 h at room temperature.
2. Wash the samples twice for 30 min each in water.

3.1.3. Dehydration and Embedment

1. Dehydrate the samples by immersion for 1 h each in 70, 100, and 100% ethanol.
2. Prepare embedding resin according to the manufacturer's instructions and mix thoroughly.
3. Infiltrate samples with embedding resin by immersing in 1/3 resin and 2/3 ethanol for 4 h, 2/3 resin and 1/3 ethanol for 4 h, and 100% resin for 8–16 h twice.
4. After two cycles in 100% resin, transfer the samples to resin-filled embedding capsules or other suitable molds designed for curing at elevated temperature. Orient the samples for the optimal position and angle of sectioning. For centrifuged material, this is typically not an issue due to the random orientation of cells and bacteria within the pellets. For adherent cells on Thermanox® cover slips, Aclar®, or silicon chips, orient the substrates such that the cells face toward the resin at the preferred angle for sectioning, usually either parallel or perpendicular to the cell layer. It is helpful to place a small paper label within the resin block with identifying information written in pencil or printed in a small font (e.g., 5 point font) for later reference.
5. Carefully place the samples into an oven for curing at 65 °C or as recommended by the resin supplier (*see* **Note 4**). Curing is temperature dependent. At 65 °C, blocks are usually hardened within 24–48 h.

3.1.4. Block Face Preparation and Sectioning

1. Remove the cover slip or any other substrate from the block (*see* **Note 5**). The cells should remain in the block, and exposed on the surface.
2. Secure the block in a microtome chuck or small vice. Using single-edge razor blades, carefully trim the block down to a suitable block face for sectioning. This is typically a trapezoid ranging in size from about 0.2–1.0 mm on each side.

3. Using an ultramicrotome and glass or diamond knife, cut sections of suitable thickness (*see* **Note 6**).
4. Pick up the sections on clean EM grids. The grids can either be coated with thin films such as carbon, Parlodion® or Formvar®, or left uncoated. Wick off excess water by gently touching absorbant filter paper to the edge of the grid. Place the grid into a grid storage box and record any necessary identifying information.

3.1.5. Post-Section Staining (Optional).

Specialized devices are available for post-section staining that facilitate handling of multiple samples and help provide consistency and reproducibility, for example, Ted Pella, Inc., item number 22510. Traditionally, grids have been stained with droplets as follows.

1. On a freshly uncovered surface of ParaFilm® or other laboratory film, pipette 30-μL droplets of Anotop®-filtered 1% uranyl acetate in water.
2. Carefully place one grid on each droplet, with the sections facing the droplet. The grids will normally float on the droplet unless a surfactant is present. Cover the droplets and samples to protect against dust and drying. Plastic lids from pipette tip racks work well for this purpose. Allow 10 min for staining. In low-humidity environments, it is prudent to supply a source of hydration, such as a wet gauze sponge or laboratory wipe within the covered space to reduce evaporation.
3. Transfer the grids to 100-μL droplets of filtered deionized or distilled water. Thoroughly wash each grid with water by passing 5–10 mL of filtered water over both grid surfaces dropwise. Carefully wick excess water from each grid using filter paper as above. Allow to dry.
4. Prepare fresh 1% lead citrate stain by adding 0.1 g of lead citrate to a mixture of 9.5 mL of water and 0.5 mL of 10 *N* sodium hydroxide. Vortex the mixture until dissolved. Pass through Anotop® filters before use. Alternative preparative methods are also available.
5. Repeat **steps 1–3** above using 1% filtered lead citrate instead of uranyl acetate. The grids are then ready to examine.

3.2. Conventional SEM

The methods described in this section are applicable to a wide variety of samples.

As with the TEM methods in **Subheading 3.1**, the time periods for each step are sufficient for most samples of individual cells or cell layers with associated Gram-negative or non-spore-forming Gram-positive bacteria. For bulk samples such as biopsied tissues or bacterial colonies on solid medium, incubation periods should be adjusted to ensure thorough diffusion of reagents into the samples.

Samples for SEM examination are often cells, cell layers, or bulk tissues mounted on rigid or semi-rigid substrates (*see* **Note 7**). Several preparative steps are similar to those listed above for TEM, in **Subheading 3.1**. The methods diverge to prepare the samples for viewing exposed surfaces rather than internal structures.

3.2.1. Fixation

1. Apply suspended cells to substrate if necessary. Place the desired number of silicon chips into wells of a 24-well tissue-culture plate, shiny side up. Pipette 30 μL of cell and/or bacterial suspensions that have been washed and resuspended at 10^6 or 10^7/mL of HBSS, respectively, onto the shiny surface of silicon chips. Cover and place in a hydrated CO_2 incubator. Allow to settle for 10–15 min.
2. Gently add 1 mL of fixative containing 2.5% glutaraldehyde, 4% paraformaldehyde, and 0.1 *M* sodium cacodylate buffer, pH 7.2, to each well. Allow 30–60 min for primary fixation. Mammalian cells, cell monolayers, and Gram-negative bacteria fix relatively rapidly. Bulk tissue samples, Gram-positive bacteria (particularly spore-forming bacteria), and hydrophobic mycobacteria may require extended fixation time periods.
3. Wash the samples twice for 15–30 min with 1–2 mL of cacodylate buffer.
4. Post-fix the samples with 1 mL of 1% OsO_4 in cacodylate for 30–60 min.
5. Wash the samples once with cacodylate buffer and twice with water.

3.2.2. Dehydration, Coating, and Mounting

1. Dehydrate the samples by replacing the water with hour-long series of 70, 100, and 100% ethanol.
2. Dry the samples using a critical point dryer and carbon dioxide.
3. Mount the samples, oriented with the material of interest facing upward, onto sample stubs manufactured for the available scanning EM, using conductive material such as silver, gold, or graphite paint, or double-adhesive carbon disks.
4. (Optional) Sputter-coat the samples with suitable material such as chromium, iridium, gold-palladium, or carbon. The samples are ready for examination.

3.3. Immune Labeling with Quantum Dots

The methods described in this section pertain to labeling antigens located on the surface of cells or thin sections for examination by either transmission or scanning electron microscopy (SEM). Despite being similar to colloidal gold for detection by electron microscopy, quantum dots offer the ability to observe labeling and associated biological processes by fluorescence microscopy prior to fixation for electron microscopy. Although use of quantum dots for biological studies is relatively new, a significant number of primary and secondary conjugates of quantum dots with varying sizes and emission wavelengths are commercially available. Furthermore, quantum dots with reactive linkers are also available for producing direct conjugates of interest, making the dots a convenient, adaptable, and flexible choice as probes for molecules and molecular interactions. As above, the protocol described applies to bacteria and cells adsorbed to a substrate such as silicon chips or Thermanox® cover slips. If used with pellets or suspended material, the methods can be modified accordingly. For some applications, it may be advantageous to pre-fix the samples to prevent progression of biological processes during labeling. When labeling live material, active processes such as internalization, intracellular trafficking, and cell division can be controlled as needed with temperature or suitable chemicals such as azide or other metabolic inhibitors. Using quantum dots, common steps such as titrating antibodies, conjugates, and blocking reagents to optimize labeling and detection of targets by fluorescence microscopy generally provide optimal electron imaging as well. Thus, fluorescence microscopy is useful for rapidly assessing the specificity and sensitivity of the labeling experiment before the samples are processed for electron imaging. However, this is not necessarily the case when detergents or solvents are used to permeabilize the cells or bacteria, as cellular ultrastructure can be affected adversely and dramatically by such agents, without noticeable detriment in a light microscope.

3.3.1. Pre-Fixation (Optional)

1. Wash samples twice with an appropriate physiological buffer such as PBS or HBSS to remove residual medium components.
2. Immerse the samples into PBS containing 4% paraformaldehyde and 0.1% glutaraldehyde. Incubate at room temperature for 15–30 min.
3. Wash sample twice for 15 min with buffer.

3.3.2. Immune Labeling

1. Block the samples by immersing in PBS containing 2% (w/v) globulin-free BSA. Incubate at room temperature for 30 min.
2. Replace the blocking solution with primary antibody such as immune or pre-immune mouse serum diluted to an appropriate concentration in blocking buffer. Incubate the mixture for 30 min at room temperature or overnight at 4 °C. Determining the optimal concentration and length of incubation may require empirical experimentation.
3. Wash the samples at least twice for 15 min in blocking buffer.

4. Label the samples by immersion into 5 n*M* secondary quantum dot conjugate, such as goat anti-mouse 655 nm quantum dots, in blocking buffer. Incubate for 30 min at room temperature.

5. Wash the samples at least twice for 15 min in blocking buffer. Wash the samples two more times in PBS to remove residual protein from the blocking buffer. The samples are now ready to examine by fluorescence microscopy and to proceed with processing for TEM or SEM (*see* **Subheadings 3.1** and **3.2**).

3.3.3. Imaging by TEM

Quantum dots are readily detectable by TEM in thin section and negatively stained preparations at voltages commonly used for biological material between 60 and 120 kV. No special requirements are needed for imaging.

3.3.4. Imaging by SEM

Quantum dots are detectable by SEM by topographical contrast with secondary electron imaging (SEI), compositional contrast with backscattered electron imaging (BEI), and scanning transmission electron imaging. By SEI, the quantum dots appear as particles on exposed surfaces that can resemble normal biological structures. However, the CdSe core and ZnS coating impart

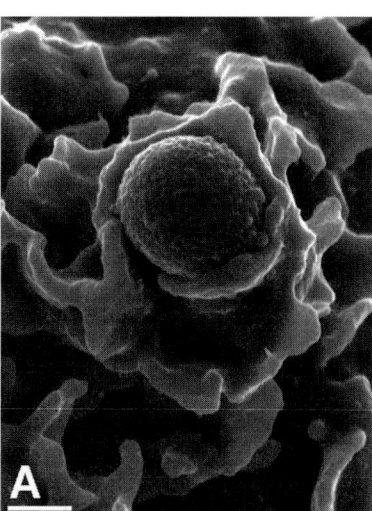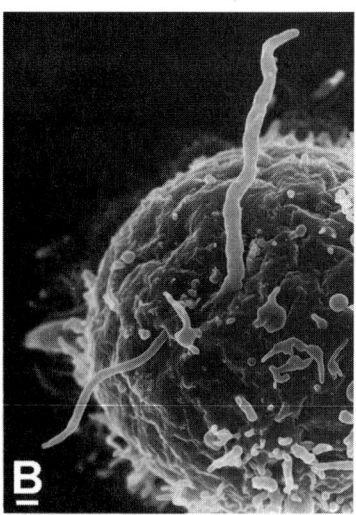

Fig. 2. Scanning electron micrographs showing interactions between bacteria and host cells. (**A**) Digital image of methicillin-resistant *S. aureus* enwrapped within lamellapodia on the surface of a primary human neutrophil. (**B**) Photographic image of *B. burgdorferi* and a human lymphocyte. The spirochete appears to simultaneously penetrate and emerge from the B cell, lymphocyte (unpublished data). Bars, 500 nm.

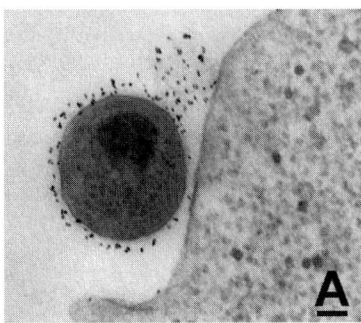

Fig. 3. Immune electron micrographs of quantum dot-labeled bacteria. Mixtures of cultured HeLa cells and *Chlamydia trachomatis* elementary bodies (EB) were pre-fixed, probed with anti-EB rabbit serum, and labeled with goat anti-rabbit IgG/655 nm quantum dot conjugates. Thin sections examined by TEM (**A**) and cell layers examined by SEM (**B**) showed numerous quantum dots on bacterial surfaces, but not on host cells (unpublished data). Bars, 50 nm.

clear compositional contrast against carbon-rich biological material when imaged by BEI or scanning transmitted electron detectors. Such contrast is maximized using relatively low accelerating voltages and carbon sputter coating.

1. Follow **steps 1–5** in **Subheading 3.2.1** and **steps 1–3** in **Subheading 3.2.2**.
2. Lightly sputter coat samples with carbon. The coating thickness should be determined empirically as the minimum thickness that enables viewing without static charge artifact in the microscope. A thickness setting of 20 Å was used for the image in **Fig. 3B**.
3. Choose an initial accelerating voltage of 1–3 kV (*see* **Note 8**). Check and adjust beam alignment as necessary. Examine the sample in backscatter mode at ×25,000 or greater magnification. Quantum dots should be visible as relatively bright particles on exposed surfaces.

4. Notes

1. Sample processing for EM involves multiple fluid exchanges. Such exchanges must be sufficiently rapid to prevent samples from drying. For example, fluids can be removed from and immediately replaced for each sample in an experiment rather than removing fluids from all samples before replacement.
2. It is the practice in this laboratory to centrifuge suspended cells at 600–800 × g for 1–5 min as necessary. After removing the supernatant, pellets are gently disaggregated into the residual liquid by tapping the tube prior to adding the next volume of liquid. Vortexing and rapid passage through pipette tips are not recommended.

3. Sodium phosphate provides excellent buffering for fixatives at 0.1 M and is often used in place of cacodylate. Each buffer system has certain advantages and limitations. Cacodylate is an organic arsenic compound, dimethylarsenate, which is poisonous. Whereas it may provide additional TEM contrast in biological samples, it is a relatively weak buffer at pH 7.2. Alternatively, phosphate requires fewer safety precautions. However, our laboratory has observed precipitants in some biological samples, such as avian erythrocytes, that were attributed to the phosphate buffer.

4. Epoxy resins emit noxious and potentially carcinogenic fumes. Therefore, curing ovens should be located within a fume hood or designed to be vented away from occupied spaces as with vacuum ovens.

5. Removal of Thermanox® cover slips can be facilitated by immersing the cover slip into liquid nitrogen for about 5 s. Prolonged exposure of the resin to liquid nitrogen should be avoided, as cracks can develop within the block. Similarly, silicon chips are removed cleanly after pressing the chip onto dry ice for 10–15 s.

6. The quality of sections are affected by many variables including personal ability, type of microtome, type of knife, flaws in the knife, cutting speed, sample and block integrity, ambient temperature, air circulation, humidity, and others. Section thickness settings on ultramicrotomes are approximate values. Typically, sections are cut between 50–70 nm in thickness. At this thickness, the sections appear silver to light-gold in color and provide good characteristics of stability, contrast, and resolution under typical biological imaging conditions of 80–100 kV. Furthermore, compression artifact during sectioning can be minimized by orienting the block face such that parallel edges of the block face are aligned parallel with the knife blade, and that the wider edge advances across the knife first.

7. In some cases, the SEM substrate can be a TEM grid. More typically, ideal substrates are flat or relatively featureless or both. Pre-cleaned glass or plastic cover slips are commonly used for cell layers and cells that can adhere. Use of pre-cut silicon chips as a substrate is gaining popularity, as these are very flat and featureless and sufficiently conductive to minimize charge buildup in samples. In many cases, cells and bacteria adhere to cover slips and chips without further treatment. If necessary, the substrates can be coated with polypeptides or other agents such as collagen, fibronectin, serum, or polylysine to promote adherence.

8. In our laboratory, backscatter detection of quantum dots has been achieved at accelerating voltages ranging from 0.5 to 10 kV. However, during testing, backscatter contrast decreased steadily as the accelerating voltage was increased beyond 2 kV, presumably due to reduced scattering by the nanocrystal or increased scattering by the substrate or both. If available on the SEM, mixing signals from the secondary and backscattered electron detectors, or adjusting for detection of optimum electron energy levels may enhance image quality with respect to compositional contrast and sample topography.

Acknowledgments

This work was supported by the Intramural Research Program of the National Institute of Allergy and Infectious Diseases.

References

1. Hawkes, P. W., and Spence, J. C. H. (eds) (2006) *Science of Microscopy*. Springer, New York, NY.
2. Hayat, M. A. (ed.) (2000) *Principles and Techniques of Electron Microscopy: Biological Applications*. Cambridge University Press, Cambridge, UK.
3. Flegler, S. L., Heckman, J. W., and Klomparens, K. L. (eds) (1993) *Scanning and Transmission Electron Microscopy: An Introduction*. Oxford University Press, New York, NY.
4. Giberson, R. T., and Demaree, R. S. (eds) (2001) *Microwave Techniques and Protocols*. Humana Press, Totowa, NJ.
5. Ubero-Pascal, N., Fortuno, J. M., and de Los Angeles Puig, M. (2005) New application of air-drying techniques for studying Ephemoptera and Plecoptera eggs by scanning electron microscopy. *Micros. Res. Tech.* **68**, 264–271.
6. McIntosh, J. R. (2001) Electron microscopy of cells: a new beginning for a new century. *J. Cell Biol.* **153**, F25–F32.
7. Giepmans, B. N., Deerinck, T. J., Smarr, B. L., Jones, Y. Z., and Ellisman, M. H. (2005) Correlated light and electron microscopy imaging of multiple endogenous proteins using quantum dots. *Nat. Methods* **2**, 742–749.
8. van Dam, G. J., Bogitsh, B. J., Fransen, J. A., Kornelis, D., van Zeyl, R. J., and Deelder, A. M. (1991) Application of the FITC-anti-FITC-gold system to ultrastructural localization of antigens. *J. Histochem. Cytochem.* **39**, 1725–1728.
9. Plowman, S. J., Muncke, C., Paxton, R. G., and Hancock, J. F. (2005) H-ras, K-ras, and inner plasma membrane raft proteins operate in nano clusters with differential dependence on the actin cytoskeleton. *Proc. Nat. Acad. Sci. U.S.A.* **102**, 15500–15550.
10. Gaietta, G., Deerinck, T. J., Adams S. R., Bouwer, J., Tour, O., Laird, D. W., Sosinsky, G. E., Tsien, R. Y., and Ellisman, M. H. (2002) Multicolor and electron microscopic imaging of connexin trafficking. *Science* **296**, 503–507.
11. Rodriguez-Viejo, J., et al. (1997) Cathodoluminescence and photoluminescence of highly luminescent CdSe/ZnS quantum dot composites. *Appl. Phys. Lett.* **70**, 2132–3134.
12. Rodriguez-Viejo, J., et al. (2000) Evidence of photo-and electrodarkening of (CdSe)ZnS quantum dot composites. *J. Appl. Phys.* **87**, 8526–8534.
13. Leon, R., Nadeau, J., Evans, K., Paskove, T., and Monemar, B. (2004) Electron irradiation effects on nanocrystal quantum dots used in bio-sensing applications. *IEEE Trans. Nucl. Sci.* **51**, 3186–3192.

15

Infection of Human Monocyte-Derived Macrophages With *Coxiella burnetii*

Jeffrey G. Shannon and Robert A. Heinzen

Summary

Coxiella burnetii, the agent of Q fever, is an obligate intracellular bacterium that has a tropism for cells of the mononuclear phagocyte system. Following internalization, *C. burnetii* remains in a phagosome that ultimately matures into a vacuole with lysosomal characteristics that supports pathogen replication. Most in vitro investigations of *Coxiella* – macrophage interactions have employed continuous cell lines. Although these studies have been informative, genetic alterations of immortalized cells may result in attenuated biological responses to infection relative to primary cells. Consequently, primary macrophages are preferred as in vitro model systems. Here, we describe procedures for propagation and isolation of *C. burnetii* from cell culture and the use of these preparations to infect primary macrophages derived from human peripheral blood monocytes. Both virulent phase I and avirulent phase II *C. burnetii* productively infect human monocyte-derived macrophages (MDMs) and replicate with approximately the same kinetics, thereby providing a more physiologically relevant in vitro model system to study the infectious process of this pathogen.

Key Words: *Coxiella*; macrophage; purification; quantitative PCR; Q fever; phagolysosome; cell culture.

1. Introduction

Coxiella burnetii is a bacterial obligate intracellular pathogen and the etiologic agent of human acute and chronic Q fever. The organism has an impressively broad host range that includes arthropods (primarily ticks), fish, birds, and a variety of mammals *(1)*. *C. burnetii* has a worldwide distribution with the exception of Antarctic regions and New Zealand *(2,3)*. In both wild and domestic animals,

From: *Methods in Molecular Biology, vol. 431: Bacterial Pathogenesis*
Edited by: F. DeLeo and M. Otto © Humana Press, Totowa, NJ

C. burnetii does not appear to cause overt disease with the exception of sheep and goats, where it can cause abortions *(4)*. Unlike other obligate intracellular bacteria, *C. burnetii* is remarkably resistant to desiccation and therefore can persist in contaminated soils for extended periods *(2)*. Inhalation of contaminated aerosols is the primary route of human infection with *C. burnetii (5)*, and disease can be initiated in both humans and animals by less than 10 organisms *(6,7)*. Consequently, *C. burnetii* can represent an occupational hazard, particularly for individuals involved in animal husbandry operations where large numbers of organisms can be shed into the environment during parturition *(8)*.

As an obligate intracellular bacterium, *C. burnetii* requires a viable eucaryotic host cell for propagation. Laboratory cultivation of *C. burnetii* has been accomplished in animals (primarily guinea pigs and mice), embryonated hen's eggs, and tissue culture cells *(5)*. Regardless of the host and cell type, *C. burnetii* replicates in a membrane-bound vacuole with lysosomal characteristics *(9)*. Cell culture is currently the method of choice for propagation of *C. burnetii*. A variety of primary and continuous cell lines support vigorous growth of the organism including primary chick and mouse embryo fibroblasts and continuous cell lines, like Vero (African green monkey kidney epithelial), BHK-21 (hamster kidney fibroblast), L-929 (murine fibroblast), J774.1 and P388D1 (murine macrophage-like), and THP-1 (human monocyte-like) *(1,10–12)*. Organisms are typically harvested when large vacuoles filled with *C. burnetii* are observed throughout the cell culture (~6–8 days post-infection). Cells are scraped from culture flasks and mechanically disrupted to release intracellular bacteria *(13)*. Purification of *C. burnetii* from host cell lysates involves a series of differential centrifugation steps followed by density gradient centrifugation through sucrose or RenoCal-76 *(13–15)*. The hydrophobic, truncated LPS of phase II *C. burnetii* results in adherence of host material that consequently results in lower yields of this strain when compared to yields of phase I organisms producing full-length LPS *(15)*.

In vivo, *C. burnetii* primarily targets cells of the mononuclear phagocyte system such as alveolar macrophages of the lung and Kupffer cells of the liver *(16)*. Thus, primary human macrophages represent the most physiologically relevant in vitro system to study *Coxiella*–host interactions. Here, we describe protocols for quantifying infection of human peripheral blood monocyte-derived macrophages (MDMs) by purified *C. burnetii*.

2. Materials
2.1. Coxiella burnetii Propagation and Purification

1. RPMI medium with Glutamax I (Invitrogen, Carlsbad, CA) supplemented with 10 or 2% fetal bovine serum (FBS) (Invitrogen). Store at 4 °C.

2. African green monkey kidney epithelial (Vero) cells (CCL-81; American Type Culture Collection, Manassas, VA).
3. Phosphate-buffered saline (PBS): Na_2HPO_4, 53.9 mM; KH_2PO_4, 12.8 mM; and NaCl, 72.6 mM. Sterilize by autoclaving and store at room temperature.
4. PBS-sucrose (PBSS): PBS containing 0.25 M sucrose. Sterilize by passage through a 0.22-µm Millipore filter and store at room temperature.
5. 30% (v/v) RenoCal-76 (Bracco Diagnostics Inc, Princeton, NJ) in PBSS. Store at 4 °C.
6. 150-cm² cell-culture flasks (Corning, Corning, NY).
7. Cell scrapers (39-cm handle, 3-cm blade) (Corning).
8. 50-mL disposable conical centrifuge tubes (Corning).
9. Sterile 50 mL O-ringed screw cap centrifuge tubes (Nalge Nunc International, Rochester, NY).
10. Sterile 250-mL O-ringed screw cap centrifuge bottles (Nalge Nunc International).
11. Ultra-Clear centrifuge tubes (25 mm × 89 mm) (Beckman Coulter, Fullerton, CA).
12. Ethanol (70%).
13. 10-mL Luer-lock Tip disposable syringes (Becton Dickinson, Franklin Lakes, NJ).
14. 14-gauge, 4-inch metal cannulas (Popper and Sons, Inc., New Hyde Park, NY).
15. Sterile 8-ounce disposable plastic beakers (Daigger and Company, Vernon Hills, IL).
16. Sterile 2-mL O-ring screw cap microfuge tubes (Sarstedt AG and Co., Numbrecht, Germany).

2.2. PBMC Isolation and MDM Culture

1. Ficoll-Paque Plus (Amersham Pharmacia, Piscataway, NJ). Store at room temperature and protect from light.
2. PBS supplemented with 2% FBS.
3. RosetteSep Monocyte Enrichment Cocktail (StemCell Technologies, Vancouver, Canada).
4. RPMI medium with Glutamax I, supplemented with 10% FBS (Invitrogen). Store at 4 °C.
5. Macrophage colony-stimulating factor (M-CSF) (Peprotech, Rocky Hill, NJ) dissolved in PBS plus 2% FBS to make a 50 µg/mL stock. Stable for weeks at –20 °C or months at –80 °C.
6. 150-cm² cell-culture flasks (Corning).
7. Cell scrapers (39-cm handle, 3-cm blade) (Corning).
8. 25-mL pipettes (BD Falcon, Balford, MA).

2.3. Quantitative PCR Coxiella burnetii Growth Assay

1. Ultraclean Microbial DNA isolation kit (MoBio, Carlsbad, CA).
2. TaqMan Master Mix solution (Applied Biosystems, Foster City, CA).
3. PCR primers and probe corresponding to *C. burnetii dotA*: dotA-F (5′-GCGCAA TACGCTCAATCACA-3′), dotA-R (5′-CCATGGCCCCAATTCTCTT-3′), and

dotA-Probe (5′-CCGGAGATACCGGCGGTGGG-3′ plus FAM/TAMRA) (Applied Biosystems).
4. Control template DNA consisting of *C. burnetii* *dotA* amplified by PCR using the forward primer 5′-ATGAATAAATTAGTGTCATCGCTG-3′ and reverse primer 5′-AGGACCCAAGTGTAGTTGTTTCCC-3′ and cloned into the TA-cloning vector pCR2.1-TOPO (Invitrogen).
5. Applied Biosystems Prism 7000 Sequence Detection System (Applied Biosystems).

3. Methods

Vero cells, a widely used cell line for animal virus propagation *(17)*, have also been routinely employed as a host cell for amplification of *C. burnetii* *(18)* and other obligate intracellular bacteria *(19)*. Vero cells are highly permissive for *C. burnetii* infection and support robust replication of the pathogen. Moreover, they are a hardy, fast growing cell line that does not require special growth conditions or media. We have recently established that the stationary phase of the *C. burnetii* growth cycle begins approximately 6 days post-infection in Vero cells following an approximately 100-fold increase in *C. burnetii* genome equivalents *(20)*. Therefore, optimal yields of the organism are obtained when purification is conducted after this time point.

All protocols for purifying *C. burnetii* from host cells include a lysis step to release intracellular organisms. Lysis methods include subjecting infected cells to osmotic stress with distilled water *(21)*, passing cells through a syringe needle *(22)*, and disrupting cells by sonication *(23)*, tissue homogenization *(24)*, or by vortexing in the presence of glass beads *(25)*. We find that gentle sonication effectively breaks host cells without damaging the more fragile large cell morphological form of *C. burnetii* *(20)*. *C. burnetii* preparations reasonably free of host cell material can be obtained simply by pelleting unbroken nuclei and intact/partially lysed cells, followed by high-speed centrifugation of the post-nuclear supernatant to pellet released bacteria. An additional step of pelleting bacteria through a 30%-RenoCal-76 pad will lessen the amount of host cell carryover, as most insoluble and buoyant host cell membrane material will remain at the pad/buffer interface.

Coxiella burnetii infection is acquired through the aerosol route, and alveolar macrophages are considered the initial target cell of the organism *(16)*. While macrophage or monocyte-like cell lines have been widely employed to study *Coxiella*–host interactions *(1,10,11)*, the biological response of these cells to infection may not accurately reflect the in vivo situation. Indeed, included among the deficiencies of cell lines is attenuated oxidative killing *(11,26)* and cytokine production *(26)*. To more closely mimic the in vivo growth conditions of *C. burnetii*, we developed primary human MDMs as an in vitro model system. Treatment of monocytes with M-CSF results in their differentiation into cells with typical macrophage morphology that efficiently ingest and support the growth of *C. burnetii*.

3.1. Infection of Vero Cells with Coxiella burnetii

1. Passage confluent Vero cells in a single T-150 flask 1:12 into 12 T-150 flasks. This split will provide confluent monolayers in new flasks for *C. burnetii* infection in 24–48 h. Grow cells in RPMI medium with 10% FBS at 37 °C in 5% CO_2.
2. Quick thaw a *C. burnetii* seed stock in a 37 °C water bath, then place on ice. In a 50-mL conical tube, make 48 mL of inoculum consisting of a 1:50 dilution of the seed stock in RPMI with 2% FBS. Discard the tissue culture media and add 4 mL of inoculum per T-150 flask. Incubate for 2 h at room temperature with slow rocking (*see* **Notes 1** and **2**).
3. Add 40 mL of RPMI with 2% FBS directly to flasks without removing inoculum (*see* **Note 3**). Incubate for approximately 7–10 days at 37 °C in 5% CO_2. At this time point, *C. burnetii* will be in its stationary growth phase, and the yield of organisms will be maximized *(20)*. Large, usually single replicative vacuoles become apparent beginning at approximately 2 days post-infection (*see* **Fig. 1**).

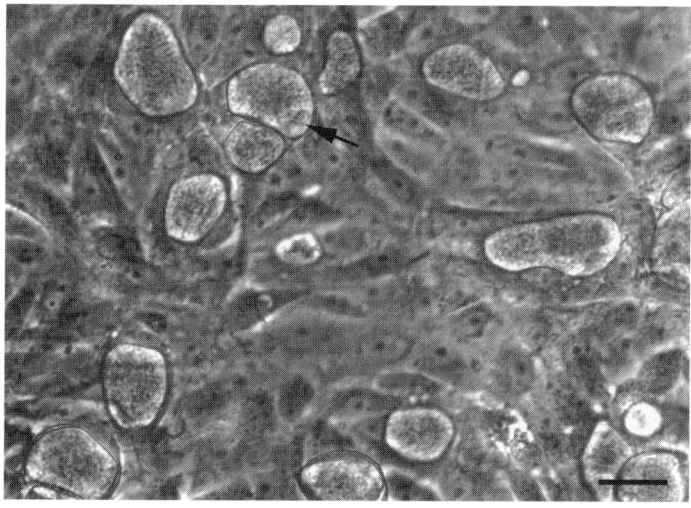

Fig. 1. Vero cells infected with the Nine Mile phase I strain (RSA493) of *Coxiella burnetii*. Vero cells in a 24-well plate were infected for 5 days, then viewed by phase contrast light microscopy using a Nikon TE-2000E inverted microscope equipped with a CoolSNAP HQ digital camera (Roper Scientific, Tuscon, AZ). Images were acquired using Metamorph software (Universal Imaging, Dowingtown, PA) and processed using Adobe Photoshop (Adobe Systems, San Jose, CA). Vero cells containing large and usually singular replicative vacuoles harboring *C. burnetii* are evident (arrow). Bar, 25 µm.

3.2. Purification of Coxiella burnetii from Vero Cells

1. Detach infected monolayer into culture medium by scraping with a cell scraper Transfer equal amounts of medium containing detached cells into two sterile 250-mL centrifuge bottles. Serially, rinse flasks with 10 mL of PBSS and add an equal amount of rinse to each centrifuge bottle. Centrifuge for 15 min at 21,000 × *g* at 4 °C. This centrifugation step pellets not only infected host cells but also *C. burnetii* present in the culture medium (*see* **Note 4**).

2. Discard supernatant. Combine pellets in approximately 30 mL of cold PBSS. Use a 10-mL syringe and sterile cannula to resuspend and disperse the pellet. Transfer resuspended organisms to a sterile disposable plastic beaker and place on ice.

3. Sonicate for 15 s at medium setting to lyse infected Vero cells. Cool on ice for 30 s and then sonicate for an additional 15 s (*see* **Note 5**).

4. Transfer material to a 50-mL conical centrifuge tube. Centrifuge at 900 × *g* for 5 min at 4 °C to pellet unbroken cells and nuclei. Transfer post-nuclear supernatant containing *C. burnetii* to a 50-mL O-ringed screw cap centrifuge tube.

5. Centrifuge sample for 15 min at 31,000 × *g* at 4 °C to pellet *C. burnetii* (*see* **Note 6**). Discard supernatant. Resuspend pellet in 20 mL of cold PBSS using a 10-mL syringe and cannula.

6. Gently layer 10 mL of the *C. burnetii* suspension equally over two 8-mL 30%-RenoCal-76 cushions in 25 × 89 Ultra-Clear centrifuge tubes (*see* **Note 7**). Fill tubes with PBSS. Centrifuge for 30 min at 58,400 × *g* at 4 °C. *C. burnetii* will move through the pad to form a whitish pellet while most host cell material will remain at the RenoCal-76/PBSS interface (*see* **Note 8**).

7. Gently pour off the supernatant. Resuspend each pellet with 10 mL of cold PBSS using a 10-mL syringe and cannula. Transfer each 10-mL resuspension to a 50-mL O-ringed screw cap centrifuge tube. Fill tubes with PBSS. Centrifuge sample for 15 min at 31,000 × *g* at 4 °C to pellet *C. burnetii* (*see* **Note 9**).

8. Discard supernatant. Resuspend and combine pellets in 6 mL of PBSS using a 10-mL syringe and cannula. Aliquot purified *C. burnetii* in volumes of 0.1–1.0 mL in 2-mL microfuge tubes. Store at –80 °C (*see* **Note 10**).

3.3. PBMC Isolation and Culture of Monocyte-Derived Macrophages

1. Prior to starting, set aside at least 3 mL of whole blood from the donor for use in the next step. Dilute whole blood or buffy coat (highly enriched for leukocytes but still containing a large number of erythrocytes) 1:2 with PBS containing 2% FBS (*see* **Note 11**). Carefully layer 25 mL of diluted blood onto 20 mL of Ficoll-Paque Plus in a 50-mL conical tube. Centrifuge for 30 min at 500 × *g* with the brake off. After centrifugation, a layer of PBMC will be found at the Ficoll-diluted blood interface. Carefully remove this layer with a 25-mL pipette and transfer to a clean 50-mL conical tube (*see* **Note 12**). To remove platelets, wash PBMC twice with PBS containing 2% FBS by centrifuging at 200 × *g* in between each wash step.

2. Resuspend PBMC pellet in PBS with 2% FBS, adjusting the cell concentration to 2–5 ×10^7 cells/mL. For every 25 mL of PBMC solution, add 3.3 mL of

whole blood from the same donor and 2 mL of RosetteSep Monocyte Enrichment cocktail (*see* **Note 13**). Incubate on a tube rotator for 20 min at room temperature.

3. Dilute mixture 1:2 with PBS containing 2% FBS and carefully layer 25 mL onto 25 mL of Ficoll-Paque Plus in a 50-mL conical tube. Centrifuge for 20 min at 1200 × *g* with the brake off. After centrifugation, a layer of highly enriched monocytes will be found at the Ficoll–PBS interface. Carefully remove this layer with a pipette and transfer to a new 50-mL conical tube. Wash monocytes twice with PBS containing 2% FBS, centrifuging at 500 × *g* in between each wash step. Resuspend cells in RPMI plus 10% FBS and 50 ng/mL M-CSF to a final concentration of 0.5–1.0 × 10^6 cells/mL (should require approximately 40 mL of media). Transfer to a T-150 tissue culture flask and incubate at 37 °C in 5% CO_2. Add fresh M-CSF to the flask after 3 days of incubation.

4. Note the monocytes will differentiate into adherent macrophages by day 7 of incubation. To harvest cells, discard the culture medium and wash once with 25 mL of cold PBS without Ca^{++} or Mg^{++}. Add 20 mL of fresh cold PBS to the flask and incubate on ice until cells begin to round-up and detach (usually 15–20 min) (*see* **Note 14**). Gently scrape cells into the PBS with a cell scraper and transfer the cell suspension to a 50-mL conical centrifuge tube. Pellet the cells by centrifugation at 500 × *g* for 5 min. Resuspend cells in 5–10 mL of RPMI plus 10% FBS and 50 ng/mL M-CSF. Count cells, add medium to adjust to the desired cell concentration, and re-plate in a tissue culture plate (*see* **Note 15**). Incubate at 37 °C in 5% CO_2 for at least 6 h to allow macrophages to adhere.

3.4. Infection of Monocyte-Derived Macrophages with Coxiella burnetii

1. A good infection of macrophages can be achieved by direct addition of purified *C. burnetii* to the culture medium. To achieve near 100% infection, multiplicities of infection of approximately 10 and 100 are required for avirulent phase II and virulent phase I strains, respectively (*see* **Note 1**). Add the appropriate amount of bacteria to each well and mix by gently rocking and swirling the plate. Incubate the plate at 37 °C in 5% CO_2 4–24 h.

2. Gently wash the cells three times with PBS or culture medium to remove extracellular bacteria.

3. Add fresh culture medium and incubate at 37 °C in 5% CO_2 to allow *C. burnetii* replication and vacuole development. Replication vacuoles should be visible by phase contrast microscopy within 2–3 days post-infection. At 6 days post-infection, growth of *C. burnetii* should have reached stationary phase and essentially all cells will contain large bacteria-filled vacuoles (*see* **Fig. 2**).

3.5. Enumeration of Coxiella burnetii Replication

1. Conduct quantification of *C. burnetii* replication in macrophages by measuring the number of bacterial genome equivalents by quantitative PCR. Total (host

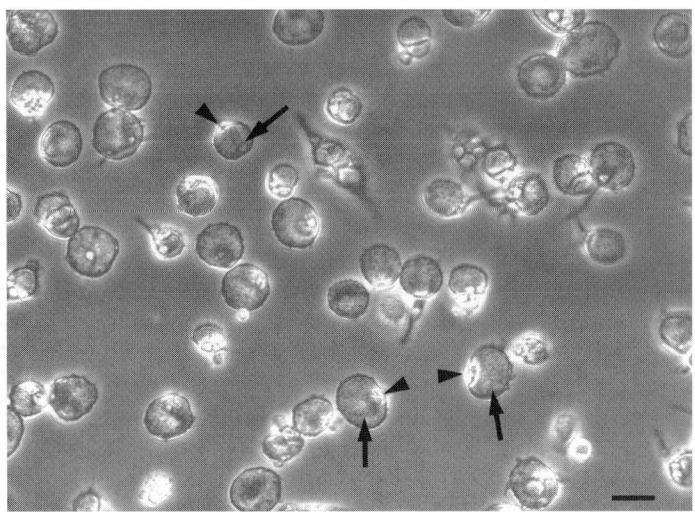

Fig. 2. Human monocyte-derived macrophages (MDMs) infected with the Nine Mile phase I strain (RSA493) of *Coxiella burnetii*. MDMs in a 24-well plate were infected for 2 days, then viewed by phase contrast light microscopy using a Nikon TE-2000E inverted microscope equipped with a CoolSNAP HQ digital camera (Roper Scientific, Tuscon, AZ). Images were acquired using Metamorph software (Universal Imaging, Dowingtown, PA) and processed using Adobe Photoshop (Adobe Systems, San Jose, CA). In macrophages, the replicative vacuole (arrow) enlarges to encompass nearly all the intracellular space that causes the cells to round up and pushes the phase-bright nucleus (arrow heads) to the cell periphery. Bar, 25 μm.

and bacterial) DNA is extracted from infected cells in a 24-well plate using an Ultraclean Microbial DNA Isolation Kit.

2. Remove the culture medium to a 2-mL microfuge tube and centrifuge at $20,000 \times g$ for 15 min to pellet any extracellular bacteria or cells that have detached.

3. To retrieve adherent macrophages, add 300 and 50 μL of microbead and MD1 kit solutions, respectively, directly to well. Pipette up and down to lyse cells and transfer the lysate to the tube containing the pellet from **step 2** (*see* **Note 16**).

4. From this point, follow kit instructions verbatim to obtain purified DNA from infected cells.

5. Set up reactions to determine the number of *C. burnetii* genome equivalents in each DNA sample. For each unknown DNA sample, make 80 μL of TaqMan Universal PCR Master Mix containing 10 μ*M* of dotA-F and dotA-R primers, and 333 n*M* of dotA-probe, as per kit instructions. Mix 10 μL of purified DNA from **step 4** with Master Mix for a final volume of 90 μL. To generate a standard curve of 10^3–10^8 dotA copies, add 10 μL of 10-fold dilutions of purified pCR2.1-TOPO DNA containing the *C. burnetii* dotA gene to 80 μL of Master Mix (*see* **Note 17**). To perform PCR reactions in triplicate, add 3–25 μL aliquots of the 90 μL Master

Mix/DNA sample to the reaction plates. Perform real-time PCR detection of *dotA* using an ABI Prism 7000 Sequence Detection System (Applied Biosystems) or comparable instrument. Extrapolate the number of *C. burnetii* genomes present in the DNA sample from the standard curve.

4. Notes

1. Assuming a stock of approximately 3×10^9 *C. burnetii* genome equivalents per milliliter (*see* **Subheading 3.5.**), this will result in a multiplicity of infection of approximately 10 if using phase II organisms. [Phase I organisms are roughly 10-fold less infectious for cultured cells than phase II organisms *(7)*.] The low inoculum volume with rocking facilitates adherence and internalization of *C. burnetii*.
2. The Nine Mile, phase II, clone 4 isolate (RSA439) can be worked with under biosafety level 2 laboratory conditions *(27)*. All other *C. burnetii* strains or isolates are considered biosafety level 3 organisms. (*See* CDC/NIH Biosafety in Microbiological and Biomedical Laboratories, 4th edition.)
3. Supplementation of culture medium with 2% FBS rather than 10% retards Vero cell growth without significantly affecting yields of *C. burnetii*.
4. A centrifugal force of 21,000 × g is attained using a Beckman JA14 or JLA16.250 rotor at 12,000 rpm.
5. Sonication is conducted at approximately 45 w with a 0.5-inch horn. The sonicator horn can be sterilized by spraying with 70% ethanol. Let air dry before use.
6. A centrifugal force of 31,000 × g is attained using a Beckman JA20 or JA25.15 rotor at 16,000 rpm.
7. RenoCal-76 is chemically similar to Renografin-76 that has been historically used to purify obligate intracellular bacteria *(15)*. Both contain 10% diatrizoate and 66% diatrizoate meglumine. Bacteria retain significant viability when centrifuged through this medium which has a lower osmotic pressure than some commonly used density gradient materials used to purify *C. burnetii* *(28)*. We have also found that the Ultra-Clear centrifuge tubes are sterile if carefully handled when removed from the box.
8. A centrifugal force of 58,400 × g is attained using a Beckman SW28 rotor at 18,000 rpm.
9. This wash step will rid organisms of residual RenoCal-76.
10. Yields of approximately 1.5×10^{10} *C. burnetii* genome equivalents (*see* **Subheading 3.5.**) per infected T-150 flask are common.
11. A buffy coat obtained from 450 mL of blood will typically yield between 0.7 and 1.2×10^8 monocytes.
12. After removing the PBMC layer to a new tube, dilute the cells at least 1:2 with PBS containing 2% FBS before centrifugation.
13. Two milliliters of RosetteSep cocktail is typically sufficient to process PBMC from 450 mL of blood.
14. Wait until the cells begin to take on a rounded appearance and detach before scraping.

15. Plate $1–2 \times 10^5$ cells/well in 24-well plates or 1×10^6 cells/well in 6-well plates. The volume of the culture medium in each well should be kept low. Volumes of 0.5 and 2 mL are sufficient for individual wells of 24- and 6-well plates, respectively.

16. At this point, the cell lysate can be stored at $-20\,°C$ for months before continuing on with the rest of the protocol.

17. The *dotA*/pCR2.1-TOPO plasmid is 6349 bp in size (2418 bp *dotA* PCR product plus 3931 pCR2.1-TOPO vector). One microgram of this plasmid equals 0.24 pmoles which equals 1.44×10^{11} plasmid molecules (i.e., *dotA* copies). Determine the concentration of a stock solution of plasmid spectrophotometrically. Make a dilution series of the plasmid such that a 10 μL aliquot of each diluted stock results in a range of $10^3–10^8$ plasmid copies.

Acknowledgments

We thank Dale Howe and Diane Cockrell for critical review of this manuscript. This research was supported by the Intramural Research Program of the National Institutes of Health, National Institute of Allergy and Infectious Diseases.

References

1. Baca, O. G. and Paretsky, D. (1983) Q fever and *Coxiella burnetii*: a model for host-parasite interactions. *Microbiol Rev.* **47**, 127–149.
2. Babudieri, C. (1959) Q fever: a zoonosis. *Adv. Vet. Sci.* **5**, 81–84.
3. Hilbink, F., Penrose, M., Kovacova, E., and Kazar, J. (1993) Q fever is absent from New Zealand. *Int. J. Epidemiol.* **22**, 945–949.
4. Palmer, S. R. and Key, D. W. (1983) Placentitis and abortion in goats and sheep in Ontario caused by *Coxiella burnetii*. *Can. Vet. J.* **24**, 60–63.
5. Maurin, M. and Raoult, D. (1999) Q fever. *Clin. Microbiol. Rev.* **12**, 518–553.
6. Benenson, A. S. and Tigertt, W. D. (1956) Studies on Q fever in man. *Trans. Assoc. Am. Physicians.* **69**, 98–104.
7. Moos, A. and Hackstadt, T. (1987) Comparative virulence of intra- and interstrain lipopolysaccharide variants of *Coxiella burnetii* in the guinea pig model. *Infect. Immun.* **55**, 1144–1150.
8. Hatchette, T. F., Hudson, R. C., Schlech, W. F., Campbell, N. A., Hatchette, J. E., Ratnam, S., et al. (2001) Goat-associated Q fever: a new disease in Newfoundland. *Emerg. Infect. Dis.* **7**, 413–419.
9. Heinzen, R. A., Scidmore, M. A., Rockey, D. D., and Hackstadt, T. (1996) Differential interaction with endocytic and exocytic pathways distinguish parasitophorous vacuoles of *Coxiella burnetii* and *Chlamydia trachomatis*. *Infect. Immun.* **64**, 796–809.
10. Ghigo, E., Honstettre, A., Capo, C., Gorvel, J. P., Raoult, D., and Mege, J. L. (2004) Link between impaired maturation of phagosomes and defective *Coxiella burnetii* killing in patients with chronic Q fever. *J. Infect. Dis.* **190**, 1767–1772.

11. Brennan, R. E., Russell, K., Zhang, G., and Samuel, J. E. (2004) Both inducible nitric oxide synthase and NADPH oxidase contribute to the control of virulent phase I *Coxiella burnetii* infections. *Infect. Immun.* **72**, 6666–6675.

12. Miller, J. D., Curns, A. T., and Thompson, H.A. (2004) A growth study of *Coxiella burnetii* Nine Mile Phase I and Phase II in fibroblasts. *FEMS Immunol. Med. Microbiol.* **42**, 291–297.

13. Weiss, E., Coolbaugh, J. C., and Williams, J. C. (1975) Separation of viable *Rickettsia typhi* from yolk sac and L cell host components by renografin density gradient centrifugation. *Appl. Microbiol.* **30**, 456–463.

14. Samuel, J. E., Frazier, M. E., Kahn, M. L., Thomashow, L. S., and Mallavia, L. P. (1983) Isolation and characterization of a plasmid from phase I *Coxiella burnetii*. *Infect. Immun.* **41**, 488–493.

15. Williams, J. C., Peacock, M. G., and McCaul, T. F. (1981) Immunological and biological characterization of *Coxiella burnetii*, phases I and II, separated from host components. *Infect. Immun.* **32**, 840–851.

16. Stein, A., Louveau, C., Lepidi, H., Ricci, F., Baylac, P., Davoust, B., et al. (2005) Q fever pneumonia: virulence of *Coxiella burnetii* pathovars in a murine model of aerosol infection. *Infect. Immun.* **73**, 2469–2477.

17. Govorkova, E. A., Murti, G., Meignier, B., de Taisne, C., and Webster, R. G. (1996) African green monkey kidney (Vero) cells provide an alternative host cell system for influenza A and B viruses. *J. Virol.* **70**, 5519–5524.

18. Burton, P. R., Stueckemann, J., Welsh, R. M., and Paretsky, D. (1978) Some ultrastructural effects of persistent infections by the rickettsia *Coxiella burnetii* in mouse L cells and green monkey kidney (Vero) cells. *Infect. Immun.* **21**, 556–566.

19. Policastro, P. F., Peacock, M. G., and Hackstadt, T. (1996) Improved plaque assays for *Rickettsia prowazekii* in Vero 76 cells. *J. Clin. Microbiol.* **34**, 1944–1948.

20. Coleman, S. A., Fischer, E. R., Howe, D., Mead, D. J., and Heinzen, R. A. (2004) Temporal analysis of *Coxiella burnetii* morphological differentiation. *J. Bacteriol.* **186**, 7344–7352.

21. Zamboni, D. S. and Rabinovitch, M. (2003) Nitric oxide partially controls *Coxiella burnetii* phase II infection in mouse primary macrophages. *Infect. Immun.* **71**, 1225–1233.

22. Beron, W., Gutierrez, M. G., Rabinovitch, M., and Colombo, M. I. (2002) *Coxiella burnetii* localizes in a Rab7-labeled compartment with autophagic characteristics. *Infect. Immun.* **70**, 5816–5821.

23. Meconi, S., Jacomo, V., Boquet, P., Raoult, D., Mege, J. L., and Capo, C. (1998) *Coxiella burnetii* induces reorganization of the actin cytoskeleton in human monocytes. *Infect. Immun.* **66**, 5527–5533.

24. Baca, O. G., Roman, M. J., Glew, R. H., Christner, R. F., Buhler, J. E., and Aragon, A. S. (1993) Acid phosphatase activity in *Coxiella burnetii*: a possible virulence factor. *Infect. Immun.* **61**, 4232–4239.

25. Veras, P. S., de Chastellier, C., Moreau, M. F., Villiers, V., Thibon, M., Mattei, D., et al. (1994) Fusion between large phagocytic vesicles: targeting of yeast and

other particulates to phagolysosomes that shelter the bacterium *Coxiella burnetii* or the protozoan *Leishmania amazonensis* in Chinese hamster ovary cells. *J. Cell Sci.*,**107**, 3065–3076.

26. Baca, O. G., Akporiaye, E. T., Aragon, A. S., Martinez, I. L., Robles, M. V., and Warner, N. L. (1981) Fate of phase I and phase II *Coxiella burnetii* in several macrophage-like tumor cell lines. *Infect. Immun.* **33**, 258–266.
27. Hackstadt, T. (1996) Biosafety concerns and *Coxiella burnetii*. *Trends Microbiol.* **4**, 341–342.
28. Wiebe, M. E., Burton, P. R., and Shankel, D. M. (1972) Isolation and characterization of two cell types of *Coxiella burnetii* phase I. *J. Bacteriol.* **110**, 368–377.

16

Infection of Epithelial Cells With *Salmonella enterica*

Olivia Steele-Mortimer

Summary

Salmonella enterica serovars cause a variety of diseases ranging from self-limiting gastroenteritis to severe systemic infections. Virulence of these facultative intracellular pathogens is dependent on their ability to invade and replicate within non-phagocytic cells, and cultured epithelial cell systems have been used extensively to dissect the molecular mechanisms involved. For efficient invasion *in vitro*, the bacterial cell growth conditions as critical since the invasion associated type III secretion system (T3SS1) must be expressed and functional. The ability of *Salmonella* to invade, and replicate within, epithelial cells can be easily assessed using a gentamicin protection assay or immunofluorescence microscopy. Here, the protocols used in our laboratory are described in detail.

Key Words: Bacteria; immunofluorescence; intracellular; invasion; type III secretion.

1. Introduction

The more than 2500 serovars of *Salmonella enterica* are genetically highly similar yet tend to have highly specific host and virulence characteristics *(1)*. For example, *S. enterica* serovar Typhimurium is one of the most common causes of gastroenteritis in humans but causes a systemic "typhoid-like" disease in susceptible mice and is used as an animal model for typhoid fever. Multiple virulence factors contribute to pathogenesis including two sets of bacterial effector proteins that are delivered directly into the host cell by the type III secretion systems, T3SS1 and T3SS2 *(2)*. T3SS1 effectors are translocated across the plasma membrane and act cooperatively to induce actin rearrangements and membrane ruffling, resulting in the internalization of *Salmonella* into a membrane-bound vacuole known as the *Salmonella*-containing vacuole

From: *Methods in Molecular Biology, vol. 431: Bacterial Pathogenesis*
Edited by: F. DeLeo and M. Otto © Humana Press, Totowa, NJ

or SCV *(3)*. In contrast, T3SS2 effectors are translocated across the vacuole membrane and mediate SCV biogenesis and intracellular survival *(4)*. Recent evidence suggests that there is overlap in the temporal expression and function of T3SS1 and T3SS2 although this remains largely uncharacterized *(5–8)*. The T3SS1 translocon or injectosome, as well as the regulatory components and some effector proteins are encoded on *Salmonella* pathogenicity island 1 (SPI1). Significantly, other T3SS1 effectors, encoded elsewhere on the chromosome, are also included in the SPI1 regulon *(9,10)*. As invasion is mediated by T3SS1, the SPI1 regulon must be induced extracellularly; however, the cues involved in this process remain incompletely characterized. We have found that SPI1 induction is strongly growth phase dependent such that bacteria grown to late log/early stationary phase are "hyper-invasive" (*see* **Fig. 1**). Therefore, to synchronize invasion of cultured epithelial cells, we use SPI1-induced bacteria at a high multiplicity of invasion (MOI) for a short infection time (2–10 min) followed by removal of the extracellular bacteria. Gentamicin is routinely added to kill any remaining extracellular bacteria, as this antibiotic does not cross the plasma membrane and should not kill intracellular bacteria. However, gentamicin can be taken up by fluid phase endocytosis and could be delivered to the SCV; adding antibiotic 15–20 min after removing the inoculum should minimize the chance of killing intracellular bacteria.

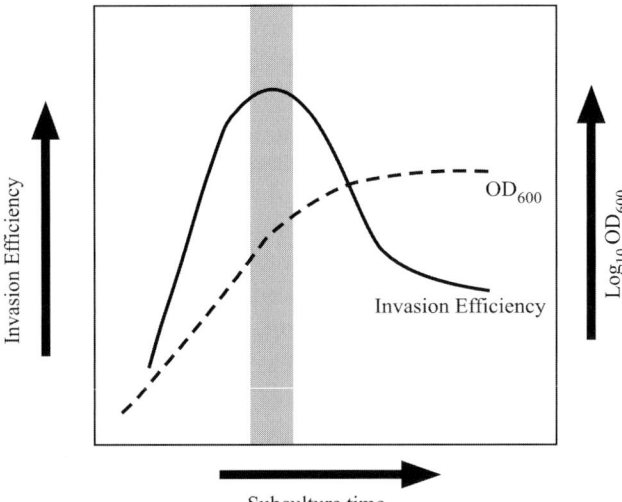

Fig. 1. *Salmonella enterica* serovar Typhimurium invasiveness is growth phase dependent. Salmonella grown to late log/early stationary phase (dotted line) are "hyper-invasive" (solid line). The gray bar illustrates the subculture time required to obtain "hyper-invasive bacteria.

Internalized bacteria can be assayed by recovery of colony-forming units, in a classical gentamicin assay or by microscopical analysis after fixation and staining using *Salmonella*-specific antibodies. For both assays, epithelial cells are grown in monolayers and infected with *Salmonella* for various lengths of time. Because of the time lag that occurs before the onset of intracellular replication, invasion is usually monitored at 1–1.5 h post-infection and replication at 4.5–8 h or later. Both invasion and replication can be affected by numerous factors—for example, host cell type, viability, confluency, and growth conditions. Adhering to a standardized routine can minimize most experimental variability. The gentamicin assay, which involves solubilization of the cells and plating the released bacteria, yields information on the total number of colony-forming units (i.e., viable bacteria) in a host cell monolayer. In contrast, the microscopical assay determines the number of bacteria inside individual cells but cannot differentiate between live and dead bacteria. Together these assays are indispensable tools for any lab studying *Salmonella*–host cell interactions.

2. Materials

1. Glycerol stocks of *S. enterica*. This protocol has been optimized for serovar Typhimurium SL1344 *(11)* (*see* **Note 1**).
2. Luria-Bertani (LB) Broth, Miller: 10 g/L Bacto tryptone and 5 g/L Bacto yeast extract 10 g/L NaCl. For plates add 15 g/L of Bacto-agar. Sterilize by autoclaving.
3. Antibiotic stocks (stored at –20 °C): streptomycin, 100 mg/mL in H_2O and gentamicin sulfate, 50 mg/mL in H_2O.
4. Minimal essential medium (MEM) with Earle's salts without L-glutamine (Invitrogen, Carlsbad, CA).
5. Sodium pyruvate, 100 mM in H_2O.
6. L-glutamine, 200 mM in H_2O.
7. Trypsin EDTA, 1×: solution of 0.25% Trypsin and 2.21 mM EDTA in HBSS without sodium bicarbonate, calcium, and magnesium. Available pre-made from suppliers of cell-culture materials.
8. Fetal bovine serum (FBS): Performance tested, mycoplasma tested, virus tested, bacteriophage tested, and endotoxin tested. To heat inactivate, thaw the serum slowly at 37 °C and mix the contents of the bottle thoroughly. Place the thawed bottle of serum into a 56 °C water bath containing enough water to immerse the bottle to just above the level of the serum. Swirl the serum every 5–10 min to ensure uniform heating and to prevent protein coagulation at the bottom of the bottle. After 30 min at 56 °C, cool the serum immediately. Store at –20 °C.
9. Growth medium (GM): MEM supplemented with 2 mM L-glutamine(*see* **Note 2**), 1 mM sodium pyruvate, and 10% heat-inactivated FBS.

10. Phosphate-buffered saline (PBS) without divalent cations: 1.50 g/L dibasic sodium phosphate ($Na_2HPO_4 \cdot 7H_2O$), 0.2 g/L anhydrous monobasic potassium phosphate (KH_2PO_4), 8 g/L sodium chloride (NaCl), and 0.2 g/L potassium chloride (KCl), pH to 7.4.

11. Tissue culture-treated plastic flasks, dishes, and 24-well plates.

12. HeLa cells (human adenocarcinoma cervix epithelial, CCL-2) obtained from the American Type Culture Collection, Manassas, VA (*see* **Note 3**).

13. Hank's-buffered salt solution (HBSS): 0.14 g/L anhydrous calcium chloride ($CaCl_2$), 0.10 g/L magnesium chloride ($MgCl_2 \cdot H_2O$), 0.10 g/L magnesium sulfate ($MgSO_4 \cdot 7H_2O$), 0.40 g/L potassium chloride (KCl), 0.60 g/L monobasic potassium phosphate (KH_2PO_4), 0.35 g/L sodium bicarbonate ($NaHCO_3$), 8.0 g/L sodium chloride (NaCl), 48 mg/L anhydrous dibasic sodium phosphate (Na_2HPO_4), and 1 g/L D-glucose (dextrose).

14. Solubilization buffer: PBS containing 1% Triton X-100 and 0.1% SDS.

15. Sterile glass coverslips, 12-mm diameter (#1).

16. Stock formaldehyde solution (25%) (*see* **Note 4**): Dissolve 10 g paraformaldehyde in 30 mL water (in capped 50 mL tube). Heat to 75 °C in water bath with frequent mixing. Add 10*N* NaOH to clear (usually 5–20 drops). Add H_2O to 40 mL. Store in 1-mL aliquots at −20 °C.

17. Working formaldehyde fixative (2.5%): Thaw aliquot of 25% formaldehyde and heat to clear. Do not heat formaldehyde above 75 °C. Mix with approximately 6 mL of ddH_2O and 1 mL 10× PBS. Adjust pH to 7.2–7.4 with HCl (Use pH strips, not pH meter). Add ddH_2O to 10 mL.

18. SS-PBS: PBS, 10% (v/v) normal goat serum (*see* **Note 5**), and 0.2% (w/v) saponin.

19. Antibodies to lipopolysaccharide (LPS) of *S. enterica* serovar Typhimurium (Group B), such as mouse monoclonal antibody, clone 1E6 (BioDesign International, Saco, ME) or rabbit polyclonal antibody, and BBL *Salmonella* O Group B antisera (Becton Dickinson, Sparks, MD).

20. Fluorescently conjugated antibodies that have been highly cross-absorbed, such as Alexa Fluor 488 goat anti-mouse IgG (H + L), Alexa Fluor 568 goat anti-mouse IgG (H + L) (from Invitrogen™ Corporation), or Cy5™-conjugated Affinipure Goat Anti-Mouse IgG (H + L) (Jackson ImmunoResearch Laboratories, Inc., West Grove, PA).

21. Fluorescently conjugated phallotoxins such as Alexa Fluor 488-phalloidin (Invitrogen™). Dissolve in methanol (200 U/mL) and store at −20 °C.

22. Mowiol™ mounting media, stock solution: Add 2.4 g of Mowiol™ 4-88 (*see* **Note 6**) to 6 g of glycerol. Stir to mix. Add 6 mL water and leave stirring at room temperature for several hours. Add 12 mL of 0.2 *M* Tris–HCl (pH 8.5) and heat to 50 °C for 10 min with occasional mixing. After Mowiol™ dissolves (never 100%), clarify by centrifugation at 5000 × *g* for 15 min. Aliquot into airtight screw capped 1.5-mL tubes and store at −20 °C. Mowiol™ stock is stable for a few weeks at room temperature or 4 °C if in an airtight tube. Optional: Add 1,4,-diazobicyclo-[2.2.2]-octane (DABCO; Sigma-Aldrich) to 2.5% (w/v) to reduce fading of fluorophores.

3. Methods

3.1. Preparation of HeLa Cells for Infection

1. Grow HeLa cells in a humidified 37 °C, 5% CO_2 tissue-culture incubator. Low passage number (<15 after receipt from ATCC) cells should be used (*see* **Note 7**).
2. Passage cells when they are actively growing (*see* **Note 8**).
3. For cells grown in a 75-cm^2 flask, decant media and rinse monolayer with 5 mL PBS.
4. Immediately add 5 mL Trypsin/EDTA and gently swirl over the monolayer. Remove 4 mL of the Trypsin/EDTA and leave cells at room temperature for up to 5 min. The cells should be easily detached by tapping the flask firmly, but gently, against an open hand.
5. Immediately add 5 mL of GM and resuspend the cells by pipetting up and down with a 5-mL pipette.
6. Count the cells using a hemocytometer. For infection experiments, seed in 24-well (5×10^4 per well) or 6-well dishes (2×10^5 per well) and grow overnight (14–20 h) before infection.
7. Cells should still be actively growing, in logarithmic phase, at the time of infection.

3.2. Preparation of SPI1-Induced Salmonella

1. Grow *S. enterica* serovar Typhimurium overnight (16–18 h) in 1.5–2 mL LB in a 15-mL tube with loose cap. Incubate at 37 °C in a shaking incubator (225 rpm).
2. Subculture *Salmonella* by transferring 300 μL of the overnight culture into 10 mL of LB in a loosely capped 125-mL Erlenmeyer flask. Incubate at 37 °C in a shaking incubator (25 rpm) for 3.5 h or to late log phase (*see* **Note 9**). At this time, there should be approximately 3×10^9 CFU/mL (*see* **Note 10**).
3. Pellet 1 mL of the *Salmonella* subculture by centrifugation at $1000 \times g$ in a microfuge for 2 min at room temperature.
4. Remove 900 μL of supernatant and gently resuspend the pellet in 900 μL HBSS.
5. Use immediately.

3.3. Invasion

1. Inoculate cells with *Salmonella* (MOI 10–100) by adding bacteria directly to the cell-culture supernatant (*see* **Note 11**).
2. Incubate for 2–10 min at 37 °C in 5% CO_2.
3. Aspirate media and rinse the monolayer twice with PBS, or HBSS, to remove extracellular bacteria.
4. Add fresh GM and incubate for 20 min at 37 °C in 5% CO_2.
5. Replace GM with fresh GM containing 50 μg/mL gentamicin.
6. Incubate for 40 min at 37 °C in 5% CO_2.
7. Replace GM with fresh GM containing 5 μg/mL gentamicin for remainder of experiment.

3.4. Gentamicin Protection Assay for Quantification of Intracellular Bacteria

1. Aspirate media and rinse the monolayer once with PBS to remove gentamicin.
2. Solubilize the monolayer in 1 mL solubilization buffer.
3. Immediately transfer to microfuge tubes and dilute in PBS (1:10, 1:100, and 1:1000).
4. Spot 10 μL of each dilution on dry LB plates (*see* **Fig. 2**).
5. Incubate overnight at 37 °C.
6. Count colonies as follows:

   a. if three colonies in 10 μL of undiluted, then total cfu/well = 3 × 100 = 300.
   b. if three colonies in 10 μL of 1:1000, then total cfu/well = 3 × 100 × 1000 = 3×10^5.

3.5. Immunofluorescence "Inside-Outside" Assay

1. Passage cells as described above, except grow in 24-well plates with a glass coverslip in the bottom of each well. Make sure that the coverslips are not floating by gently pushing each one to the bottom with a pipette tip.
2. Infect cells 16–20 h after passaging as above.
3. Rinse monolayer once with HBBS.
4. Fix in fresh 2.5% formaldehyde (warmed to 37 °C), 300–500 μL/well, for 10 min at room temperature. Discard the formaldehyde appropriately.
5. Rinse twice with PBS (use "squirt bottle" and direct jet onto side of well).
6. Wash three times for 10 min each with PBS.
7. Remove PBS thoroughly by aspiration and make sure coverslip is not touching sides of well.

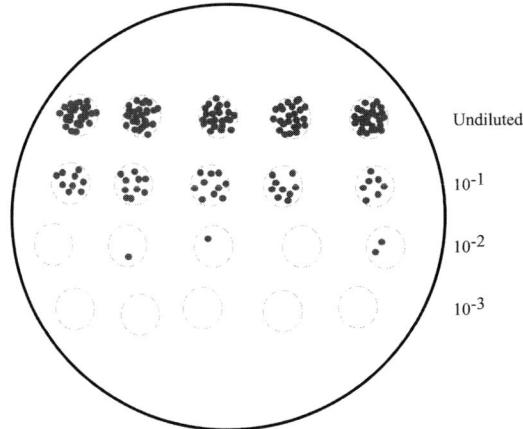

Fig. 2. Schematic diagram of an LB plate used to estimate colony-forming units in a cell lysate. See text for explanation.

8. Immediately add 12–30 µL of primary (1°) antibody (e.g., rabbit anti-LPS, diluted 1:200–1:1,000 in PBS) directly onto the coverslip (*see* **Note 12**). The antibody solution must stay on the coverslip. Leave at room temperature for 15–30 min depending on antibody.
9. Rinse twice with PBS.
10. Wash three times for 10 min each with PBS.
11. Remove PBS thoroughly by aspiration and make sure coverslip is not touching sides of well.
12. Immediately add 12–30 µL of secondary (2°) antibody, for example, AlexaFluor568-conjugated goat anti-rabbit antibody diluted in PBS (*see* **Note 13**). The antibody solution must stay on the coverslip. Leave at room temperature for 15–30 min depending on antibody.
13. Rinse twice with PBS.
14. Wash three times for 10 min each with PBS.
15. Permeabilize cells and block by incubating with SS-PBS, 250–400 µL/well, for 15–20 min at room temperature.
16. Remove SS-PBS thoroughly by aspiration and make sure coverslip is not touching sides of well.
17. Immediately add 12 µL of primary (1°) antibody (e.g., mouse anti-LPS, diluted 1:200–1:1000 in SS-PBS) (*see* **Note 14**). The antibody solution must stay on the coverslip. Leave at room temperature for 20–60 min depending on antibody.
18. Rinse once with PBS.
19. Wash three times with PBS, 10 min each.
20. Incubate briefly with SS-PBS for 2–5 min.
21. Immediately add 12 µL of appropriate secondary (2°) antibody, for example, if AlexaFlur568 conjugated antibodies were used in step #12, then use AlexaFluor488-conjugated goat anti-rabbit antibody diluted in PBS (*see* **Note 13**). The antibody solution must stay on the coverslip. Leave at room temperature for 20–60 min depending on antibody.
22. Rinse twice with PBS.
23. Wash three times for 10 min each with PBS. It is crucial to remove all unbound antibody before proceeding to the next step.
24. Optional: Rinse once with distilled H_2O.
25. Mount coverslips, cells down, on clean glass slides using 5–10 µL Mowiol (*see* **Note 6**) per coverslip.
26. Allow Mowiol™ to harden overnight at room temperature in the dark.
27. Observe cells using a fluorescence microscope equipped with the appropriate filters/laser lines for the fluorophores used.

4. Notes

1. Glycerol stock cultures should be kept at –80 °C. Bacteria for experiments should be picked from colonies grown on LB plates not more than 1 week old and stored at 4 °C. The SL1344 strain is streptomycin resistant, so can be grown on LB plates containing 100 µg/mL streptomycin.

2. L-glutamine is an essential component of cell-culture media, but it is unstable and decomposes spontaneously in aqueous solutions producing ammonia, a toxic product of degradation. Buy MEM without glutamine and add L-glutamine immediately before use. Alternatively there are now some commercially available, stabilized forms of L-glutamine such as GlutaMAX™ (Invitrogen).

3. When working with mammalian cells, it is crucial to obtain them from a reliable source and to keep good frozen stocks and records. The American Type Culture Collection is the best source in North America.

4. Paraformaldehyde or formaldehyde? Formaldehyde (HCHO) is a colorless, flammable gas with a pungent, suffocating odor. Formalin is a 37% solution of formaldehyde dissolved in water. Normally, an additive such as methanol is used to stabilize formalin making it unsuitable for microscopy studies. Paraformaldehyde, a white powder, is a solid polymer (*n* up to 100) of formaldehyde. Fixative solution is made from paraformaldehyde but must contain monomeric formaldehyde as its major solute. Dilution with water breaks up the small polymers but heating and an added source of hydroxide ions is required to break up the large polymers. As formaldehyde in solution rapidly forms polymers, it must be made fresh although this can be avoided by storing a stock (25%) solution at $-20\,°C$ and making the working solution immediately before use. Technically, the fixative we describe here is "2.5% formaldehyde made from paraformaldehyde"; it is usually incorrectly referred to as 2.5% paraformaldehyde (PFA) solution *(12)*.

5. Normal serum is used for blocking non-specific binding sites before adding primary antibodies in immunofluorescence staining protocols. Ideally, it should be of the same species as the secondary antibodies, but if good quality highly cross-absorbed antibodies are used, any normal serum will suffice.

6. Mowiol™ is the trade name of polyvinyl alcohol, a group of water-soluble polymers manufactured by the alcoholysis of polyvinyl acetate. Mowiol™ comes in different grades and can be partially or fully hydrolyzed, with partially hydrolyzed ones being more water soluble. Mowiol™ 4-88 is a partially hydrolyzed polyvinyl alcohol with low to medium viscosity. Glycerol is added as a plasticizer and can be added up to 30% (v/v). When thawing Mowiol™, ensure that it has come to room temperature before using, otherwise small bubbles will form under the coverslip. Mowiol™ mounting media hardens to form a stable mount so that, in contrast to liquid mounting media such as glycerol, nail varnish is not required to seal the coverslips. Alternatively a variety of pre-made mounting media are available from companies such as invitrogen or sigma.

7. The routine addition of antibiotics such as streptomycin and/or penicillin can mask low-level contamination and should be avoided. For experiments involving *Salmonella*, the cells will in any case have to be cultured in antibiotic-free media for several hours before the experiment.

8. To achieve consistency and reproducibility, it is essential to adhere to a routine protocol when passaging and maintaining cell cultures. The following points should be considered. Even if the original stock is of unknown passage number

staying within a certain range, for example, 10–20 passages from stock cultures, always using well-documented stocks is important for consistency. Cells should be passaged on a routine schedule and always while in the logarithmic growth phase. For each experiment plate, the same number of cells at the same time. Observe the monolayers on an inverted phase contrast microscope and aim for absolute consistency in confluency between different experiments.

9. It is crucial that the incubator remains at 37 °C throughout the subculture. Incubators that are opened and closed frequently may not maintain steady temperature and will result in changes in the growth rate of the *Salmonella*.

10. The subculture conditions may need to be optimized to get hyper-invasive bacteria. This depends on *Salmonella* strain as well as other less tangible variables that differ from laboratory to laboratory. The most important factor in our hands is that the bacteria should be at late logarithmic phase of growth just before going into stationary phase. Remember that mutant strains or strains containing expression plasmids may grow at different rates and this should be taken into account. They should be extremely motile and will rapidly induce ruffles in HeLa cells that can be observed on a phase contrast tissue culture microscope using a ×20 or ×40 objective.

11. It is often assumed that low MOI should be used for all experiments. However, it is important to consider each individual experiment and the question that is being asked. Often it may be better to use a high MOI but a short invasion time to synchronize invasion. Consider also whether it is better to use low MOI and centrifugation rather than high MOI and no centrifugation.

12. The antibody solution must cover the whole coverslip. At this step, as the cells have not yet been permeabilized, the surface is extremely hydrophobic and the antibody solution will tend to bead up and retract from the edges of the coverslip. If the antibodies are readily available, this step can be carried out using larger volumes, for example, 250–350 µL. Alternatively, the coverslips can each be placed "cell-side down" on a drop of antibody solution on some parafilm. For permeabilized cells (after detergent treatment) on 12-mm coverslips, 12–14 µL will form a meniscus and surface tension will keep it on the coverslip as long as there is no sudden movement. The coverslips should be incubated in a humid environment (a plastic box with wetted tissue in the bottom works fine), especially if incubations longer than 30 min are being carried out. It is essential that the coverslips do not dry out at any stage during the process or the background fluorescence will increase dramatically.

13. To reveal the host cells, a fluorescent marker such as phalloidin can be added at this step. Phalloidin is a toxin from the toadstool "Death Cap" (*Amanita phalloides*) that binds F-actin and is used to reveal the distribution of actin filaments in permeabilized cells.

14. For this protocol, two 1° and 2° antibodies are added sequentially: first 1° (e.g., rabbit polyclonal), followed by fluorescently conjugated first 2° antibody (e.g., anti-rabbit IgG), followed by second 1° (e.g., mouse monoclonal), followed by fluorescently conjugated second 2° (e.g., anti-mouse IgG). The first 1° and 2°

antibodies will stain only the extracellular bacteria as they are added before the cells are permeabilized. The second 1° and 2° antibody pair will stain *both* extracellular and intracellular bacteria as they are added after permeabilization. Thus, extracellular bacteria will be detected in two channels, whereas intracellular bacteria will be detected in only one. Partially internalized bacteria may be stained at only one end (the exposed end) with the "extracellular" marker.

Acknowledgments

The research in this laboratory is supported by the Intramural Research Program (DIR) of the NIAID and NIH.

References

1. Edwards, R. A., Olsen, G. J., and Maloy, S. R. (2002) Comparative genomics of closely related salmonellae. *Trends Microbiol.* **10**, 94–99.
2. Galan, J. E. (2001) *Salmonella* interactions with host cells: type III secretion at work. *Annu. Rev. Cell Dev. Biol.* **17**, 53–86.
3. Patel, J. C. and Galan J. E. (2005) Manipulation of the host actin cytoskeleton by *Salmonella*–all in the name of entry. *Curr. Opin. Microbiol.* **8**, 10–15.
4. Waterman, S. R. and Holden, D. W. (2003) Functions and effectors of the *Salmonella* pathogenicity island 2 type III secretion system. *Cell. Microbiol.* **5**, 501–511.
5. Drecktrah, D., Knodler, L. A., Galbraith, K., and Steele-Mortimer, O. (2005) The *Salmonella* SPI1 effector SopB stimulates nitric oxide production long after invasion. *Cell. Microbiol.* **7**, 105–113.
6. Drecktrah, D., Knodler, L. A., Ireland, R., and Steele-Mortimer, O. (2006) The mechanism of *Salmonella* entry determines the vacuolar environment and intracellular gene expression. *Traffic* **7**, 39–51.
7. Brown, N. F., Vallance, B. A., Coombes, B. K., Valdez, Y., Coburn, B. A., and Finlay, B. B. (2005) *Salmonella* pathogenicity island 2 is expressed prior to penetrating the intestine. *PLoS Pathog.* **1**, e32.
8. Coburn, B., Li, Y., Owen, D., Vallance, B. A., and Finlay, B. B. (2005) *Salmonella enterica* serovar Typhimurium pathogenicity island 2 is necessary for complete virulence in a mouse model of infectious enterocolitis. *Infect. Immun.* **73**, 3219–3227.
9. Ahmer, B. M, van Reeuwijk, J., Watson, P. R., Wallis, T. S., and Heffron, F. (1999) *Salmonella* SirA is a global regulator of genes mediating enteropathogenesis. *Mol. Microbiol.* **131**, 971–982.
10. Knodler, L. A., Celli, J., Hardt, W. D., Vallance, B. A., Yip, C., and Finlay, B. B. (2002) *Salmonella* effectors within a single pathogenicity island are differentially expressed and translocated by separate type III secretion systems. *Mol. Microbiol.* **43**, 1089–1103.

11. Hoiseth, S. K. and Stocker, B. A. (1981) Aromatic-dependent *Salmonella typhimurium* are non-virulent and effective as live vaccines. *Nature* **291**, 238–239.
12. Manoonkitiwongsa, P. S. and Schultz, R. L. (2002) Proper nomenclature of formaldehyde and paraformaldehyde fixatives for histochemistry. *Histochem. J.* **34**, 365–367.

17

Determining the Cellular Targets of Reactive Oxygen Species in *Borrelia burgdorferi*

Julie A. Boylan and Frank C. Gherardini

Summary

The response of *Borrelia burgdorferi* to the challenge of reactive oxygen species (ROS) is a direct result of its limited biosynthetic capabilities and lack of biologically significant levels of intracellular Fe. In other bacteria, the major target for oxidative damage is DNA as a consequence of the reaction of "free" intracellular with ROS through the Fenton reaction. Therefore, cellular defenses in these bacteria are focused on protecting this essential cellular component. This does not seem to be the case for *B. burgdorferi*. In this chapter, we describe methods that were used to analyze the potential targets for ROS in *B. burgdorferi*. Surprisingly, membrane lipids (e.g., linoleic and linolenic acids) derived from host are the major target of ROS in the Lyme disease spirochete.

Key Words: Oxidative stress; DNA damage; lipid damage.

1. Introduction

One imposing challenge faced by organisms living in an oxygen-rich environment is that of dealing with toxic products resulting from incomplete reduction of oxygen. This assault primarily comes from reactions involving endogenous metalloproteins, flavoproteins, and so on as a consequence of exposure to or metabolism of oxygen *(1)*. Pathogenic bacteria face a similar metabolic test but also are challenged by exogenous reactive oxygen generated by host immune cells for the express purpose of eliminating invading microbes. These compounds constitute a formidable challenge that must be overcome for bacterial pathogens to successfully survive, colonize, and cause disease in mammalian hosts.

From: *Methods in Molecular Biology, vol. 431: Bacterial Pathogenesis*
Edited by: F. DeLeo and M. Otto © Humana Press, Totowa, NJ

The biological targets for ROS are DNA, RNA, proteins, and lipids. In bacteria, it has been shown that the most damaging effects of ROS result from its interactions with reduced iron (and to a lesser extent, copper) centers in proteins. These initial oxidation reactions cause the release of Fe^{3+} from Fe-S proteins contributing to the "free" Fe pool. This Fe can then react with H_2O_2 and generate $OH^{\cdot-}$ through the Fenton reaction *(2)*. $OH^{\cdot-}$ reacts with most biomolecules and because it is so reactive, will react with them within a cell at diffusion-limited rates. The average diffusion distance is only a few nanometers, and thus, its effect on any given biomolecule will depend largely upon proximity to the target. Because Fe can localize along the phosphodiester backbone of nucleic acid, DNA is a major target of ROS. Active species can attack both the base and sugar moieties producing single- and double-strand breaks in the backbone and crosslinks to other molecules. As some of the base damage can result in miscoding, lesions formed by endogenous oxidants may be a significant or even preponderant source of spontaneous mutagenesis in aerobically growing cells. Because the reaction requires Fe^{3+}/Cu^{3+}, the amount of DNA damage that results from Fenton chemistry depends upon the metal metabolism of the bacteria. In the case of *Borrelia burgdorferi*, intracellular Fe concentrations are estimated to be <10 atoms per cell *(3)*. At those levels, it is unlikely that *B. burgdorferi* DNA is a target for ROS.

Lipids are a major target during oxidative stress in eukaryotes. Free radicals can attack polyunsaturated fatty acids in membranes and initiate lipid peroxidation. A primary effect of this is a decrease in membrane fluidity that affects the properties of the membrane and alters the function of membrane-associated proteins. Once lipid peroxides have formed, they react with adjacent polyunsaturated lipids causing an amplification of the damage. Lipid peroxides degrade into a variety of products, including aldehydes, which can subsequently damage membrane proteins and affect membrane fluidity. Unlike reactive free radicals, aldehydes are rather long-lived and can therefore diffuse from the site of origin and attack targets that are distant from the initial reaction. Until recently, it was assumed that bacterial lipids were not subject to oxidative damage observed in eukaryotic cells. Only certain polyunsaturated lipids, such as linoleic and linolenic acid, are subject to attack *(4),* and it is clear that most bacteria do not synthesize or contain these types of lipids in their membranes. Interestingly, linoleic acid and linolenic acid is incorporated into the membranes of some pathogenic bacteria including *B. burgdorferi* *(5,6)*. As these bacteria cannot synthesize their own lipids and must scavenge them, their membrane composition often reflects the host's lipid profile or that of their growth medium *(7–9)*. Approximately, 10% of *B. burgdorferi's* total lipid content is linoleic acid, suggesting that their membranes could undergo peroxidation when exposed to ROS.

In this chapter, we describe the methods used to determine the targets of ROS in *B. burgdorferi*. DNA damage is determined using spontaneous DNA mutation rate or by calculating the number of DNA lesions per unit DNA. The effect of ROS on lipids is determined using fluorescence microscopy and HPLC analysis.

2. Materials

1. BSK II: 9.8g/L 10× CMRL-1066 (Sigma, St. Louis, MO), 0.4% neopeptone, 0.16% yeastolate, 0.6% HEPES (Na), 0.4% dextrose, 0.153% sodium carbonate, 0.056% sodium citrate, 0.064% sodium pyruvate, 0.032% *N*-acetylglucosamine, 60 mL/L rabbit serum, and 4% BSA, pH 7.6.
2. Side-arm bottles and serum bottles, stoppers, and crimp tops.
3. P-BSK plating medium: 0.35% neopeptone, 0.42% HEPES, 0.05% sodium citrate, 0.35% glucose, 0.06% sodium pyruvate, 0.028% *N*-acetylglucosamine, 0.16% sodium bicarbonate, 0.18% Yeastolate, 35 g/L BSA, 7.5% 10× CMRL-1066, and 0.5% agarose, pH 7.5.
4. Anaerobic gas jar with an $H_2 + CO_2$ generator envelope.
5. HEPES/NaCl buffer: 20 mM NaCl and 50 mM HEPES, pH 7.6.
6. DNeasy tissue kit (Qiagen, Valencia, CA).
7. Expand Long Template PCR system (Roche, Applied Science, Indianapolis, IN).
8. Biotinylated aldehyde reactive probe (ARP) (Oxford Biomedical Research, Oxford, MI).
9. Tris–EDTA: 10 mM Tris–HCl, pH 7.5, and 1 mM EDTA.
10. Glycogen.
11. Ethanol.
12. Optically clear 96-well microplate.
13. TPBS: 137 mM NaCl, 2.7 mM KCl, 10mM Na_2HPO_4, 2 mM KH_2PO_4, and 0.5% Tween-20, pH 7.4.
14. HRP-streptavidin conjugate (Invitrogen, Carlsbad, CA).
15. Assay buffer: 0.15 M NaCl, 10 mM Na_2HPO_4, 1.5 mM KH_2PO_4, 2.5 mM KCl, 5 mg/mL BSA, and 0.1% Tween, pH 7.5.
16. TMB single solution (Invitrogen).
17. 1 M H_2SO_4.
18. ARP standards (Oxford Biomedical Research).
19. PKH26 Red Flourescent Cell Linker Kit (Sigma): PKH26 linker dye, diluent C.
20. 2.5 mM diphenyl-1-pyrenylphosphine (DPPP) stain, resuspend in DMSO.
21. Paraformaldehyde.
22. Microscope slides and cover slips.
23. 70% perchloric acid.
24. 6.1N trichloroacetic acid (TCA).
25. 97% tetraethoxypropane (MDA).
26. 98% thiobarbituric acid (TBA).
27. Methanol.
28. C18 4.6 × 150 mm HPLC column and guard.
29. Mobile phase 72:17:11 50 mM KH_2PO_4, methanol, and ACN.

3. Methods

The methods described outline treatment (1) of cells with oxidants, (2) identification of DNA damage, (3) quantitation of DNA damage, (4) identification of lipid damage, and (5) identification of lipid peroxidation intermediates.

3.1. Treatment of Borrelia Cells with Oxidants

1. Degas (1) BSK II liquid medium and (2) sterile side-arm and 25 mL serum bottles overnight under an atmosphere of 5% CO_2, 4% H_2, and 91% N_2 to achieve 0% oxygen.
2. Transfer 5 mL of anaerobic BSK II to a sealed 25-mL serum bottle and inoculate with 5×10^5 cells of low-passage *Borrelia*. Incubate at 34 °C for 4–6 days or until cells reach approximately 5×10^7 cells/mL. Enumerate cells using a dark-field microscope.
3. To expand the culture, transfer fresh anaerobic BSK II to a sealed side-arm bottle and inoculate to a final concentration of 1×10^5 cells/mL. Incubate at 34 °C for 2–3 days or until cells reach approximately 5×10^7 cells/mL.
4. To test the sensitivity of *Borrelia* to oxidants, such as *t*-butyl peroxide, split the culture into sterile, sealed side-arm bottles, and treat 1 bottle with 2 m*M* *t*-butyl peroxide. Incubate all bottles at 34 °C for 4 h.
5. During incubation, prepare P-BSK solid-plating medium and pour the 15-mL bottom layer.
6. After the incubation, prepare four serial 10-fold dilutions of the cultures into fresh BSK II.
7. For each dilution, add 100 µL of cells to 15 mL of the P-BSK-plating medium and pour onto the previously prepared bottom layers. Allow to solidify at room temperature.
8. Incubate in an anaerobic gas jar with an $H_2 + CO_2$ generator anaerobic envelope at 34 °C for 7–14 days.
9. To determine sensitivity, count the number of colonies on treated plates versus the number of colonies on untreated plates.

3.2. Identification of DNA Damage

This protocol is a quick way to determine whether DNA damage has occurred. If DNA damage is present, DNA polymerase will not be able to extend the reactions, and therefore, no product is generated. However, to quantitate the damage, the next protocol should be used.

1. Treat cells as described above.
2. After the incubation, return the cultures to the anaerobic chamber, and transfer to centrifuge bottles.
3. Harvest the cells at $3000 \times g$ for 20 min at 4 °C. In the anaerobic chamber, remove the supernatant and wash the pellets thrice in degassed HEPES/NaCl buffer.

4. Isolate total genomic DNA using DNeasy tissue kit (Qiagen) and quantify.
5. Use primers that will generate fragments 5–10 kb in length and perform PCR using an Expand Long Template PCR system (Roche). Briefly, the 25-μL PCR reaction mixture should contain 0.1 ng of genomic DNA as a template, 300 nM of each primer, 350 μM of dNTPs, 1× PCR buffer containing 0.175 mM MgCl$_2$, and 0.75 μL of DNA polymerase.
6. Separate products on a 1% agarose gel, stain with ethidium bromide, and visualize using UV.

3.3. Quantitation of DNA Damage

1. Treat cells and isolate DNA as described above.
2. Dilute DNA to 0.1 mg/mL in PBS.
3. Combine 5 μL of diluted DNA with 5 μL of 10 mM ARP reagent (Oxford Biomedical Research) and incubate for 1 h at 37 °C.
4. To precipitate the DNA, combine 88 μL of Tris–EDTA, 2 μL of 10 mg/mL glycogen, and the DNA-ARP mixture. Add 300 μL of cold ethanol and mix well. Cap and store at –20 °C for 10 min. Centrifuge the samples at 14,000 × g for 10 min and discard the supernatant. Wash the pellet thrice with 0.5 mL of 70% ethanol. Allow the pellet to dry for 3–5 min. Dissolve DNA pellet in Tris–EDTA. Determine the concentration using a spectrophotometer and add Tris–EDTA to make a final concentration of 0.5 μg/mL.
5. In triplicate, pipet 60 μL of sample into one well of a 96-well microplate. The microplate should be prewashed with TPBS 4×, followed by two washes with dH$_2$O.
6. Cover the plate and incubate at 37 °C overnight.
7. After the incubation, wash the plate 4× with TPBS.
8. Dilute the HRP-streptavidin conjugate (Invitrogen) to 0.5 μg/mL in assay buffer and pipet 100 μL to each well. Shake plate at 100 rev/min on a platform shaker for 1 h at room temperature.
9. Wash wells five times with TPBS.
10. Add 100 μL of TMB single solution (Invitrogen) to each well and incubate for 1 h at 37 °C.
11. To quench, add 100 μL of 1 M H$_2$SO$_4$ and read at 450 nm. Quantitate using ARP standards (Oxford Biomedical Research). Graph the standard curve by plotting ARP versus OD.

3.4. Identification of Lipid Damage

1. Grow and treat cells as described above.
2. Centrifuge 2 × 10^7 cells at 1000 × g for 5 min at room temperature.
3. Wash cells 2× with HEPES/NaCl buffer.
4. Resuspend cells in 1 mL of diluent C.
5. Add 4 μL of red flourescent dye to 1 mL diluent C.
6. Add 1 mL of cell suspension to 1 mL of diluted dye. Incubate for 5 min at room temperature.

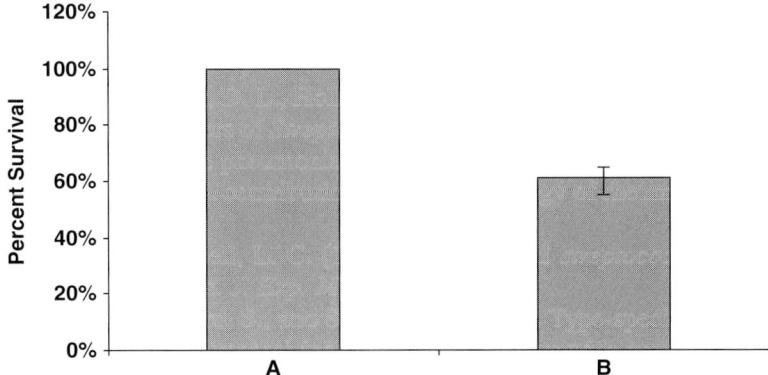

Fig. 1. Percent survival. *Borrelia burgdorferi* strain B31A3 was grown to a cell density of 5×10^7 cells/mL in BSK II under anaerobic conditions, the culture divided in half and each was treated with 0 (**A**) or 2 m*M* (**B**) *t*-butyl peroxide at 34 °C for 4 h. After treatment, cells were diluted in fresh BSK II, plated on BSKII plating media, and incubated for 7–14 days at 34 °C in an anaerobic gas jar. Percent survivability was calculated as the number of colonies on the treated plates versus the number of colonies on the untreated plates.

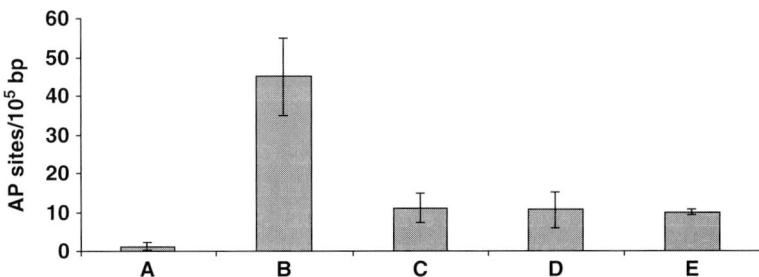

Fig. 2. Quantitation of DNA damage. *Borrelia burgdorferi* B31A cells were grown in BSK-II under anaerobic conditions to a cell density of 5×10^7 cells/mL, treated with 0, 10, or 50 m*M* *t*-butyl peroxide (**C**, **D**, and **E**, respectively) for 4 h and DNA isolated. The DNA was then mixed with an aldehyde-reactive probe (Oxford Biomedical Research, Inc) labeled with biotin and detected with an HRP-streptavidin conjugate. The color development was monitored at 450 nm. The number of aldehyde-reactive probe (DNA base lesions)/10^5 bp DNA was determined using a standard curve. *Escherichia coli* TA4315 cells were grown in minimal media to OD_{600} of 0.4, treated with 0 or 100 m*M* H_2O_2 for 30 min (**A** and **B**, respectively) and DNA isolated. The number of base lesions was determined as described above.

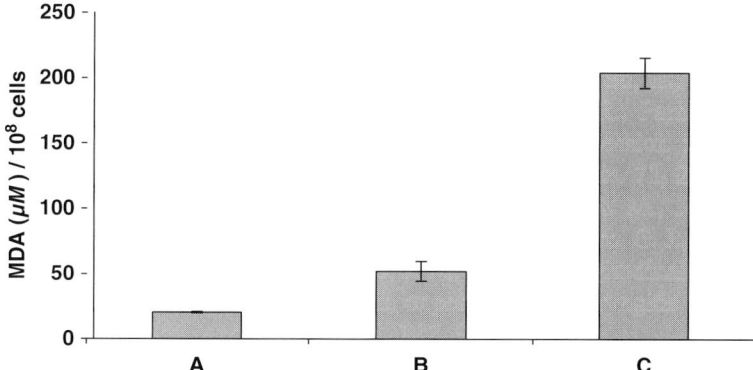

Fig. 3. Quantitation of lipid damage. *Borrelia burgdorferi* strain B31A3 was grown to a cell density of 5×10^7 cells/mL in BSK II under anaerobic, microaerophilic, or aerobic conditions (**A, B** and **C**, respectively) at 34 °C for 3–5 days. Cells were harvested, lysed, and reacted with TBA. The resulting thiobarbituric acid reacting substances (TBARS) are then injected onto a C18 HPLC column and the absorbance at 532 nm monitored. Standards made from pure MDA were used as controls. The data is calculated as microliters MDA per 10^8 cells.

7. Add 2 mL rabbit serum to stop the reaction. Incubate for 1 min at room temperature.
8. Add 4 mL HEPES/NaCl buffer.
9. Centrifuge mixture $1000 \times g$ for 10 min at room temperature. Wash cells thrice with HEPES/NaCl buffer.
10. Resuspend cells in 1 mL of HEPES/NaCl buffer and incubate at 34 °C for 5 min.
11. Add 30 μL of 2.5 m*M* DPPP.
12. Incubate for 5 min at 34 °C in the dark.
13. Centrifuge the mixture $1000 \times g$ for 10 min at room temperature. Wash pellet thrice with HEPES/NaCl buffer.
14. Resuspend cells in 100 μL of HEPES/NaCl buffer.
15. Before microscopy, the cells must be fixed by adding paraformaldehyde to 2%.
16. Pipet 5 μL of fixed cells onto a microscope slide.
17. Add 5 μL of embedding solution. Cover with coverslip.
18. To visualize

   a. Red flourescent cell linker, excitation $\lambda = 551$ nm and emission $\lambda = 567$ nm.
   b. DPPP, excitation $\lambda = 351$ nm and emission $\lambda = 380$ nm.

3.5. Identification of Lipid Peroxidation Intermediates

1. Grow and treat cells as described above.
2. Centrifuge 10^{10} cells at $3000 \times g$ for 15 min. Wash the cells thrice in HEPES/NaCl buffer. Resuspend pellet in 50 μL HEPES/NaCl buffer.

Table 1
Preparation of MDA Standards

Concentration	Third dilution	dH$_2$O
0.1 μM	1.23 μL	498.8 μL
1.0 μM	12.3 μL	481.7 μL
5.0 μM	61.7 μL	438.3 μL
10 μM	123 μL	377 μL

3. Prepare MDA standards: Dilute the stock MDA by performing three 100-fold serial dilutions in methanol (*see* **Table 1**).
4. Combine 50 μL of sample/standard, 150 μL of 0.1N PCA, and 150 μL of 40 mM TBA in a 1.5-mL tube.
5. Vortex and incubate at 97 °C for 60 min in a fume hood.
6. Incubate at –20 °C for 20 min to cool the reaction.
7. Add 300 μL of methanol and 100 μL of 20% TCA.
8. Vortex and centrifuge at 13,000 × g for 6 min.
9. Inject 10 μL onto C18 4.6 × 150-mm column with mobile phase 72:17:11 and elute isocratically at 1 mL/min over 10 min. Monitor absorbance at 532 nm.

Acknowledgments

This research was supported by the Intramural Research Program of the National Institutes of Health, National Institute of Allergy and Infectious Diseases.

References

1. Storz, G. and Imlay, J. A. (1999) Oxidative stress. *Curr. Opin. Microbiol.* **2,** 188–94.
2. Imlay, J. A. (2003). Pathways of oxidative damage. *Annu. Rev. Microbiol.* **57,** 395–418.
3. Posey, J. E., Hardham, J. M., Norris, S. J., and Gherardini, F. C. (1999). Characterization of a manganese-dependent regulatory protein, TroR, from *Treponema pallidum. Proc. Natl. Acad. Sci. U.S.A.* **96,** 10887–10892.
4. Gutteridge, J. M. and Halliwell, B. (1990) The measurement and mechanism of lipid peroxidation in biological systems. *Trends Biochem. Sci.* **15,** 129–135.
5. Beermann, C., et al., (2000) The lipid component of lipoproteins from *Borrelia burgdorferi*: structural analysis, antigenicity, and presentation via human dendritic cells. *Biochem. Biophys. Res. Commun.* **267,** 897–905.
6. Ben-Menachem, G., et al. (2003) A newly discovered cholesteryl galactoside from *Borrelia burgdorferi. Proc. Natl. Acad. Sci. USA* **100,** 7913–7918.

7. Barbour, A.G. (1984) Isolation and cultivation of Lyme disease spirochetes. *Yale J. Biol. Med.* **57,** 521–525.

8. Fraser, C. M., et al. (1997) Genomic sequence of a Lyme disease spirochaete, *Borrelia burgdorferi. Nature* **390,** 580–586.

9. Matthews, H. M., Yang, T. K., and Jenkin, H. M. (1979) Unique lipid composition of *Treponema pallidum* (Nichols virulent strain). *Infect. Immun.* **24,** 713–719.

III

ANIMAL MODELS OF BACTERIAL INFECTION

18

Bioluminescent Imaging of Bacterial Biofilm Infections In Vivo

Jagath L. Kadurugamuwa and Kevin P. Francis

Summary

Whole body biophotonic imaging (BPI) is a technique that has contributed significantly to the way researchers study bacterial pathogens and develop pre-clinical treatments to combat their ensuing infections *in vivo*. Not only does this approach allow disease profiles and drug efficacy studies to be conducted non-destructively in live animals over the entire course of the disease, but in many cases, it enables investigators to observe disease profiles that could otherwise easily be missed using conventional methodologies. The principles of this technique are that bacterial pathogens engineered to express bioluminescence (visible light) can be readily monitored from outside of the living animal using specialized low-light imaging equipment, enabling their movement, expansion and treatment to be seen completely non-invasively. Moreover, because the same group of animals can be imaged at each time-point throughout the study, the overall number of animals used is dramatically reduced, saving lives, time, and money. Also, as each animal acts as its own control over time, the issues associated with animal-to-animal variation are circumvented, thus improving the quality of the biostatistical data generated. The ability to monitor infections *in vivo* in a longitudinal fashion is especially appealing to assess chronic infections such as those involving implanted devices. Typically, bacteria grow as biofilms on these foreign bodies and are reputably difficult to monitor with conventional methods. Because of the non-destructive and non-invasive nature of BPI, the procedure can be performed repeatedly in the same animal, allowing the biofilm to be studied in situ without detachment or disturbance. This ability not only allows unique patterns of disease relapse to be seen following termination of antibiotic therapy but also *in vivo* resistance development during prolonged treatment, both of which are common occurrences with device-related infections. This chapter describes the bioluminescent engineering of both Gram-positive and Gram-negative bacteria and overviews their use in device-associated infections in several anatomical sites in a variety of animal models.

From: *Methods in Molecular Biology, vol. 431: Bacterial Pathogenesis*
Edited by: F. DeLeo and M. Otto © Humana Press, Totowa, NJ

Key Words: Bioluminescent imaging; biophotonic imaging; animal modeling; *in vivo* methods; UTI; cvs; catheter; biofilm; chronic; *lux*; luciferase; antibiotic; real time; non-invasive.

1. Introduction

In 1995, Contag et al. *(1)* first demonstrated the ability to monitor biolumines-cently engineered bacteria in live mice using whole body biophotonic imaging (BPI). Subsequent to this publication, a range of both Gram-positive and Gram-negative bacteria were bioluminescently transformed and tested *in vivo* in a number of standard animal models *(2–5)*, so establishing this technique as a platform methodology for basic research and drug development. Most of these initial studies were focused upon acute models of infection such as soft tissue, sepsis, and pneumonia *(3,4,6)*. It was quickly realized, however, that this technique was especially useful for monitoring chronic and complex infections, such as biofilm infections *(7–13)*. These biofilm methodologies are the key focus of this chapter.

Modern medical and surgical practice has come to rely increasingly on various types of prosthetic and implanted devices. Despite sterilization, aseptic procedures, and improvements in the materials and design of devices, microbial colonization and biofilm formation on these devices is a common occur-rence. These biofilm infections represent a serious medical problem, as mature biofilms are resilient to the host immune response and conventional antibiotic treatments. Most often such complications lead to failure of the implant and the need for its complete removal *(14,15)*. To better understand and control biofilm infections on indwelling medical devices, new approaches are essential, especially in *in vivo* settings. To this end, BPI methodologies have been developed to allow rapid, continuous real-time monitoring of biofilm infections in a number of different animal models *(7–10,12,13)*. The use of bioluminescent reporters to tag bacterial pathogens allows for the sensitive detection of only live, metabolically active cells by BPI. Because of its non-destructive and non-invasive nature, the imaging procedure can be performed repeatedly without disturbing the integrity of the biofilm structure, as is the case with procedure that involves detachment of the bacteria. Additionally, BPI experiments can be performed without the loss of contextual influences of the animal's host defense mechanisms.

Many of the biofilm models described in this chapter resemble characteristics commonly found in human foreign-body infections, such as the establishment of infection with low inoculum, persistence of infection without spontaneous healing, and relapse of disease soon after treatment. This sensitive and quanti-tative longitudinal monitoring approach is ideally suited to assess disease progression and allows both response and/or relapse to antimicrobial agents to be seen directly from intact live animals in real time, offering an excellent preclinical strategy to assess disease response. Furthermore, this approach is ideally suited to assess novel therapeutic agents or medical devices. This will accelerate efforts to optimize lead compounds, especially at the animal modeling step with respect to formulation, dose, or optimizing materials used in medical implants.

2. Materials

1. Strains of bacteria: *Staphylococcus aureus* strain ATCC (American Type Culture Collection, Manassas, VA) 12600 capsule type 3 strain, *Pseudomonas aeruginosa* strain ATCC 19660, *Proteus mirabilis* strain ATCC 51286, and *Escherichia coli* strain S17-1 λ *pir* pUT Amp^R mini-Tn5 *luxCDABE* Tc^R (M.K. Winson, University of Nottingham, UK).
2. Plasmid pXen-5 (Xenogen Corp, Alameda, CA).
3. Ampicillin, tetracycline, erythromycin, and kanamycin (Sigma-Aldrich, St. Louis, MO).
4. MgSO$_4$, glycerol, glucose, Luria-Betani (LB) broth, trypticase soy broth (TSB) (Difco, Detroit, MI), and brain heart infusion (BHI) media (Becton Dickinson, Franklin Lakes, NJ).
5. Ketamine and xylazine (Burns Vet Supply, Vancouver, WA).
6. GenePulser II and cuvettes (Bio-Rad, Richmond, CA).
7. Teflon intravenous catheter, 14-gauge (Abbocath-T; Burns Vet Supply).
8. Polyethylene PE 50 (OD 0.965 mm, ID 0.58 mm; Braintree Scientific, Inc, Braintree, MA).
9. IVIS® Imaging System and LivingImage® software (Xenogen Corp).
10. Ultrasonic bath (VWR, San Francisco, CA).
11. Surgical staples, scissors, and blades (SPI Supplies, West Chester, PA).
12. Balb/C, CF-1, and CD-1 female mice weighing 18–30 g (Charles River, Wilmington, MA).

3. Methods
3.1. Generation of Bioluminescent Bacteria

To maintain bioluminescence stably, all bacteria used for BPI studies were transformed so that an optimized *Photorhabdus luminescense lux* operon is integrated into their chromosome. These operons encode the genes necessary for both luciferase and substrate production, enabling bioluminescence to be produced by the transformants without the addition of exogenous products. Two methodologies for stably transforming the particular Gram-negative and Gram-positive bacteria used in the biofilm studies to a bioluminescent phenotype are described below.

3.2. Transformation Procedure for Gram-negative Bacteria (General Methodology)

Gram-negative bacteria are made bioluminescent (e.g., *P. aeruginosa* Xen-5 and *P. mirabilis* Xen-44) through the stable integration of an unaltered *P. luminescense luxCDABE* cassette onto their chromosome as first described by Winson ct al. *(16)*.

1. Mate the Gram-negative strain to be transformed (the recipient) with the *E. coli* strain S17-1 λ *pir* pUT Amp^R mini-Tn5 *luxCDABE* Tc^R(the donor) (*see* **Note 1**).
2. To allow the recipient strain to be more readily distinguished from the donor strain, initially transform the recipient with a blank vector carrying a unique antibiotic resistance gene (primary resistance marker) using standard electroporation procedures.
3. Use a single resistant transformant (carrying the primary resistance marker) to inoculate 10 mL of LB containing the primary selective antibiotic and incubate overnight at 37 °C. Similarly, use a single colony of *E. coli* S17-1 λ *pir* pUT mini-Tn5 *luxCDABE* Tc^R to inoculate 10 mL of LB containing 100 µg/mL ampicillin.
4. Individually, pellet and resuspend both bacterial cultures in 10 mL of LB.
5. Inoculate a fresh 10-mL volume of LB containing 10 m*M* MgSO$_4$ with 100 µL volumes of each of the suspensions and incubate this mixture overnight at 37 °C to allow mating to occur.
6. Plate 100 µL volumes of the mating onto LB containing both the primary selective antibiotic and 20 µg/mL tetracycline.
7. After an overnight incubation at 37 °C, screen the plates for bioluminescent transposants using a highly sensitive optical imaging system, such as an IVIS® Imaging System.
8. Select a single highly bioluminescent colony and characterize to ensure that it is stable [e.g., bioluminescent photon counts correspond to colony-forming units (CFUs)] and behaves similarly to the parental strain, both *in vitro* and *in vivo* (e.g., not attenuated).

3.3. Transformation Procedure for Gram-positive Bacteria (Staphylococcus aureus Methodology)

Gram-positive bacteria are made bioluminescent (e.g., *S. aureus* Xen-29) through the stable integration of an optimized *P. luminescense luxABCDE* cassette onto their chromosome as first described by Francis et al. *(3)*. This broad host range plasmid can be used to transform a variety of Gram-positive bacteria to a stable bioluminescent phenotype. However, the procedure described below is optimized for the transformation of *S. aureus*.

1. Grow-up an overnight culture of *S. aureus* in 10 mL of BHI.
2. Dilute the overnight culture 1:100 in 50 mL of fresh BHI broth in a 500-mL flask and incubate at 37 °C in an orbital shaker at 200 rpm until the OD$_{600}$ reached 0.8 (exponential growth phase).

3. Centrifuge and then sequentially wash cells with 50 mL, 25 mL and 5 mL ice-cold 10% glycerol solution (centrifuging between washes).

4. Resuspend the electrocompetent cells in 500 μL of ice-cold 10% glycerol.

5. Add one 1 μL of 1 μg/μL plasmid pXen-5 (pAUL-A Tn*4001 luxABCDE* Km^R) to 200 μL of electrocompetent *S. aureus* and mix (*see* **Note 2**).

6. Transfer the bacteria/plasmid mixture into a 1-cm ice-cold cuvette (Bio-Rad) and electroporate at 25 μF, 100 Ω, and 2.5 kV (GenePulser II; Bio-Rad).

7. Immediately add 0.8 mL of BHI, gently invert cuvette and incubate statically at 37 °C for 1h.

8. Plate 200 μL volumes of the transformation mix on BHI agar plates containing 0.3 μg/mL of erythromycin and incubate plates at 37 °C for 24 h.

9. Patch colonies (~1 cm²) onto BHI agar plates containing 0.3 μg/mL of erythromycin and incubated at 37 °C overnight.

10. Uniformly streak a quantity of each patch (10 μL loopful of cell growth consisting of ~10^8–10^9 cells) over the entire area of a BHI agar plate containing 200 μg/mL of kanamycin and incubated plates at 37 °C for 24 h.

11. Screen plates for bioluminescent transposants using a highly sensitive low-light optical imaging system.

12. Select a single highly bioluminescent colony and characterize to ensure that it is stable (e.g., bioluminescent photon counts correspond to CFUs) and behaves similarly to the parental strain, both *in vitro* and *in vivo* (e.g., not attenuated).

3.4. In vitro *Catheter-Associated Biofilm*

1. Grow the bioluminescent biofilm-forming bacterial strain overnight in TSB supplemented with 0.25% glucose (TSBG) in an orbital shaker at 37 °C.

2. Dilute the culture 1:10 in fresh TSBG and further incubate for 1.5 h at 37 °C.

3. Adjust OD at 600 nm to reach a standardized cell suspension of 10^5 CFUs/mL in TSBG (*see* **Note 3**).

4. Cut teflon intravenous catheter 14-gauge into 1-cm segments (soft tissue model), polyethylene catheter PE 50 into 6-mm segments (UTI) and sterilize each piece with 70% ethanol. Air-dry catheters overnight.

5. Develop bacterial biofilms on the cut, sterile catheters by placing individual segments into Eppendorf tubes containing 1.0 mL of bacterial cell suspension in TSBG with 10^5 CFUs/mL in the exponential phase of growth.

6. Vortex well to ensure the removal of air bubbles from inside catheter, allowing it to remain immersed during incubation (*see* **Note 4** for incubation times for each pathogen).

7. After incubating for appropriate time intervals at 37 °C recover colonized catheters aseptically and rinse once gently in TSBG to remove unbound bacteria. Flush out lumen of catheter with sterile TSBG. Image catheters for bioluminescence prior to implant to ensure proper colonization.

3.5. Soft-tissue Biofilm Infection

This device-related infection is established in the flanks of mice by subcutaneous implantation *(17)* of either pre- or *in vivo*-colonized catheter material with biofilm forming Gram-positive or Gram-negative bioluminescent bacteria *(7)*. The model is ideally suited for screening the development of biofilms on medical device material and the effects of therapeutic agents to treat them non-invasively without the need for exogenous sampling or culturing of the pathogen (*see* **Fig. 1**). Infections could be established with clinically relevant doses as low as 10^3 CFUs pathogen per device with either Gram-positive or Gram-negative pathogens. Moreover, subsequent disease progression could be followed for weeks within the same animal.

This model is especially appealing for the analysis of *in vivo* efficacy of antibiotics against young as well as mature biofilms, post antibiotic effects, *in vivo* resistance development, or *in vivo* fitness of resistant organisms *(8,9,11,13)*. As the imaging procedure can be repeatedly performed over time without killing the animal or removal of device for detachment of biofilm (*see* **Fig. 2** and Color Plate 7, following p. 46), this model, with several of the clinically relevant features, has proven to have significant advantages over conventional methods for studying foreign-body-related infections and treatment *in vivo*.

3.6. Experimental Model of Infection

1. Anesthetize Balb/C female mice weighing 18–22 g with Ketamine (100 mg/kg) and Xylazine (5 mg/kg).
2. Shave their flanks and clean the skin with Betadyne and alcohol.
3. Make a 4–5 mm skin incision (as small as possible) and dissect to create a subcutaneous tunnel.
4. Push a 1-cm segment of intravenous catheter with its associated bioluminescent bacterial biofilm (e.g., Xen) through the incision into the tunnel (subcutaneous tunnel usually created is ~1.5 cm in length). One catheter segment is inserted on each side of each animal.
5. Cover the incisions with intact skin, close with surgical staples, and disinfect the skin.
6. Following surgery, place the mice on a warming pad until fully recovered.
7. Image animals using a highly sensitive low-light optical imaging system. (see imaging procedure 3.12).
8. If post-implant infection is preferred, implant segments of sterile catheters as above, then approximately 1 h after the implantation procedure, introduce a defined quantities of bacterial suspension in 50 µL of PBS by injection into the catheter lumen (10^3–10^5 CFUs/catheter) via a 31-gauge needle.

Delaying the infection until after device implantation is useful to study factors such as post-operative colonization of implanted device or the role of host

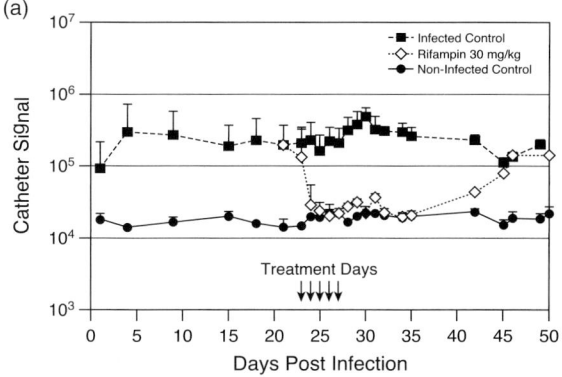

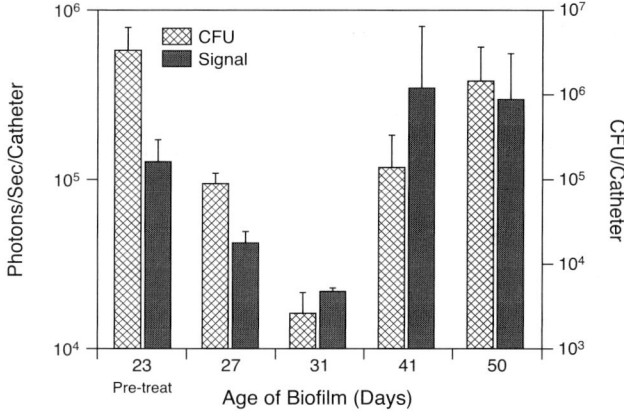

Fig. 1. (A) *In vivo* bioluminescence monitoring of *S. aureus* Xen29 in the mouse model of biofilm infection during treatment with rifampin. Effect of 4 days of rifampin treatment on a 23-day-old biofilm. Total number of photons detected per second over the infected catheter was determined using an IVIS® Imaging System and LivingImage® software, plotted with respect to time. Each mouse was implanted with two catheters and each data point is the mean ± SD of 2–3 mice. Arrow indicates the day of antibiotic administration. Data are averages of results from three to four separate experiments. (**B**) Correlation between bioluminescence and traditional colony-forming units (CFUs) for 4-day rifampin treatment on a 23-day old biofilm. Viable counts are reported as CFUs per catheter, and bioluminescence is represented as Photons/Sec/Catheter using IVIS® Imaging System. Each data point is the mean± SE of three to four catheters. Mice were first imaged for bioluminescence followed by removing and extracting the bacteria from catheters for CFU determinations. Reproduced with permission from ASM Publication Department.

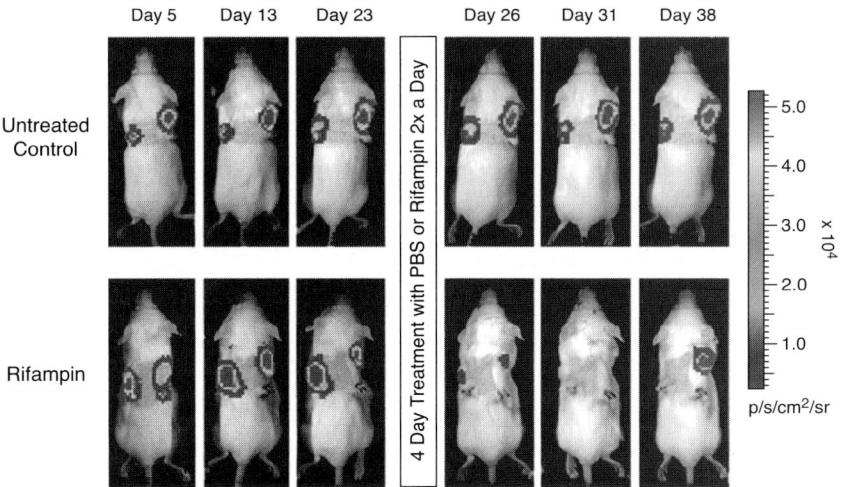

Fig. 2. Real-time *in vivo* bioluminescence monitoring of *S. aureus* Xen-29 in the mouse model of soft-tissue biofilm infection during treatment with rifampin. Effect of 4 days of rifampin treatment (twice a day with 30 mg/kg) on a 3-week-old biofilm. A representative animal from the group receiving antibiotic or the untreated infected group is shown. Note the response to treatment including relapse that can be monitored non-invasively within the same animal throughout the study period. (*See* Color Plate 7, following p. 46.)

proteins in promoting bacterial adhesion to biomaterials in initial phase of blood–material interaction. An inoculum of $\sim 10^3$–10^5 CFUs of *S. aureus* Xen-29 and *P. aeruginosa* Xen-5 per catheter results in a reproducible, localized persistent infection surrounding the catheter until the termination of the experiment on day 25–50. Doses above 10^6 CFUs/catheter for *P. aeruginosa* and 10^8 CFUs/catheter for *S. aureus* result in 100% mortality. An inoculum of $\leq 10^5$/catheter produces chronic infections.

3.7. Catheter-Associated Urinary Tract Infections

UTIs are one of the most common bacterial infections in humans. Most of these infections follow instrumentation of the urinary tract, mainly urinary catheterization (*see* **Fig. 3**), with the development of catheter-associated bacteriuria directly related to the duration of catheterization (*18*). BPI easily allows spatial information to be monitored sequentially throughout the entire disease process, from cystitis to ascending UTIs (*see* **Fig. 4** and Color Plate 8, following p. 46), as well as treatment efficacy and relapse in diseased or asymptomatic animals all without exogenous sampling (*10*).

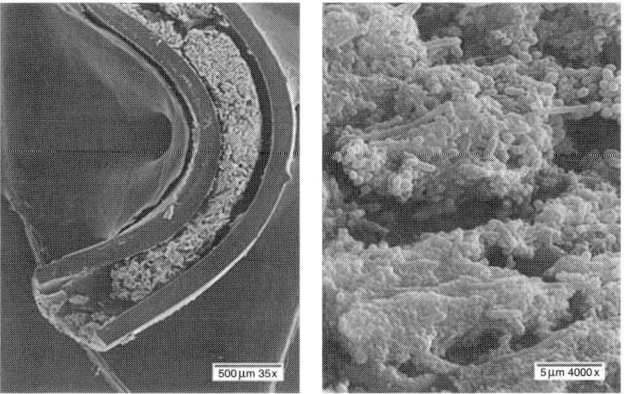

Fig. 3. Scanning electron micrograph of catheter from the bladder of a mouse infected with *P. mirabilis* Xen 44. Cross-section of catheter at low magnification (left) shows the lumen of catheter filled with a thick biofilm, and numerous rod-shaped bacteria embedded within the polymeric substance is clearly seen at higher magnification (right).

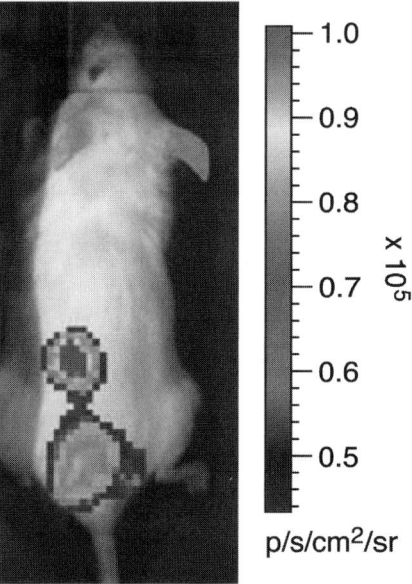

Fig. 4. *In vivo* bioluminescent monitoring of the progression of *P. mirabilis* Xen 44 in the mouse model of catheter-induced urinary track infection (UTI). The ascending nature of uropathogen from the bladder up the ureters and to the kidney can be clearly visualized from outside intact live animals using luciferase tag pathogen and bioluminescence imaging. (*See* Color Plate 8, following p. 46.)

3.8. Experimental Model of Infection

The experimental catheter-associated UTI in a rat model developed by Kurosaka et al. *(19)* was established in mice with modifications for BPI.

1. Cut polyethylene catheter material PE 50 into 6-mm long segments and fit onto a flexible metal wire.
2. Wrap this around a similar metal wire in a spiral and place in boiling water for 1 min.
3. Remove the segment and sterilize each piece with 70% ethanol and air dry overnight.
4. Bacterial biofilm is grown on the catheter material as described above in **Subheading 3.4**.
5. Cut the metal stylet of a 24-gauge Teflon catheter up to 5 cm and fit either sterile PE tubing or tubing that has been colonized with bacteria.
6. Anesthetize CF-1 female mice weighing 26–30 g with 2.5% isoflurane gas and place in the supine position.
7. Clean the periurethral area with 70% ethanol.
8. Grasp the papilla with light pressure using forceps and lift away from the abdomen.
9. While maintaining the papilla in this position, place the end of the stylet bearing the 6-mm-long tubing into the urethral opening and advance until the leading end is in the bladder.
10. Push the 5-cm Teflon catheter until the shorter segment is pushed off the stylet, thereby leaving the 6-mm-long PE tubing free in the bladder lumen (the PE tubing will revert spontaneously to its spiral form thus remain in place until the end of the monitoring period).
11. Remove the Teflon catheter with the stylet.
12. Image animals using a highly sensitive low-light optical imaging system (see imaging procedure below).
13. Remove Teflon catheter post-inoculation.

To induce post-catheterize infection, place a sterile tubing in the bladder as described above and inoculate the bladder with defined quantity of pathogen in 50 μL phosphate-buffered saline over 30 s by insertion of Teflon catheter into the bladder through the urethra (*see* **Note 5** for challenge doses required to establish chronic UTI for each pathogen).

3.9. Central Venous Catheter-Associated Infection in Mice

Long-term implantation of intravascular devices is one of the major advances in modern medical and surgical practice. A common complication with these implanted foreign bodies is their susceptibility to pyogenic infections, especially due to *Staphylococci* *(20)*. Establishment of such infections is due to either perioperative bacterial contamination or postoperative colonization of the implanted device through local or hematogenous seeding *(21)*. Little is known about the pathogenesis and treatment of this condition, mainly due to lack of in situ methods to monitor such infections *in vivo* in real time. Thus, we have applied the *in vivo* imaging technology to monitor and visualize the establishment and spread of staphylococcal infection non-invasively in mice with indwelling jugular vein catheter (*see* **Fig. 5** and Color Plate 9, following p. 46).

3.10. Experimental Model of Infection

To avoid high operative mortality rate associated with small animal vascular catheterization, it is advisable to purchase animals that are surgically prepared and ready to use from a commercial vendor such as Charles River Laboratories.

1. After 5–7 days post-catheterization, infect the catheterized mice intravenously with 50 µL of bioluminescent bacterial suspension (e.g., *S. aureus* Xen 29) in exponential growth phase in PBS through the tail vein with inocula of 10^4CFUs.
2. Inoculate a group of non-catheterized mice with a similar dose as control group.
3. Image animals using a low-light imaging system (see imaging procedure below).
4. To optimize capture of bioluminescence, shave over the heart (ventral), as well as kidneys and spleen areas (dorsal).

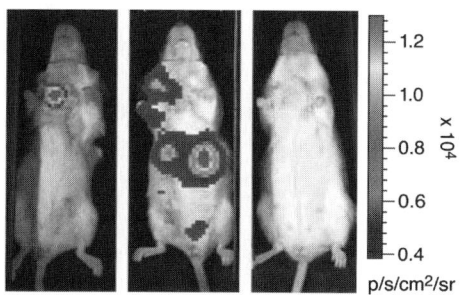

Fig. 5. Daily imaging of mice inoculated by tail vein injection with *S, aureus* Xen 29 reveals the colonization of *S. aureus* around the indwelling catheter in jugular vein catheterized mice as early as one day after infection (left). Six days later, evidence of metastatic disease with staphylococci spreading to heart, kidney, and liver as a result of seeding of bacteria from the infected vascular catheter (center). In contrast, whole-body imaging of non-catheterized mice challenged with a similar dose of bacterial suspension showed no sign of infection around heart or any other organ, even after a week into the hematogenously induced infection (right). This shows the implanted device that constitutes a particularly attractive surface for bacterial colonization. The difference in the degree of infection seen between catheterized and non-catheterized mice clearly indicates the importance of the indwelling vascular catheter in potentiating the establishment of infection. (*See* Color Plate 9, following p. 46.)

3.11. Separation of Bacteria From Catheter

1. After the final imaging time-point, kill mice humanely (in accordance with the guidelines of the Institutional Animal Care and Use Committee) and surgically remove the infected catheters for enumeration of bacteria by both bioluminescence imaging and conventional plate count method.
2. Image the harvested catheters for bioluminescence.
3. Transfer the catheters to tubes containing 1 mL of TSB, place tubes in an ultrasonic bath (VWR), and sonicate for 5–10 min (38.5–40.5 kHz).
4. Vortex the tubes for 1 min to remove the biofilm bacteria from the support surface and image catheters to confirm the loss of bioluminescence and thus the complete removal of the biofilm.
5. Repeat sonicating-vortexing procedure if necessary. Note, repeating the procedure for >10 min may affect the viability of certain pathogens.
6. Dilute the suspension of bacteria removed form the catheter, plate on trypticase soy agar, and incubate at 37 °C for colony counting.

Correlation between CFUs and bioluminescent signal (Photons/Sec) is determined by plotting Photons/Sec/Catheter versus CFUs/Catheter.

3.12. Imaging

1. Prior to and during imaging, mice are anesthetized with 2–2.5% isoflurane gas.
2. The whole animal is imaged (dorsal and ventral) for a maximum of 5 min at various times during the experiment using a highly sensitive optical imaging system, such as an IVIS® Imaging System.
3. Quantify total photon emission from defined regions of interest (e.g., bladder, kidney, flanks, abdomen, or heart) within the images of each animal (e.g., using Xenogen's LivingImage® software).
4. Following imaging, return mice to their cages until the next imaging session.

4. Notes

1. pUT is a conjugative suicide vector that is only maintained in strains of *E. coli* expressing the π protein, such as S17-1 λ *pir* and CC118 λ *pir* (*22,23*). Thus, recipient bacteria that do not express the π protein can only become stably bioluminescent and tetracycline resistant if the mini-Tn5 *luxCDABE* Tc^R cassette transposes onto their chromosome. Furthermore, as the *lux* operon does not have its own promoter, the strength of bioluminescence in the recipient bacterium is dependent upon the endogenous promoter directly upstream of the site at which the *lux* transposon integrates. Derivatives of this donor strain carrying alternative antibiotic markers are also available (e.g., chloramphenicol, kanamycin, and streptomycin).
2. The promoterless *luxABCDE* km^R operon is optimized from Gram-positive expression and is silent on pXen-5. When transposition occurs, the strength of

bioluminescence and the degree of kanamycin resistance in the bacterium is dependent upon the endogenous promoter directly upstream of the site at which the *lux* transposon integrates.

3. The following conversion factors are useful in adjusting cell suspensions to the desired levels: *S. aureus* Xen 29 OD_{600} 1.0 = 2.8 × 10^8 CFUs/mL, *P. aeruginosa* Xen 5 OD_{600} 1.0 = 7.5 × 10^8, and *P. mirabilis* Xen 44 OD_{600} 1.0 = 5.4 × 10^8 CFUs/mL.

4. Incubation time for *S. aureus* Xen 29 to reach 10^3 CFUs/Catheter = 30 min, 10^4 CFUs/Catheter = 1 h, and 10^5 CFUs/Catheter = 2 h in 10^5 CFUs/mL of suspension. For *P. aeruginosa* Xen 5 to reach 10^3 CFUs/Catheter = 10 min, 10^4 CFUs/Catheter = 1.5 h, and 10^5 CFUs/Catheter = 3 h in 10^5 CFUs/mL of suspension. The above values are valid for 14-gauge, 1-cm segments of Teflon intravenous catheter. The concentration of cell suspension and time required to reach desired inoculum varies with surface area and the type of material.

5. The challenge doses required to establish chronic UTI with *P. aeruginosa* Xen 5 is 1.6 × 10^6 CFUs/catheter or 2.0 × 10^6 CFUs/50-μL of suspension and with *P. mirabilis* Xen 44, 7.1 × 10^3 CFUs/catheter or 1.3 × 10^4 CFUs/50-μL of suspension.

References

1. Contag, C. H., Contag, P. R., Mullins, J. I., Spilman, S. D., Stevenson, D. K., and Benaron, D. A.. (1995) Photonic detection of bacterial pathogens in living hosts. *Mol. Microbiol.* **18**, 593–603.
2. Burns-Guydish, S. M., Olomu, I. N., Zhao, H., Wong, R. J., Stevenson, D. K., and Contag, C. H. (2005) Monitoring age-related susceptibility of young mice to oral *Salmonella enterica* serovar Typhimurium infection using an *in vivo* murine model. *Pediatr. Res.* **58**, 153–158.
3. Francis, K. P., Yu, J., Bellinger-Kawahara, C., Joh, D., Hawkinson, M. J., Xiao, G., Purchio, T. F., Caparon, M. G., Lipsitch, M., and Contag, P. R. (2001) Visualizing pneumococcal infections in the lungs of live mice using bioluminescent *Streptococcus pneumoniae* transformed with a novel gram-positive lux transposon. *Infect. Immun.* **69**, 3350–3358.
4. Rocchetta, H. L., Boylan, C. J., Foley, J. W., Iversen, P. W., LeTourneau, D.L., McMillian, C. L., Contag, P. R., Jenkins, D. E., and Parr, T. R., Jr. (2001) Validation of a noninvasive, real-time imaging technology using bioluminescent *Escherichia coli* in the neutropenic mouse thigh model of infection. *Antimicrob. Agents Chemother.* **45**, 129–137.
5. Francis, K. P., Joh, D., Bellinger-Kawahara, C., Hawkinson, M. J., Purchio, T. F., and Contag, P. R. (2000) Monitoring bioluminescent *Staphylococcus aureus* infections in living mice using a novel *luxABCDE* construct. *Infect. Immun.* **68**, 3594–3600.

6. Orihuela, C. J., Gao, G., McGee, M., Yu, J., Francis, K. P., and Tuomanen, E. (2003) Organ-specific models of *Streptococcus pneumoniae* disease. *Scand J Infect. Dis.* **35**, 647–652.

7. Kadurugamuwa, J. L., Sin, L., Albert, E., Yu, J., Francis, K., DeBoer, M., Rubin, M., Bellinger-Kawahara, C., Parr, T. R., Jr., and Contag, P. R. (2003) Direct continuous method for monitoring biofilm infection in a mouse model. *Infect. Immun.* **71**, 882–890.

8. Kadurugamuwa, J. L., Sin, L.V., Yu, J., Francis, K. P., Kimura, R., Purchio, T., and Contag, P. R. (2003) Rapid direct method for monitoring antibiotics in a mouse model of bacterial biofilm infection. *Antimicrob. Agents Chemother.* **47**, 3130–3137.

9. Kadurugamuwa, J. L., Sin, L. V., Yu, J., Francis, K. P., Purchio, T. F., and Contag, P. R. (2004) Noninvasive optical imaging method to evaluate postantibiotic effects on biofilm infection *in vivo. Antimicrob. Agents Chemother.* **48**, 2283–2287.

10. Kadurugamuwa, J. L., Modi, K., Yu, J., Francis, K. P., Purchio, T., and Contag, P. R. (2005) Noninvasive biophotonic imaging for monitoring of catheter-associated urinary tract infections and therapy in mice. *Infect. Immun.* **73**, 3878–3887.

11. Kuklin, N. A., Pancari, G.D., Tobery, T.W., Cope, L., Jackson, J., Gill, C., Overbye, K., Francis, K. P., Yu, J., Montgomery, D., Anderson, A. S., McClements, W., and Jansen, K. U. (2003) Real-time monitoring of bacterial infection *in vivo*: development of bioluminescent staphylococcal foreign-body and deep-thigh-wound mouse infection models. *Antimicrob. Agents Chemother.* **47**, 2740–2748.

12. Xiong, Y. Q., Willard, J., Kadurugamuwa, J. L., Yu, J., Francis, K. P., and Bayer, A. S. (2005) Real-time *in vivo* bioluminescent imaging for evaluating the efficacy of antibiotics in a rat *Staphylococcus aureus* endocarditis model. *Antimicrob. Agents Chemother.* **49**, 380–387.

13. Yu, J., Wu, J., Francis, K. P., Purchio, T. F., and Kadurugamuwa, J. L. (2005) Monitoring *in vivo* fitness of rifampicin-resistant *Staphylococcus aureus* mutants in a mouse biofilm infection model. *J. Antimicrob. Chemother.* **55**, 528–534.

14. Davies, D. (2003) Understanding biofilm resistance to antibacterial agents. *Nat. Rev. Drug Discov.* **2**, 114–122.

15. Mermel, L. A., Farr, B. M., Sherertz, R. J., Raad, I. I., O'Grady, N., Harris, J. S., and Craven, D. E. (2001) Guidelines for the management of intravascular catheter-related infections. *Clin. Infect. Dis.* **32**, 1249–1272.

16. Winson, M. K., Swift, S., Hill, P. J., Sims, C. M., Griesmayr, G., Bycroft, B. W., Williams, P., and Stewart, G. S. (1998) Engineering the *luxCDABE* genes from *Photorhabdus luminescens* to provide a bioluminescent reporter for constitutive and promoter probe plasmids and mini-Tn5 constructs. *FEMS Microbiol. Lett.* **163**, 193–202.

17. Rupp, M. E., Ulphani, J. S., Fey, P. D., Bartscht, K., and Mack, D. (1999) Characterization of the importance of polysaccharide intercellular adhesin/hemagglutinin of *Staphylococcus epidermidis* in the pathogenesis of biomaterial-based infection in a mouse foreign body infection model. *Infect. Immun.* **67**, 2627–2632.

18. Warren, J. W. (2001) Catheter-associated urinary tract infections. *Int. J. Antimicrob. Agents* **17**, 299–303.
19. Kurosaka, Y., Ishida, Y., Yamamura, E., Takase, H., Otani, T., and Kumon, H. (2001) A non-surgical rat model of foreign body-associated urinary tract infection with *Pseudomonas aeruginosa. Microbiol. Immunol.* **45**, 9–15.
20. Dugdale, D. C. and Ramsey, P. G. (1990) *Staphylococcus aureus* bacteremia in patients with Hickman catheters. *Am. J. Med.* **89**, 137–141.
21. Rupp, M. E. and Archer, G. L. (1994) Coagulase-negative staphylococci: pathogens associated with medical progress. *Clin. Infect. Dis.* **19**, 231–243; quiz 244–235.
22. Herrero, M., de Lorenzo, V., and Timmis, K. N. (1990) Transposon vectors containing non-antibiotic resistance selection markers for cloning and stable chromosomal insertion of foreign genes in gram-negative bacteria. *J. Bacteriol.* **172**, 6557–6567.
23. de Lorenzo, V., Herrero, M., Jakubzik, U., and Timmis, K. N. (1990) Mini-Tn5 transposon derivatives for insertion mutagenesis, promoter probing, and chromosomal insertion of cloned DNA in gram-negative eubacteria. *J. Bacteriol.* **172**, 6568–6572.

19

The Cotton Rat as a Model for *Staphylococcus aureus* Nasal Colonization in Humans
Cotton Rat S. aureus *Nasal Colonization Model*

John F. Kokai-Kun

Summary

Staphylococcus aureus nasal colonization is a well-known risk factor for development of *S. aureus* infections in humans, but despite this established association, we are only beginning to understand the factors, both host and pathogen, that play a role in the colonization of the nares by *S. aureus*. The cotton rat is a model for many human respiratory pathogens and has proved its utility as a robust model for *S. aureus* nasal colonization. In this animal model, *S. aureus* is instilled in the nostrils of adult cotton rats, the bacteria rapidly colonize, and 7 days later *S. aureus* nasal colonization is enumerated by surgical removal of the nose and recovery of the colonizing *S. aureus*. This model is an excellent animal model to allow for the evaluation of the efficacy of various therapies, including semi-solid formulations, for determination of their ability to eradicate *S. aureus* nasal colonization. Further, the cotton rat model allows for assessment of the ability of defined genetic mutants of *S. aureus* to colonize mucosal surfaces. Finally, this model has demonstrated its utility for the assessment of various antigens as vaccine candidates to protect against *S. aureus* nasal colonization. This chapter will discuss in detail the method to establish nasal colonization, treatment and eradication of colonization, and recovery of the colonizing bacteria from the nose.

Key Words: *Staphylococcus aureus*; nasal colonization; animal model; cotton rat; therapy.

1. Introduction

The anterior nares is a primary ecologic niche for *Staphylococcus aureus* *(1–3)* with ~30% of the population being colonized at any one time *(1)*. *S. aureus* nasal colonization is a well-established risk factor for *S. aureus*

From: *Methods in Molecular Biology, vol. 431: Bacterial Pathogenesis*
Edited by: F. DeLeo and M. Otto © Humana Press, Totowa, NJ

infection *(4–7)*, and a significant correlation has been demonstrated between the colonizing strains of *S. aureus* and the strains of *S. aureus* in infected individuals. As the incidence of community-associated methicillin-resistant *S. aureus* infections continues to rise *(8)*, there appears to be a parallel increase in nasal colonization with methicillin-resistant *S. aureus* (MRSA) *(1)*. Strategies to reduce *S. aureus* nasal colonization and thus reduce infection rates have met with mixed success *(9–13)*, but this type of prophylaxis may still prove to be an important intervention that can effectively reduce overall *S. aureus* infection rates. This chapter details an animal model for studying *S. aureus* nasal colonization, which allows for both investigation of staphylococcal factors that play a role in nasal colonization as well as allowing for screening of potential therapies to eradicate *S. aureus* nasal colonization.

The cotton rat *(Sigmodon hispidus)* is a well-established model for a number of human pathogens, especially viral respiratory pathogens *(14,15)*. The model detailed in this chapter is an adaptation of a model of *S. aureus* nasal colonization originally described in mice *(16)* but now adapted to cotton rats. The cotton rat's nasal histology is comparable to that of humans *(17)* and pretreatment of cotton rats with antibiotics like streptomycin is not required as it is in mice *(16)* to establish nasal colonization by *S. aureus*. We have successfully used the cotton rat *S. aureus* nasal colonization model to demonstrate the efficacy of lysostaphin as a therapy for *S. aureus* nasal colonization *(18)* as well as to study the roles of wall teichoic acid *(19)* and IsdA and IsdH *(20)* of *S. aureus* in nasal colonization.

This chapter will describe in detail the materials and methods necessary for studying this cotton rat model of *S. aureus* nasal colonization. These steps include (1) *instillation* of the *S. aureus* to establish nasal colonization, (2) *treatment* of the nares with various therapeutic agents, and (3) *recovery* and enumeration of colonizing *S. aureus*.

2. Materials

2.1. Instillation

1. Cotton rats *(Sigmodon hispidus)* *(14)* may be purchased from Harlan (Hsd:Cotton Rat, Indianapolis, IN). There are also several breeding colonies that may be able to provide animals, including colonies at Virion Systems (Gaithersburg, MD) and Baylor College of Medicine (Houston, TX). Cotton rats should be 6 weeks old or older for use in this model (*see* **Note 1**).
2. Frozen stocks of *S. aureus* (*see* **Note 2**).
3. Tryptic soy broth (TSB): As per the manufacturer's instructions, 30 g/L in ddH$_2$O.

4. Columbia salt agar (CSA): Columbia broth (as per the manufacturer's instructions, 35 g/L in ddH_2O), 2% (w/v) Bacto Agar, and 2% (w/v) NaCl (store at 4 °C).
5. Tryptic soy agar (TSA): As per the manufacturer's instructions 40 g/L in ddH_2O or blood agar—TSA with 5% sheep's blood (store agar plates at 4 °C).
6. Phosphate-buffered saline (PBS) without calcium or magnesium: 5.6 m*M* Na_2HPO_4, 0.94 m*M* KH_2PO_4, pH 7.4, and 155 m*M* NaCl.
7. Anesthesia: Ketamine 33 mg/kg, Xyla-Ject (xylazine) 2.5 mg/kg, and Acepromezine Maleate 2.5 mg/kg (see **Table 1A**) (make anesthesia fresh for each use).
8. Small volume (20 µL) disposable pipette tips.
9. Syringes, 1 mL with 27-gauge needles.
10. A spectrometer capable of reading light transmission at 650 nm.

2.2. Treatment

1. Anesthesia: As above (freshly made).
2. Syringes, 1 mL with 27-gauge needles.
3. BD Angiocath flexible FEP polymer catheters, 22-gauge (for delivery of semisolid formulations if desired).
4. Syringes, 1 mL.
5. Small volume (20 µL) disposable pipette tips.

2.3. Recovery

1. Ethanol: 70% (v/v) in water (or equivalent for surface sanitation) and 100% for flaming instruments.
2. PBS + Tween-20: PBS (above), 0.5% (v/v) polyoxyethylene monolaurate (Tween 20) (make fresh for each harvest, Tween 20 is generally sterile as received).
3. TSA + 7.5% NaCl: TSA (above) and 7.5% (w/v) NaCl (store at 4 °C).

Table 1A
Anesthesia Preparation and Dosing by Animal Weight

Anesthesia	Dose (mg/Kg)
Ketamine	33
Xylazine	2.5
Acepromezine	2.5

Preparation of standard anesthesia solution: Ketamine solution at 100 mg/mL, use 20 µL/animal dose; Xyla-Ject (xylazine) solution at 20 mg/mL, use 7.5 µL/animal dose; acepromezine solution at 10 mg/mL, use 15 µL/animal dose; phosphate-buffered saline, use 157.5 µL/animal dose. Make ~50% more animal doses than needed.

Table 1B
Administration Volume of Standard
Anesthesia Preparation by Animal Weight
(i.m. with 1-mL syringe, 27-guage needle)

Cotton rat weight (g)	Dose (µL)
50	150
52	160
54	170
56	180
58	190
60	200
62	210
64	220
66	230
68	240
70	250

4. Chromagar Staph aureus (Chromoagar Microbiology, Paris, France): As per the manufacturer's instruction, 82.5 g/L (agar is light sensitive and can be stored for 1 month at 4 °C after preparation).
5. Sterile alcohol prep pads: 70% isopropyl alcohol.
6. Sterile scalpel blades, #21.
7. Sterile disposable snap cap test tubes, 15 mL.

3. Methods

3.1. Instillation

1. Inoculate a CSA plate from a thawed TSB stock of *S. aureus* using a sterile swab. Cover the agar surface completely and allow any liquid to absorb into the agar.
2. Incubate the inoculated CSA plate at 35–37 °C for 16–24 h.
3. Harvest the *S. aureus* from the inoculated CSA plate into PBS.
4. Aliquot 1 mL of PBS per animal to be instilled into a sterile glass test tube, using a separate test tube for each different *S. aureus* strain (*see* **Note 3**).
5. Determine the percent transmittance (%*T*) of the blank PBS at 650 nm on a spectrometer and set to *T* = 100% (*see* **Note 4**).
6. Harvest the *S. aureus* from the CSA plate using a sterile swab resuspending the bacteria in the PBS and vortex briefly to disperse the bacteria. Check the %T at 650 nm.
7. Add bacteria to the PBS until the %T is ∼10% for a 1-cm path length (*see* **Note 5**).
8. Pellet the *S. aureus* by centrifugation at 7750 × *g* for 7.5 min.

9. Decant or remove the supernatant by pipette and discard as biohazardous waste.
10. Resuspend the bacterial pellet in 10 µL of PBS per animal to be instilled (e.g., if original volume was 6 mL, use 60 µL of PBS to resuspend bacteria) and vortex the resuspension on maximum speed for 3 s.
11. Anesthetize the cotton rats by intramuscular injection in either hind leg muscle with a 27-gauge needle using volumes appropriate for various body weights as indicated in **Table 1B** (*see* **Notes 1** and **6**).
12. Briefly revortex the *S. aureus* suspension on maximum speed prior to instillation.
13. Hold the unconscious cotton rat on its back in the palm of your hand and restrain the chin with your thumb.
14. Using a small (20-µL size) disposable pipette tip, evenly distribute a 10-µL volume of *S. aureus* suspension between each nostril drop-wise (usually 2–3 drops per nostril) without touching the nose of the animal. Allow the rat to draw the drops into its nose by natural respiration (*see* **Note 7**).
15. Hold the rat on its back until it is breathing normally and has cleared its nose (*see* **Note 8**).
16. Return the rat to the cage, laying the rat on its back (*see* **Note 9**).
17. Repeat the instillation with each cotton rat in the group and each group of cotton rats until all animals have been instilled.
18. Monitor the animals until they are demonstrating purposeful movements.
19. Return food and water ad libitum and cover the cage with a barrier cover (*see* **Note 10**).
20. Perform serial dilutions in PBS of the remaining *S. aureus* inoculum and plate the serial dilutions on blood agar or TSA to determine the actual number of *S. aureus* instilled per animal (*see* **Note 2**).

3.2. Treatment

If desired, the nasally colonized cotton rats can be treated intranasally to affect nasal colonization (*see* **Note 11**).

3.2.1. Nasal Installation of Liquid Formulations

1. Liquid formulations of various compounds can be instilled into the cotton rat nares in the same manner as the initial bacterial instillation (*see* **Note 12**).

3.2.2. Nasal Installation of Semi-Solid Formulations

1. Remove the metal needle from a flexible catheter and affix the flexible catheter to a 1-mL syringe.
2. Load the semi-solid formulation into a 1-mL syringe fitted with the 22-gauge flexible catheter for nasal delivery (*see* **Note 13**).
3. Anesthetize the cotton rats as described above (*see* **Note 14**).
4. Once the cotton rats have succumbed to the anesthesia, gently insert the catheter tip 2–3 mm into one nostril (*see* **Note 15**) and dispense the desired volume

(usually $\sim$20 μL) by applying continuous gentle pressure on the syringe plunger and slowly withdrawing the catheter tip from the nostril (*see* **Note 16**).

5. Gently message the nose of the cotton rat to disperse the semi-solid and open the nares to allow normal breathing.
6. Lay the cotton rat down on its back and do not proceed to the second nostril until normal breathing is resumed.
7. Once the animal is breathing normally, repeat **steps 3** and **4** to instill the semi-solid in the second nostril (*see* **Note 17**).
8. Return animals breathing normally to their cages laying them on their backs.
9. Repeat the treatment with all animals and groups to be treated.
10. Monitor the animals until they are demonstrating purposeful movements.
11. Return food and water ad libitum and cover the cage with a barrier cover.

3.2.3. Recovery

1. The noses of colonized cotton rats can be harvested 24 h after *S. aureus* instillation (*see* **Note 18**).
2. Prepare one 15-mL snap cap test tube per animal containing 500 μL (*see* **Note 19**) PBS + Tween 20 + any appropriate neutralizer needed (*see* **Note 20**).
3. Sacrifice the animals as appropriate to the guidelines of the presiding IACUC. We use 100% CO_2 delivered at 10 psi.
4. Cleanse the outside of the nose and the area around the nose well ensuring that the bridge of the nose is also cleansed with a sterile 70% alcohol prep pad (*see* **Note 21**). This removes any *S. aureus* colonizing the skin around the nose.
5. Sanitize fine-tipped forceps and fine-bladed dissecting scissors by flaming with 100% ethanol (*see* **Note 22**) and then let them cool prior to reuse.
6. Lay the cotton rat on its back on a sanitized dissection board.
7. Hold the chin of the cotton rat between the thumb and forefinger of one hand and place the blade of a scalpel with a #21 blade just under the upper lip and against the teeth of the cotton rat (*see* **Fig. 1A**). Slide the blade down the teeth and back along the bridge of the nose removing the anterior nares and soft tissue and bony cartilage along the bridge of the nose (*see* **Fig. 1B**).
8. Grasp the cartilage with the flamed forceps, and using the scissors, bisect the nostrils laterally (*see* **Figs. 1C and 1D**) exposing the nostril lumen.
9. Place the bisected nose into an appropriately labeled recovery tube and cap the tube.
10. Vortex the tube on maximum for 10 s to release adherent bacteria (*see* **Note 23**).
11. To prepare for the processing of the next animal, rinse the forceps, scissors, and dissecting board with 70% ethanol and wipe with a clean paper towel. Dispose of scalpel blade and flame forceps and scissors as above.

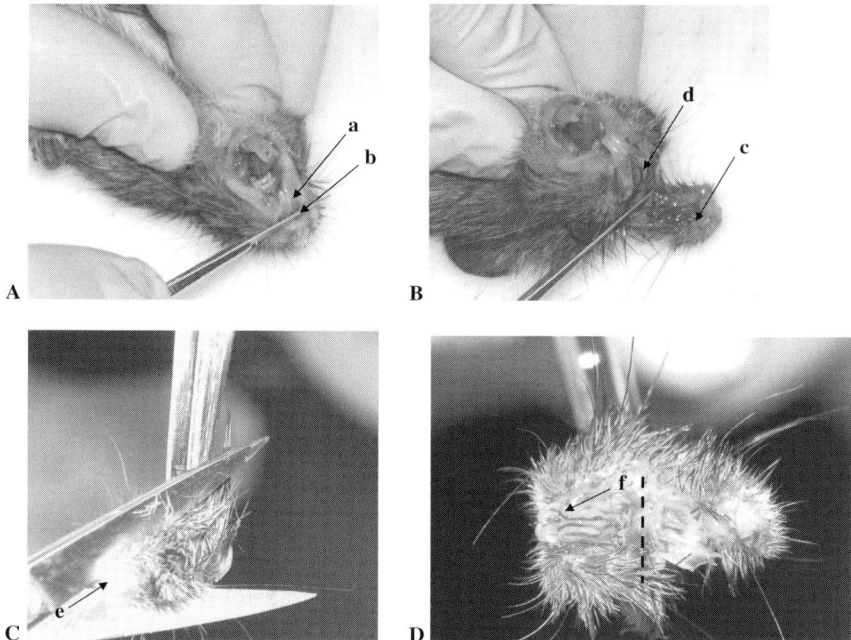

Fig. 1. Surgical processing of the cotton rat nose for recovery of colonizing *Staphylococcus aureus*. (**A**) With the sacrificed animal lying on its back, secure the chin between the thumb and forefinger and slide the scalpel blade down the prominent front teeth (**a**) until it contacts the soft tissue of the lips (**b**). (**B**) Slide the scalpel blade under the nostrils (**c**) and back along the bone of the bridge of the nose (**d**) removing the nose along with the cartilage and skin from the bridge of the nose. (**C**) Grasp the skin and cartilage of the bridge of the nose with forceps and laterally bisect the nostrils (**e**) with scissors. (**D**) The bisected cotton rat nose ready to be vortexed. The arrow (**f**) indicates the lumen of the nostril bisected by the cut. The dashed line indicates the midline of the nose where the cut was made.

12. Repeat the procedure with each animal and each group until completed.
13. Discard the sacrificed animals as appropriate for the facility (*see* **Note 24**).
14. Just prior to inoculating agar plates, vortex samples again on maximum for 10 s.
15. Inoculate an individual TSA + 7.5% NaCl (*see* **Note 25**) bacterial agar plate with 100 μL (*see* **Note 19**) of supernatant from each sample and spread the solution evenly on the agar surface with a disposable spreader.
16. Repeat **step 15** for each sample.
17. Incubate TSA + 7.5% NaCl plates for 48 h and then count the colonies (*see* **Note 26**).

18. Following the appropriate incubation time, count the *S. aureus* colonies on each plate (*see* **Note 25**) and multiply by the dilution factor (in this case by five) to determine the *S. aureus* recovered per nose (see **Note 27**).

4. Notes

1. A leather gardening glove should be worn on the grabbing hand when handling cotton rats. The animals should be pinned under a gloved hand and the rat then grabbed by its scruff for control. Cotton rats are very skittish and will attempt to escape, so open the cage lid only as far as needed to grab a single rat. Do not grab a cotton rat by its tail as the tail will deglove *(14)*.
2. Frozen stocks of *S. aureus* are stored in TSB at −70 °C. Most wild-type *S. aureus* strains examined successfully colonize the cotton rat nares (*see* **Table 2**).
3. Make an additional two or three animal doses (2–3 mL) than required to allow for extra volume and to titer the actual dose of *S. aureus* instilled per animal.

Table 2
Cotton Rat Nasal Colonization by Various *Staphylococcus aureus* strains

S. aureus strain	Animals colonized	Mean colonization per nose
ATCC 49521 (MSSA, Type 5 capsule)	5/5	$> 10,000$
ATCC 12605 (MSSA, Type 8 capsule)	5/5	5418
Sa113 (MSSA)	15/15	6011
SH1000 (MSSA)	5/5	2553
Newman (MSSA)	5/5	4021
SA5 StrR (Streptomycin resistant[a])	1/5	100
MBT 5040 (MRSA)	41/41	5262
MBT 5040 LysoR (Lysostaphin resistant[b])	1/5	200
MRSA 12/12 (MRSA, StrR[c])	5/5	567
Col (MRSA, StrR[c])	5/5	3825
BK 2352 (MRSA)	5/5	3481
BK 2454 (MRSA)	5/5	2323
SA 3865 (mupirocin resistant)	5/5	268
NRS 79 (VISA)	5/5	2574

All animals instilled with ∼5 × 10^8 colony-forming units (CFUs) of *S. aureus* and killed 6–7 days post-instillation.

MSSA, methicillin-sensitive *S. aureus*; MRSA, methicillin-resistant *S. aureus*; StrR, streptomycin resistant; VISA, vancomycin intermediate susceptibility.

[a] A streptomycin-resistant variant isolated by serial passage on streptomycin agar.
[b] See reference *(21)*.
[c] Naturally streptomycin resistant.

Nephelo culture flasks (125 mL) work well for larger groups of animals (10 or greater).

4. A Spectronic 20 (Theromo Scientific, Waltham, MA) with 1-cm diameter sterile glass test tubes works well for this, allowing the %T for the whole suspension to be determined. Alternatively, the %T of an aliquot of suspended bacteria can be determined using a cuvette and a conventional visible light spectrometer set to 650 nm.

5. A *T* (transmittance) of ~10% is equivalent to ~5×10^8 *S. aureus* per milliliter of buffer at a 1-cm path length, but it is best to determine this empirically for each strain of *S. aureus*. A dose >10^8 *S. aureus* instilled in the nares will ensure consistent nasal colonization even if much of the initial dose is expelled during instillation (*see* **Note 7**). Consistent colonization can be achieved with initial inoculums of as low ~10^5 per animal, but care must be taken that the animal does not expel any of the lower inoculum volume. Instillation of <10^5 *S. aureus* have not resulted in consistent nasal colonization in our hands.

6. The injection volumes given in **Table 1** will usually render a cotton rat unconscious in ~5 min, and they will remain that way for about 20–30 min. Occasionally, an individual animal will not be rendered unconscious by the initial dose of anesthesia; in this case, additional doses of 50–100 μL depending on the weight of the animal can be administered until the animal in unconscious. It is important that the cotton rats be fully unconscious prior to instillation of the bacteria (*see* **Note 7**).

7. Even when unconscious, cotton rats retain a strong expiratory response, that is, they will often "snort-out" much of the *S. aureus* inoculum. For safety reasons, a full face mask and hair cover should be worn when instilling *S. aureus* in cotton rat nares. Expiration can be controlled to some extent by gently rolling the loose skin on the bridge of the nose over the nostrils with your forefinger after instillation. This is particularly necessary when instilling a lower inoculum of bacteria.

8. Occasionally, instilled cotton rats will appear to have difficulty in breathing demonstrating a gasping appearance because of nasal blockage by the bacterial bolus. The animal will usually clear its own airway within about a minute, but occasionally artificial respiration in the form of gentle compressions on the rib cage are needed to aid the animal's recovery from the instillation.

9. Female rats can be housed five per 18.5" × 10" cage, but after manipulation of any kind, male rats must be housed individually or they will fight. This fighting often involves biting on the nose, which will interfere with this model and can lead to active infection. The cotton rats should begin to demonstrate purposeful movement about 20–30 min after bacterial instillation.

10. All cage bedding should be considered as biohazardous and treated appropriately as *S. aureus* continues to be shed by the animals for a couple of days after instillation.

11. Cotton rats can generally tolerate anesthesia and treatment once or twice a day for up to 5 days. The more often an animal is anesthetized, the less sensitive it becomes to the anesthesia.

12. Use no more than 20 µL for intranasal administration of liquid formulations as instillation volumes of >20 µL will result in some of the liquid entering the trachea and possibly the lungs.

13. If the semi-solid formulation is heat tolerant, then the formulation can be warmed in a water bath above its melting point to allow the melted semi-solid to be drawn into the syringe without the attached catheter. For non-heat tolerant formulations, the semi-solid can be loaded into a syringe using a spatula and the syringe plunger to load the formulation in the syringe barrel from the top. Before affixing the catheter to the syringe, ensure that the syringe tip is wiped free of formulation or the catheter can "pop-off" the syringe during instillation. Using a lure lock syringe or parafilm to secure the catheter can also help prevent this.

14. For semi-solid formulations, lighter anesthesia than is used for liquid formulations is generally better than deep anesthesia. This will better allow the cotton rats to clear their nares of the semi-solid formulation and help prevent respiratory distress.

15. Care should be taken when inserting the catheter in the nostril as the tip of the catheter is sharp and can cut the inside of the nostrils. To help prevent this, smear some of the formulation on the tip of the catheter to act as a lubricant.

16. The more viscous the formulation, the more difficult it is to control the exact dosing of the semi-solid. More viscous formulations will continue to ooze from the syringe after pressure is released from the plunger making exact dosing difficult.

17. Instillation of semi-solid formulations into both nostrils will occasionally lead to respiratory distress as cotton rats are obligate nose breathers. Some distress is normal as the animals gasp to clear their nasal passages, but very labored breathing may require additional massage of the nose and possible breathing assistance in the form of gentle compressions of the rib cage to deflate and inflate the lungs. Respiratory distress mostly occurs with smaller, over-anesthetized, or sick animals. Cotton rats should be monitored until they display normal breathing following nasal instillation of semi-solids.

18. Depending on the strain of *S. aureus* instilled, cotton rats can remain nasally colonized with *S. aureus* for up to and possibly greater than 2 months *(18)*. Shedding of excess or non-adherent *S. aureus* appears to occur primarily within the first 2 or 3 days, after which point the nasal colonization appears to reach a relatively steady state *(19)*. We generally allow 4–6 days post-instillation prior to any treatments and harvest the noses between 7 and 10 days post-bacterial instillation or 24 h after the final treatment. For some defined genetic *S. aureus* mutants, a longer colonization time (2–3 weeks) is required to see a significant difference between colonization by the mutant and wild-type bacteria *(20)*.

19. A recovery buffer (PBS + Tween 20) volume of 500 μL will allow sufficient buffer to plate 100 μL on one recovery media or 2 × 75 μL on two recovery media. Greater volumes of buffer can be used if further recovery media are required, but these reduce the sensitivity of the recovery correspondingly.

20. Topical applications of compounds like antibiotics can be retained in the nose above MIC/MBC levels for >24 h in certain formulations and can lead to ex vivo killing of the bacteria and false positive results, thus an appropriate neutralizer should be used to ensure that any apparent antibacterial activity by the administered compound actually occurs in the nares and not in the recovery tube *(18)*.

21. Hold the head of the animal down so that any excess alcohol does drip into the nostrils.

22. We use an IBS Integra Biosciences Fireboy Plus with an ignition sensor instead of a constant open flame like a bunsen burner for safety around open containers of ethanol.

23. Vigorous vortexing of the nose recovers the vast majority of the colonizing *S. aureus*. We have followed the vigorous vortexing of samples with sonication to try to release more *S. aureus* and found no greater recovery of *S. aureus* following sonication.

24. The food and bedding from the instilled cotton rat cages should be considered contaminated and disinfected by autoclaving or other appropriate methods.

25. Unlike the mouse *(16)*, the cotton rat has a variety of bacteria as normal nasal flora that are recovered by this procedure (**Table 3**), so the nasal supernatant must be plated on a selective media to ensure accurate enumeration of the instilled colonizing *S. aureus*. Depending on the characteristics of the *S. aureus* strain instilled, a variety of bacterial media can be used for recovery from the cotton rat nares. Antibiotic-resistant strains, for example, streptomycin-resistant strains, MRSA, or gene knock-outs with antibiotic resistance markers, are the easiest to recover because the appropriate antibiotic can be added to the TSA + 7.5% NaCl agar to inhibit the growth of most other native flora. The cotton rat native nasal flora includes a prevalent Gram-negative organism that is naturally erythromycin resistant but which will not grow in the presence of 7.5% NaCl. The cotton rat nose also has a variety of coagulase negative staphylococci that will also grow on 7.5% NaCl. If the instilled strain of *S. aureus* is not antibiotic resistant, Chromagar Staph aureus works well for recovery of *S. aureus* with the instilled *S. aureus* growing as large purple colonies. It should be noted that a naturally occurring *Staphylococcus muscae* found in the cotton rat nose will grow as a small purple colony on Chromagar Staph aureus, so care should be taken to differentiate these bacteria from the instilled *S. aureus* that will grow as a larger purple colony.

26. The *S. aureus* colonies grow more slowly on the high salt agar than on other agars thus 48 h incubation at 37 °C is required for accurate enumeration.

27. *S. aureus* colonization varies by strain (*see* **Table 2**), but generally, the number of colonies recovered per plate from the primary sample tube are within the countable range (<2000 CFUs) for a colony counter that can count up to 2000 colonies per plate. If hand counting is done, then it may be beneficial to dilute

Table 3
**Recoverable Native Aerobic Flora from Uninstilled and *Staphylococcus*
aureus-Instilled Cotton Rats Noses**

Bacteria[a,b]	Number colonized/number instilled	Mean or individual CFUs recovered
Uninstilled cotton rats		
Gram-negative of the enterobaceriacea family[c]	5/5	13,390
Streptococcus species[c]	5/5	730
Streptococcus thoralensis	4/5	810
Staphylococcus muscae[d]	3/5	128
Staphylococcus xylosus	2/5	1000 and 5
Aerobacter viridans	2/5	160 and 10
Non-typeable Gram+[e]	1/5	80
S. *aureus*-instilled cotton rats		
Gram-negative of the enterobacteriacea family	4/5	22,512
Streptococcus species	5/5	800
Streptococcus thoralensis	5/5	56
Staphylococcus muscae	5/5	71
Staphylococcus xylosus	0/5	n/a[f]
Aerobacter viridans	4/5	26
Non-typeable Gram+	3/5	47
S. aureus (instilled)[g]	5/5	1179

[a] Bacteria was recovered on blood agar and Chromagar Staph aureus at 37 °C.
[b] Bacterial identification confirmed by comparative DNA sequence analysis Accugenix Inc.
(Newark, DE).
[c] Comparable DNA sequence not available in database.
[d] Grows as a small purple colony of Chromagar Staph aureus.
[e] Was not amenable to comparative DNA sequence analysis.
[f] Not applicable.
[g] *S. aureus* strain MBT 5040.

each nasal supernatant 1:10 in PBS by adding 75 µL of supernatant to 675 µL of
PBS and then plate 75 µL of both the primary supernatant and the 1:10 dilution
on separate agar plates to ensure that the resulting colony count is within the
countable range. This will require twice as many agar plates.

Acknowledgements

We acknowledge of the excellent technical assistance of Tanya Chanturiya
and Jimmy Mond for critical reading of this manuscript.

References

1. Kuehnert, M. J., Kruszon-Moran, D., Hill, H. A., McQuillan, G., McAllister, S. K., Forsheim, G., McDougal, L. K., Chaitram, J., Jensen, B., Fridkin, S. K., Killgore, G., and Tenover, F. C. (2006) Prevalence of *Staphylococcus aureus* nasal colonization in the United States, 2001–2002. *J. Infect. Dis.* **193**, 172–179.
2. Kluytmans, J., van Belkum, A., and Verbrugh, H. (1997) Nasal carriage of *Staphylococcus aureus*: epidemiology, underlying mechanisms, and associated risks. *Clin. Microbiol. Rev.* **10**, 505–520.
3. White, A. and Smith, J. (1963) Nasal reservoir as the source of extranasal staphylococci. *Antimicrob. Agents Chemother.* **3**, 679–683.
4. Polgreen, P. M. and Herwaldt, L. A. (2004) *Staphylococcus aureus* colonization and nosocomial infections: implications for prevention. *Curr. Infect. Dis. Rep.* **6**, 435–441.
5. Yu, V. L., Goetz, A., Wagener, M., Smith, P. B., Rihs, J. D., Hanchett, J., and Zuravleff, J. J. (1986) *Staphylococcus aureus* nasal carriage and infection in patients on hemodialysis. *N. Engl. J. Med.* **315**, 91–96.
6. von Eiff, C., Becker, K., Machka, K., Stammer, H., and Peters, G. (2001) Nasal carriage as a source of *Staphylococcus aureus* bacteremia. *N. Engl. J. Med.* **344**, 11–16.
7. Corne, P., Marchandin, H., Jonquet, O., Campos, J., and Banuls, A.-L. (2005) Molecular evidence that nasal carriage of *Staphylococcus aureus* plays a role in respiratory tract infections of critically ill patients. *J. Clin. Microbiol.* **43**, 3491–3493.
8. Kowalski, T. J., Berbari, E. F., and Osmon, D. R. (2005) Epidemiology, treatment and prevention of community-acquired methicillin-resistant *Staphylococcus aureus* infections. *Mayo Clin. Proc.* **80**, 1201–1208.
9. Kluytmans, J. A., Mouton, J. W., Vandenbergh, M., Manders, M.-J., Maat, A., Wagenvoort, J., Michel, M., and Verbrugh, H. (1996) Reduction of surgical-site infections in cardiothoracic surgery by elimination of nasal carriage of *Staphylococcus aureus*. *Infect. Control Hosp. Epidemiol.* **17**, 780–785.
10. Perl, T., Cullen, J., Wenzel, R., Zimmerman, B., Pfaller, M., Sheppard, D., Twombley, J., French, P., and Herwalt, L. (2002) Intranasal mupirocin to prevent postoperative *Staphylococcus aureus* infections. *N. Engl. J. Med.* **24**, 1871–1877.
11. Martin, J., Perdreau-Remington, F., Kartalija, M., Pasi, O., Webb, M., Gerberding, J., Chambers, H., Tauber, M., and Lee, B. (1999) A randomized clinical trial of mupirocin in the eradication of *Staphylococcus aureus* nasal carriage in human immunodeficiency virus disease. *J. Infect. Dis.* **180**, 896–899.
12. Wertheim, H. F. L., Vos, M. C., Ott, A., Voss, A., Kluytmans, J. A. J. W., Vandenbroucke-Grauls, C. M. J. E., Meester, M. H. M., van Keulen, P. H. J., and Verbrugh, H. A. (2004) Mupirocin prophylaxis against nosocomial *Staphylococcus aureus* infections in nonsurgical patients. *Ann. Intern. Med.* **140**, 419–425.
13. Yano, M., Doki, Y., Inoue, M., Tsujinaka, T., Shiozaki, H., and Monden, M. (2000) Preoperative intranasal mupirocin ointment significantly reduces postoperative infection with *Staphylococcus aureus* in patients undergoing upper gastrointestinal surgery. *Surg. Today (Japan)* **30**, 16–21.

14. Faith, R. E., Montgomery, C. A., Durfee, W. J., Anguilar-Cordova, E., and Wyde, P. R. (1997) The cotton rat in biomedical research. *Lab. Anim. Sci.* **47**, 337–345.
15. Niewiesk, S. and Prince, G. (2002) Diversifying animal models: the use of hispid cotton rats (*Sigmodon hispidus*) in infectious diseases. *Lab. Anim.* **36**, 357–372.
16. Kiser, K. B., Cantey-Kiser, J. M., and Lee, J. C. (1999) Development and characterization of a *Staphylococcus aureus* nasal colonization model in mice. *Infect. Immun.* **67**, 5001–5006.
17. Prince, G. A., Jenson, B., Horswood, R. L., Carmargo, E., and Chanock, R. M. (1978) The pathogenesis of respiratory syncytial virus infection in cotton rats. *Am. J. Pathol.* **93**, 771–783.
18. Kokai-Kun, J. F., Walsh, S. M., Chanturiya, T., and Mond, J. J. (2003) Lysostaphin cream eradicates *Staphylococcus aureus* nasal colonization in a cotton rat model. *Antimicrob. Agents Chemother.* **47**, 1589–1597.
19. Weidenmaier, C., Kokai-Kun, J. F., Kristian, S. A., Chanturiya, T., Kalbacher, H., Gross, M., Nicholson, G., Neumeister, B., Mond, J. J., and Peschel, A. (2004) Role of teichoic acids in *Staphylococcus aureus* nasal colonization, a major risk factor in nosocomial infections. *Nat. Med.* **10**, 243–245.
20. Clarke, S. R., Brummell, K. J., Horsburgh, M. J., McDowell, P. W., Mohamad, S. A. S., Stapleton, M. R., Acevedo, J., Read, R. C., Day, N. P. J., Peacock, S. J., Mond, J. J., Kokai-Kun, J. F., and Foster, S. J. (2006) Identification of *in vivo* expressed antigens of *Staphylococcus aureus* and their use in vaccinations for protection against nasal carriage. *J. Infect. Dis.* **193**, 1098–1108.
21. Kosuma, C. M. and Kokai-Kun, J. F. (2005) Comparison of four methods for determining lysostaphin susceptibility of various strains of *Staphylococcus aureus*. *Antimicrob. Agents Chemother.* **49**, 3256–3263.

20

A Non-Human Primate Model of Acute Group A *Streptococcus* Pharyngitis

Paul Sumby, Anne H. Tart, and James M. Musser

Summary

This chapter describes methods for using non-human primates as a model of group A streptococcal (GAS) pharyngitis. This model has been used successfully to study host–pathogen interactions occurring during pharyngeal GAS infections. The protocol as described will compare two different GAS strains for their ability to cause clinical symptoms of pharyngitis.

Key Words: *Streptococcus pyogenes*; GAS; pharyngitis; animal model; non-human primate.

1. Introduction

The human bacterial pathogen group A *Streptococcus* (GAS) causes a wide variety of diseases including pharyngitis and/or tonsillitis, scarlet fever, pyoderma, toxic shock syndrome, necrotizing fasciitis, poststreptococcal glomerulonephritis, and acute rheumatic fever (ARF) *(1–3)*. Annually, GAS infections account for an estimated 30 million cases of pharyngitis and 15,000 cases of invasive disease in the United States alone *(3–5)*. In addition to the extensive morbidity and mortality associated with GAS infections in the United States, the financial impact of these infections runs at over 1.5 billion dollars in direct health care costs *(3–5)*. Failure to treat GAS pharyngitis may lead to the post-GAS infection sequel ARF and subsequently to the development of rheumatic heart disease. Although antibiotic treatment of GAS pharyngitis patients has drastically reduced the incidence of these forms of heart disease in the United States and Western Europe, they remain the most common cause of

From: *Methods in Molecular Biology, vol. 431: Bacterial Pathogenesis*
Edited by: F. DeLeo and M. Otto © Humana Press, Totowa, NJ

preventable pediatric heart disease in developing countries *(1,4)*. In aggregate, GAS infections result in an estimated 500,000 deaths worldwide per year, clearly defining GAS as a major public health concern *(4)*. The mortality, morbidity, and financial burden associated with GAS infections emphasize the critical need for a more detailed understanding of the complex strategies used by this pathogen to successfully circumvent host immune defenses and cause disease.

GAS pathogenesis studies have been hampered by the absence of animal models of disease that recapitulate human infections. Although mouse models of infection are currently most frequently used, a debate about the relevance of these models with regard to the human host remains *(6,7)*. Following the reasoning that humans and non-human primates are closely related phylogenetically, several non-human primate models of GAS diseases have been developed over the last 100 years. One of the earliest studies using a non-human primate model to study streptococcal disease was published in 1907 by Beattie, who observed clinical manifestations reminiscent of rheumatic fever in infected monkeys *(8)*. About a decade later, Blake *et al.* published work in which intratracheal injections of GAS were used to study pneumonia in non-human primates *(9)*. In 1946, Watson observed that intranasal challenge of monkeys with GAS led to serotype-specific immunity to this organism *(10)*. In 1976, Köhler *et al.* reported the protective effect of M protein in an immunization study using rhesus monkeys *(11)*. Since then, several reports describe non-human primate models used to study invasive GAS diseases *(12,13)*.

The most commonly used animal model to study GAS pharyngitis is the mouse upper respiratory tract infection model *(14–16)*. However, mice do not develop a true pharyngitis and only a relatively few animals become colonized even when high concentrations of inocula are used. Moreover, in our experience the data obtained with this model are poorly reproducible, and because the blood volume of the mouse is small the ability to easily perform many standard immunologic assays is limited. In addition, several gas virulence factor do not act on mouse molecules. Therefore, the relevance, significance, and usefulness of the murine model of GAS pharyngitis is limited. Several independent studies indicate that non-human primates may provide a more suitable model to study GAS pharyngitis: animals are successfully colonized by GAS, and limited data indicate that the infections mimicked the humoral immune response characteristic for human disease *(10,17–20)*.

The inadequacies of the mouse model of infection led us to develop a non-human primate model of GAS pharyngitis that is a superb phenocopy of human infection and has become the gold-standard model for studying GAS–host molecular interactions in the upper respiratory tract *(21–24)*.

This chapter describes a non-human primate model of GAS pharyngitis that we have used successfully to gain insight into the transcriptional adaptation of GAS during upper respiratory tract infection *(21,22)*, to determine the contribution of specific GAS virulence factors to infection *(23,24)*, and to investigate possible vaccine candidates (unpublished data). The following protocol outlines a comparison study where wild-type and isogenic mutant strains of GAS are used to determine the contribution of a putative virulence gene to GAS pharyngitis.

2. Materials

2.1. Preparation of GAS inoculum

1. Wild-type and mutant GAS strains.
2. Todd-Hewitt broth (THY broth) with 0.2% yeast extract.
3. 80% glycerol, sterile.
4. THY agar plates (THY broth plus 1.5% agar).
5. Phosphate-buffered saline (PBS), sterile.
6. 50-mL falcon tubes, sterile.
7. 15-mL screw cap tubes, sterile.
8. Liquid nitrogen.

2.2. Animal Grouping

1. Purified streptolysin-O-protein (SLO, Sigma cat no. S5265).
2. Bovine serum albumin (BSA).
3. PBS with 0.1% Tween 20 (PBST).
4. Goat-anti-monkey IgG-HRP.
5. 3,3′,5,5′-tetramethylbenzidine solution (TMB).
6. 0.18 M sulfuric acid (stop solution).
7. 96-well clear bottom ELISA plates.

2.3. Non-human Primate Infections

1. Eight cynomolgus macaques (*Macaca fascicularis*).
2. 1-mL syringes.
3. PBS, sterile.

2.4. Disease Development

1. Laryngoscope (reusable handle with disposable blades).

2.5. Blood Draws

1. Eight Vacutainer™ systems (VWR).
2. Red-capped Vacutainer™ plasma tubes (VWR).
3. 2-mL red-capped tubes, sterile.

2.6. Throat Swabbing

1. 2-mL yellow-capped tubes, sterile.
2. Cotton-tipped throat swabs, sterile.
3. Selective strep agar plates with 5% sheep blood (cat.# 221780; BD Biosciences, San Jose, CA, USA).
4. 100 m*M* dithiothreitol (DTT), sterile.
5. PCR reagents and machine.
6. *emm* PCR primers (forward: CTATTSGCTTAGAAAATTAA and reverse: GCAAGTTCTTCAGCTTGTTT).
7. Agarose gel electrophoresis equipment.
8. Vortex.
9. Racks for 1.5-/2-mL tubes.
10. 1.5-mL tubes for serial dilutions, sterile.
11. PBS.
12. Ethanol.
13. Glass spreader.
14. 100 blood agar plates.
15. Light box.
16. Colony counter.
17. Disposable absorbent bench covers.
18. Liquid nitrogen.

2.7. Saliva Aspiration

1. 2-mL blue-capped tubes, sterile.
2. 1-mL syringes, sterile.

2.8. Nasal Washes

1. 2-mL white-capped tubes, sterile.
2. 1-mL syringes, sterile.
3. PBS.

3. Methods·

3.1. Preparation of Bacterial Inoculum

GAS is prepared for infection of non-human primates by first growing the wild-type and mutant strains in THY broth to mid-exponential phase as detailed below. This step should be performed 2–3 weeks prior to monkey infection to enable adequate time to generate accurate bacterial titers and repeat the culturing step if necessary. Accurate bacterial titers are essential to ensure an equal infectious dose is given to each animal.

1. Grow 5-mL overnight cultures of the wild-type and mutant GAS strains statically at 37 °C with 5% CO_2 in THY broth.
2. Into four 50-mL tubes, place 40 mL of THY broth and prewarm to 37 °C.
3. For each of the two overnight GAS cultures, perform 1/100 dilutions into two of the prewarmed 40-mL THY broth aliquots (400 μL of culture into 40 mL of THY).
4. Incubate cultures statically at 37 °C with 5% CO_2 to the mid-exponential phase of growth as previously determined by optical density at 600 nm (OD_{600}) measurement.
5. Pellet the bacterial cells by centrifugation ($3000 \times g$ for 10 min), wash each of the pellets twice with 40 mL of PBS, and resuspend each bacterial pellet in 8 mL of PBS.
6. Pool the two cultures corresponding to the same strain, giving 16 mL of washed wild-type and 16 mL of washed mutant GAS cells.
7. Add 4 mL of 80% glycerol to each 16 mL of GAS solution, mix by inversion, and aliquot in 2 mL volumes into 15-mL tubes. GAS aliquots are snap frozen in liquid nitrogen and stored at –80 °C. Approximately 10 aliquots of each strain should be generated by this protocol.
8. The following day, thaw four of the 10 aliquots of each strain on ice for $\sim$ 30 min.
9. Once thawed, perform serial dilutions of each aliquot in PBS, plate on THY agar plates, and incubate overnight at 37 °C with 5% CO_2.
10. The following morning, count colonies and determine the mean (±standard deviation) colony-forming unit (CFU) titers for the two GAS strains. Typical titers are $\sim 2 \times 10^8$ CFUs per mL.

3.2. Cynomolgus Macaque Screens for Group Selection

We commonly use group sizes of four or six cynomolgus macaques per GAS strain although experiments have been conducted with up to 20 animals *(22)*. To divide macaques into experimental groups, we use information regarding animal age, sex, weight, and concentration of serum anti-SLO antibodies. SLO is a secreted toxin of GAS such that infected individuals seroconvert to SLO following GAS infection *(24–26)*. Thus, measuring the concentration of anti-SLO antibodies provides an indicator of previous potential GAS infection. Note that all animals are tattooed to enable individual identification.

3.2.1. Animal Age, Sex, and Weight

To reduce animal-specific differences in the data, all cynomolgus macaques should ideally be of the same sex and of similar age ($\sim$2–4 years old). It should be noted, however, that mixed sex experiments and animals ranging in age from 2 to 9 years have previously been used successfully *(22)*. Similarly, differences in animal weights should be minimalized and taken into consideration during the grouping process.

3.2.2. Determination of Serum Anti-SLO Antibody Concentrations

Three-milliliter blood samples are taken on day minus-seven of the protocol to enable determination of serum anti-SLO antibody titers (*see* **Table 1**). The sera isolated will also serve as a pre-infection control for antibody measurements comparing pre- and post-infection samples. Blood samples are isolated and processed as described in **Subheading 3.4.1**. Sera samples isolated from the macaques are titered for anti-SLO antibody concentrations by ELISA as follows:

1. Add 100 µL of SLO solution [2.5 µg of Sigma-Aldrich SLO (contains 3–9% SLO protein) per milliliter of PBS] to the wells of a 96-well ELISA plate. Dry overnight at 40 °C.
2. Wash wells with PBST.
3. To each well add 100 µL of 1% BSA in PBST and incubate on a rocking platform for 30 min at room temperature to block.
4. Wash wells with PBST.
5. To triplicate wells add 100 µL of each serum sample (diluted 1:500 with PBS) and incubate for 2 h at room temperature on a rocking platform.
6. Wash wells with PBST.

Table 1
Timeline of Sample Isolations During the Non-Human Primate Experiment

Day of the experiment	Clinical scoring	Throat swab	Blood withdrawal	Nasal wash	Saliva
–7 Monday	X	X	X		
0 Monday	X	X		X	X
2 Wednesday	X	X			
4 Friday	X	X			
7 Monday	X	X	X	X	X
10 Thursday	X	X			
14 Monday	X	X	X	X	X
17 Thursday	X	X			
21 Monday	X	X	X	X	X
28 Monday	X	X	X	X	X

Throat swabs, blood withdrawals, nasal washes and saliva samples are isolated on the indicated (X) days. Blood specimens (3 mL) are taken by venipuncture from the cephalic vein. Nasal washes are collected by instilling 1.5 mL of sterile phosphate-buffered saline (PBS) into the nostril and immediately aspirating wash fluid and secretions with a sterile syringe. Saliva specimens are collected by aspiration from the cheek pouch. Prior to throat swabbing, clinical scoring of tonsillitis and pharyngeal erythema severity is performed by a trained veterinarian who is blinded to the two animal groups.

7. Add 100 µL of goat-anti-monkey IgG-HRP (diluted 1:30,000 with PBS) to each well and incubate for 1 h at room temperature on a rocking platform.
8. Wash wells with PBST.
9. Wash wells with PBS.
10. Add 100 µL of TMB solution to each well and incubate for 26 min at room temperature on a rocking platform.
11. Add 100 µL of stop solution (0.18 *M* sulfuric acid) to each well.
12. Read absorbance at 450 nm.
13. Calculate mean (±standard deviation) values for each serum sample.

3.2.3. Additional Considerations for Animal Grouping

To ensure that the macaques do not have current carriage of GAS, the animals should be cultured for the presence of GAS in the upper respiratory tract on day minus-seven. Culturing for GAS is performed by swabbing the posterior pharynx and plating to look for GAS colonies as described in **Subheading 3.4.3**. A final consideration for animal grouping should be the pre-infection scoring of pharyngeal erythema and tonsil size. Scoring is performed as described in **Subheading 3.4.2**.

3.3. GAS Non-Human Primate Infection

Non-human primates are infected on day 0 of the experiment through instillation of 1 mL PBS containing GAS into the nares of each animal. Several parameters of infection are monitored during the 5-week protocol, with samples isolated as listed in **Table 1**. All animals should be housed individually during the infection phase of the protocol to prevent cross-contamination.

1. Thaw 1 h prior to animal infection one tube each of the wild-type and mutant strains on ice and dilute to 1×10^7 CFUs/mL using PBS.
2. Fill four 1-mL syringes with GAS-PBS solution for each of the two GAS strains.
3. Perform day-0 sample isolations and clinical scorings as outlined in **Table 1** and **Subheading 4.4**.
4. Infect each animal by slowly dribbling 1 mL of GAS-PBS solution into their nares (~0.5 mL per nostril).
5. Titer dilute GAS-PBS solutions to confirm that correct infectious dose was given to the animals for each of the two GAS strains. Use THY agar or TSA blood agar plates for plating serial dilutions for titer determination.

3.4. Sample Isolations and Clinical Scoring of Disease

Samples and clinical scoring occurs on the days outlined in **Table 1**. The order of events is as follows.

1. Anesthetize animals with an intramuscular injection of ketamine (10 mg/kg) and bring them individually into the examination room.
2. Obtain rectal temperature and weight of the animal.
3. Take a blood sample (*see* **Subheading 3.4.1.**) and place the animal onto sheets of disposable absorbent bench covers.
4. Using a laryngoscope with disposable blades, score pharyngeal erythema and tonsil size (performed by a veterinarian) (*see* **Subheading 3.4.2.**). Replace bench covers and laryngoscope blades after examination of each animal.
5. Following clinical scoring, obtain five throat swabs (*see* **Subheading 3.4.3.**), collect approximately 1 mL of saliva (*see* **Subheading 3.4.4.**), and perform a nasal wash using PBS (*see* **Subheading 3.4.5.**).
6. Return animals to holding room.

3.4.1. Serum is Used to Measure Host Immune Responses to Infection

The host response to infection is monitored by measurement of serum C-reactive protein levels and antibody development against various GAS antigens *(21–24)*. C-reactive protein concentrations are determined by ELISA with commercially available anti-C-reactive protein antibodies. Antibody production is assayed by ELISA using purified streptococcal proteins (e.g., SLO; **Subheading 3.2.2.**).

1. Take 3-mL blood samples from each of the anesthetized animals by venipuncture in the cephalic vein.
2. Draw blood using a Vacutainer™ system with a red-capped collection tube and incubate on ice until all macaque manipulations are completed.
3. Incubate blood samples at 4 °C for 1 h to allow the blood to clot before centrifuging at $500 \times g$ for 20 min at 4 °C.
4. Remove aqueous (serum) phase from each sample and place into individual red-capped 2-mL tubes.
5. Freeze serum samples at –80 °C until required.

3.4.2. Clinical Scoring of Disease

Development of disease is monitored by determination of tonsillitis and pharyngeal erythema severity scores prior to throat swabbing. Briefly, score pharyngeal erythema on a 4-point scale as follows: no erythema *(0)*, mild erythema with hyperemic blood vessels (+1), more intense erythema and palatal petechiae (+2), and intense erythema with palatal petechiae and exudative tonsillitis (+3). Score macaque tonsil size on a 5-point scale using the same criteria as established by Feinstein and Levitt *(27)* for scoring tonsil enlargement during human GAS pharyngitis (*see* **Fig. 1**).

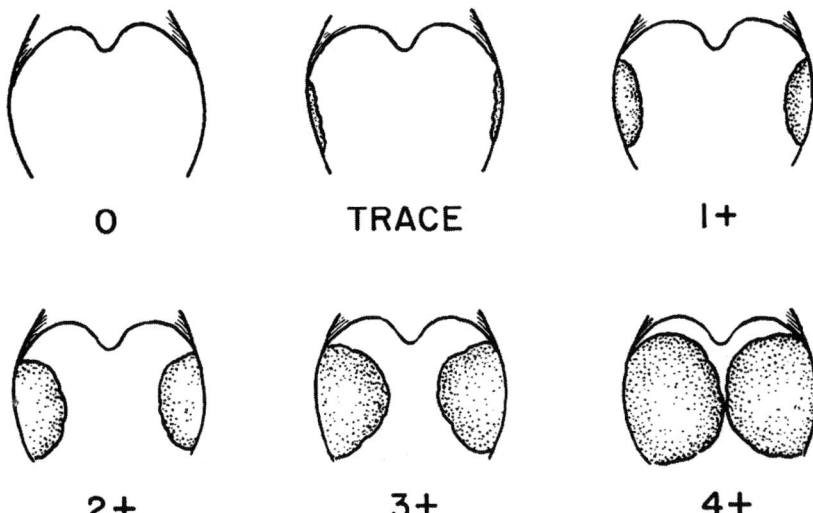

Fig. 1. Diagram of grading scheme used for denoting size of tonsils. Figure taken with permission from reference *(27)*. Copyright © 2006, Massachusetts Medical Society. All rights reserved.

3.4.3. GAS Colonization Levels

The level of GAS colonization during infection is monitored by enumeration of CFUs from two of five throat swabs taken at various intervals from each of the infected animals during the infection protocol. Throat swabs are processed to release adherent bacteria and plated on select agar to minimize growth of resident microbial flora from the macaque posterior pharynx. The three remaining throat swabs are snap frozen and stored at –80 °C. Our laboratory freezes these three throat swabs for future use in expression microarray and/or quantitative PCR analyses to monitor GAS and/or host gene expression during upper respiratory tract infection *(21,22)*.

1. Vigorously swab the throat of each anesthetized animal using five sterile cotton-tipped wood applicators sequentially.
2. Use the first and second swabs to determine GAS colonization levels. Place the first and second swabs into individual 2-mL yellow-capped tubes containing 1 mL PBS. Break off the cotton-tipped end of each swab inside the tube, close tube, and place on ice.
3. Place throat swabs three through five individually into empty 2-mL yellow-capped tubes, snap off the cotton tips, close tubes, freeze in liquid nitrogen, and place at –80 °C for future manipulation [discussion of how throat swabs are processed for gene expression studies is outside the limitation of this chapter, see Virtaneva *et al. (21)* for details].

4. To process throat swabs for GAS CFU enumeration, add DTT to 0.7 mM (7 µL of a 0.1 M stock DTT solution), vortex, and incubate for 15 min at room temperature. DTT facilitates the release of GAS from the swab into the surrounding PBS.
5. Vortex samples a second time following the 15-min DTT incubation.
6. Make serial dilutions in PBS and plate 100 µL of neat, 10^{-1}, 10^{-2}, and 10^{-3} dilutions onto select agar plates.
7. A single plate is used per throat swab per dilution, resulting in 64 plates per time-point (8 animals × 2 swabs × 4 dilutions).
8. Incubate plates at 37 °C with 5% CO_2 for 24 h.
9. Following incubation, enumerate GAS colonies and determine colonization levels for each animal. Note that a single GAS strain may produce colonies with several different morphologies. Coupled with the natural microbial flora of the macaque posterior pharynx, this complicates the task of GAS CFU enumeration.
10. To confirm that scored bacterial colonies are indeed GAS, amplify the *emm* gene by colony PCR from several colonies per throat swab per time-point. *emm* encodes the M protein, a major GAS-specific virulence factor *(28)*.

3.4.4. Saliva Isolation

Saliva samples from infected macaques enable testing of IgA production during the course of infection. Approximately 1 mL of saliva is collected by aspiration from the cheek pouch while the animals are under anesthesia. Saliva samples from each animal are placed into individual blue-capped 2-mL tubes and put on ice until all animal manipulations are completed. No processing of the saliva samples is required, with the tubes simply being stored at –80 °C until needed.

3.4.5. Nasal Washes

Nasal washes also provide a means to test IgA production during infection. Washes are collected by instilling 0.7 mL of sterile PBS into each nostril and immediately aspirating wash fluid and secretions with a sterile syringe. Nasal washes are placed into white-capped 2-mL tubes on ice until all animal manipulations are completed. Nasal wash samples are processed by centrifugation in a microfuge at 500 × g for 10 min at 4 °C, removing the supernatant to new white-capped tubes and storing at –80 °C.

3.5. Post-experiment Antibiotic Treatment of Animals

Following sample isolations on day 28 of the experiment, the macaques are each given an intramuscular injection of penicillin G (20,000 U/kg). Two additional doses of penicillin G are given during the following week. Following antibiotic treatment, throat swabs are taken of the animals and used to confirm that they are culture-negative for GAS and hence have cleared the infection.

3.6. Data Analysis

To test for significance between data from animals grouped based upon the strain of infecting GAS (wild-type or mutant), we use generalized estimating equations to analyze the repeated binary outcomes in the data *(23)*. A *P* value < 0.05 is considered significant.

4. Notes

1. Occasionally, a small amount of blood is visible in the posterior pharynx of the animal as a result of the infection; this is to be avoided if possible when swabbing. Blood on the throat swab provides a nutrient source that can enable GAS to replicate following isolation but prior to plating for CFUs, thus skewing the data.
2. The presence of blood at the posterior pharynx may serve as a useful additional clinical symptom and should be noted for each time-point for each animal.
3. Animals may require an increased dose of ketamine as the experiment progresses to overcome resistance development (increase from 10 mg/kg to 14 mg/kg).
4. Cynomolgus macaque tonsils are routinely larger in size than human tonsils, and hence it is normal for macaques to have pre-infection tonsil size scores of +1 or +2.
5. All tubes should be pre-labeled to prevent confusion during sample isolation. We also strongly recommend using color-coded tubes for each of the sample types isolated (red, blood; yellow, throat swab; blue, saliva; and white, nasal wash).
6. A small flashlight may be useful to illuminate the macaque posterior pharynx during clinical scoring of disease.
7. Additional confirmation that throat-swab recovered bacteria are GAS can be gained by using a GAS agglutination kit. We have used the BD BBL Streptocard Acid Latex Test kit with good success.

References

1. Cunningham, M. W. (2000). Pathogenesis of group A streptococcal infections. *Clin. Microbiol. Rev.* **13**, 470–511.
2. Musser, J. M. and Krause, R. M. (1998). The revival of group A streptococcal diseases, with a commentary on staphylococcal toxic shock syndrome, p. 185–218. In R.M. Krause (ed.), *Emerging Infections*. Academic Press, San Diego, CA.
3. Bisno, A. L., Rubin, F. A., Cleary, P. P., and Dale, J.B. (2005). Prospects for a group A streptococcal vaccine: rationale, feasibility, and obstacles–report of a National Institute of Allergy and Infectious Diseases workshop. *Clin. Infect. Dis.* **41**, 1150–1156.
4. Carapetis, J. R., Steer, A. C., Mulholland, E. K., and Weber, M. (2005). The global burden of group A streptococcal diseases. *Lancet Infect. Dis.* **5**, 685–694.
5. Musser, J. M. and DeLeo, F. R. (2005). Toward a genome-wide systems biology analysis of host-pathogen interactions in group A *Streptococcus. Am. J. Pathol.* **167**, 1461–1472.

6. Hollingshead, S. K., Simecka, J. W., and Michalek, S. M. (1993). Role of M protein in pharyngeal colonization by group A streptococci in rats. *Infect. Immun.* **61**, 2277–2283.

7. Husmann, L. K., Yung, D. L., Hollingshead, S. K., and Scott, J. R. (1997). Role of putative virulence factors of *Streptococcus pyogenes* in mouse models of long-term throat colonization and pneumonia. *Infect. Immun.* **65**, 1422–1430.

8. Beattie, J. M. (1907). A contribution to the bacteriology of rheumatic fever. *J. Exp. Med.* **9**, 186–207.

9. Blake, F. G. and Russell, L. C. (1920) Experimental *streptococcus haemolyticus* pneumonia in monkeys. *J. Exp. Med.* **32**, 401–436.

10. Watson, R. F., Rothbard, S., and Swift, H. F. (1946). Type-specific protection and immunity following intranasal inoculation of monkeys with group A streptococci. *J. Exp. Med.* **84**, 127–142.

11. Köhler, W., Djikidse, E. K., Kuehnemund, O., Knoell, H., Mgakjan, G. O., and Juergens, I. (1976) Untersuchungen zur Immunobiologie von M-Proteinen des *Streptococcus pyogenes*. III. Mitteilung: Immunisierung von Rhesusaffen (*Macaca mulatta*). *Allergie und Immunologie* **22**, 197–219.

12. Bryant, A. E. and Stevens, D. L. (1997) Expression of activational markers on circulating leukocytes from baboons with group A streptococcal (GAS) bacteremia. *Adv. Exp. Med. Biol.* **418**, 801–804.

13. Taylor, F. B. J., Bryant, A. E., Blick, K. E., Hack, E., Jansen, P. M., Kosanke, S. D., and Stevens, D. L. (1999) Staging of the baboon response to group A streptococci administered intramuscularly: a descriptive study of the clinical symptoms and clinical chemical response patterns. *Clin. Infect. Dis.* **29**, 167–177.

14. Lukomski, S., Hoe, N.P., Abdi, I., Rurangirwa, J., Kordari, P., Liu, M., Dou, S.J., Adams, G. G., and Musser, J. M. (2000) Nonpolar inactivation of the hypervariable streptococcal inhibitor of complement gene (*sic*) in serotype M1 *Streptococcus pyogenes* significantly decreases mouse mucosal colonization. *Infect. Immun.* **68**, 535–542.

15. Husmann, L. K., Dillehay, D. L., Jennings, V. M., and Scott, J. R. (1996) *Streptococcus pyogenes* infections in mice. *Microb. Pathog.* **20**, 213–224.

16. Park, H. S. and Cleary, P. P. (2005) Active and passive intranasal immunizations with streptococcal surface protein C5a peptidase prevent infection of murine nasal mucosa-associated lymphoid tissue, a functional homologue of human tonsils. *Infect. Immun.* **73**, 7878–7886.

17. Vanace, P. W. (1960) Experimental infection of the rhesus monkey. *Ann. N. Y. Acad. Sci.* **85**, 910–930.

18. Zimmerman, R. A., Krushak, D. H., Wilson, E., and Douglas, J. D. (1970) Human streptococcal disease syndrome compared with observations in chimpanzees. II. Immunologic responses to induced pharyngitis and effect of treatment. *J. Infect. Dis.* **122**, 280–289.

19. Taranta, A., Spagnuolo, M., Davidson, M., Goldstein, G., and Uhr, J. W. (1969) Experimental streptococcal infections in baboons. *Trans. Proc.* **I**, 992–993.

20. Ashbaugh, C. D., Moser, T. J., Shearer, M. H., White, G. L., Kennedy, R. C., and Wessels, M. R. (2000) Bacterial determinants of persistent throat colonization and

the associated immune response in a primate model of human group A streptococcal pharyngeal infection. *Cell. Microbiol.* **2**, 283–292.

21. Virtaneva, K., Graham, M. R., Porcella, S. F., Hoe, N. P., Su, H., Graviss, E. A., Gardner, T. J., Allison, J. E., Lemon, W. J., Bailey, J. R., Parnell, M. J., and Musser, J. M. (2003) Group A *Streptococcus* gene expression in humans and cynomolgus macaques with acute pharyngitis. *Infect. Immun.* **71**, 2199–2207.

22. Virtaneva, K., Porcella, S. F., Graham, M. R., Ireland, R. M., Johnson, C. A., Ricklefs, S. M., Babar, I., Parkins, L. D., Romero, R. A., Corn, G. J., Gardner, D. J., Bailey, J. R., Parnell, M. J., and Musser, J. M. (2005) Longitudinal analysis of the group A *Streptococcus* transcriptome in experimental pharyngitis in cynomolgus macaques. *Proc. Natl. Acad. Sci. U.S.A.* **102**, 9014–9019.

23. Sumby, P., Barbian, K. D., Gardner, D. J., Whitney, A. R., Welty, D. M., Long, R. D., Bailey, J. R., Parnell, M. J., Hoe, N. P., Adams, G. G., DeLeo, F. R., and Musser, J. M. (2005) Extracellular deoxyribonuclease made by group A *Streptococcus* assists pathogenesis by enhancing evasion of the innate immune response. *Proc. Natl. Acad. Sci. U.S.A.* **102**, 1679–1684.

24. Sitkiewicz, I., Nagiec, M. J., Sumby, P., Butler, S. D., Cywes-Bentley, C., and Musser, J. M. (2006) Emergence of a bacterial clone with enhanced virulence by acquisition of a phage encoding a secreted phospholipase A_2. *Proc. Natl. Acad. Sci. USA* **103**, 16009–16014.

25. Sawada, S., Shida, M., Suenaga, R., Mizuma, H., Karasaki, M., Hashimoto, M., Kawano, K., and Amaki, I. (1986) Elicited antibody nature of human monoclonal protein with anti-streptolysin O activity–analysis with monoclonal anti-idiotype antibody. *J. Clin. Lab. Immunol.* **19**, 31–35.

26. Sumby, P., Porcella, S. F., Madrigal, A. G., Barbian, K. D., Virtaneva, K., Ricklefs, S. M., Sturdevant, D. E., Graham, M. R., Vuopio-Varkila, J., Hoe, N. P., and Musser, J. M. (2005) Evolutionary origin and emergence of a highly successful clone of serotype M1 group A *Streptococcus* involved multiple horizontal gene transfer events. *J. Infect. Dis.* **192**, 771–782.

27. Feinstein, A. R. and Levitt, M. (1970) The role of tonsils in predisposing to streptococcal infections and recurrences of rheumatic fever. *N. Engl. J. Med.* **282**, 285–291.

28. McArthur, J. D. and Walker, M. J. (2006) Domains of group A streptococcal M protein that confer resistance to phagocytosis, opsonization and protection: implications for vaccine development. *Mol. Microbiol.* **59**, 1–4.

IV

IDENTIFICATION OF THERAPEUTIC TARGETS AND TYPING METHODS

21

Target-Based Antimicrobial Drug Discovery

Lefa E. Alksne and Paul M. Dunman

Summary

The continued increase in antibiotic resistance among bacterial pathogens, coupled with a decrease in infectious disease research among pharmaceutical companies, has escalated the need for novel and effective antibacterial chemotherapies. While current agents have emerged almost exclusively from whole-cell screening of natural products and small molecules that cause microbial death, recent advances in target identification and assay development have resulted in a flood of target-driven drug discovery methods. Whether genome-based methodologies will yield new classes of agents that conventional methods have been unable to is yet to be seen. At the end of the day, perhaps a synergy between old and new approaches will harvest the next generation of antibacterial treatments.

Key Words: Antibacterial; drug discovery; antibacterial target; antimicrobial chemotherapy.

1. Introduction

In what is now sometimes referred to as the post-genomic age of microbiology *(1)*, during which hundreds of sequenced bacterial genomes have been mined for putatively essential genes, antibacterial drug discovery is primarily target driven *(2)*. For better or worse, use of whole-cell assays in which the only marker of antibacterial activity is growth inhibition or death are currently rare. The multitude of gene products that have been identified as potential antimicrobial targets, along with significant advancements in detection methods, provide researchers with a new standard for attack *(3)*. The trick to successful drug discovery, however, is to identify and pursue the right target with the right method.

From: *Methods in Molecular Biology, vol. 431: Bacterial Pathogenesis*
Edited by: F. DeLeo and M. Otto © Humana Press, Totowa, NJ

It is generally accepted that a target must be essential for survival (in a laboratory setting), broadly represented among the target organisms, and preferably not homologous, or overly so, to a eukaryotic counterpart *(4)*. It must also be "druggable," that is, amenable to inhibition by a small molecule or biotherapeutic, as well as accessible to such interaction in vivo *(5)*. When deciding whether the method used to detect target inhibitors should occur in the test tube or in whole cells, one must take into account whether the target is a pathway or a single gene product, whether the target can be isolated, and whether the cellular environment needs to be considered for target activity. Until recently, there were only a handful of cellular targets that were proven to lead to successful antibacterial chemotherapy when inhibited. More recently, other types of targets have been exploited, and representatives of both sets will be given below. In addition, assay types that are amenable to use with such targets will be reviewed. Whether target-driven antimicrobial drug discovery will ultimately prove more fruitful than tried and true methods remains to be determined.

2. Methods of Target Identification

One need not look farther than any bacterial physiology text to identify potential targets for antimicrobial drug discovery. These targets include well-characterized components of essential biological processes whose usefulness in chemotherapic intervention of bacterial infection has been validated by assorted means (discussed below). In general, any enzyme that is proven to be essential for cellular survival (either in vitro or in vivo) constitutes a starting platform for drug discovery efforts. A second "class" of targets, which are also described below, includes virulence factors and their regulatory molecules. A third "class," novel targets, generically refers to gene products that are essential for cellular survival but whose function(s) is poorly understood. Benefits of focusing drug discovery efforts on a novel target include freedom of intellectual property issues and increased probability of avoiding compound inactivation by commonly encountered antibiotic resistance determinants. However, there are several drawbacks associated with work with novel targets. Perhaps, the largest obstacle includes developing a functional assay to screen for compounds that inhibit protein function and thus constitute a potential antimicrobial agent. Nonetheless, it is exciting to consider that with each deposit of bacterial genomic sequence content into publicly available databases lies potential novel targets for antimicrobial drug discovery. To one degree or another, the identification of potential novel targets includes a bioinformatics-based comparison of bacterial genomes to delineate genes that are conserved across pathogens of interest but have acceptable amino acid sequence divergence from human gene products.

Development of broad-spectrum antibiotics should minimally include a comparison of genomic content of bacteria involved in respiratory tract, surgical site, and urinary tract infections *(6)*. Bioinformatics procedures can also provide clues into a protein's function and aid in assay development to a certain extent. For instance, one can determine whether a putative target contains ATP hydrolysis domain, which can in turn be exploited when developing an assay using purified protein (i.e., ATPase activity assay). While in silico techniques provide a starting point for identifying and perhaps prioritizing putative targets for drug discovery, wet laboratory tactics are needed to definitively determine whether a gene is essential for cellular survival. This can be accomplished through both indirect or, preferably, direct approaches *(7)*. The former involves using random mutagenesis with agents such as nitrosoguanidine to identify and characterize conditional lethal mutants. Another indirect means of assigning essentiality of putative novel targets involves analyzing members of a transposon mutant library *(8)*. Transposon library members actually contain insertions into non-essential genes, which can be subtracted from the genetic composition of a bacterial genome to indicate which genes are essential for in vitro survival. Transposon subtractive analysis has successfully identified genes that are essential for the growth of Escherichia coli and Helicobacter pylori *(9,10)*. One can also directly determine whether a putative target is essential through targeted gene disruptions, through various methods the description of which are beyond the scope of this discussion [reviewed in *(7)*]. Nonetheless, the overall goal is to establish that an essential gene cannot be completely knocked out. Below, we describe drug discovery efforts in these three "classes" of targets (characterized and poorly characterized essential genes, and virulence determinants).

3. Types of Targets

3.1. Oldies but Goodies

Until very recently, antibacterial targets could be clustered into a limited number of categories. Historically, antibiotics have been natural fermentation products derived from organisms seeking to protect themselves from microbial attack and tended to be complex molecules targeting specific macromolecular structures or pathways *(11)*. Interestingly, recent analysis of essential gene products analyzed by several genomic and proteomic methods recurrently identified the same components as antimicrobial targets *(5,12)*. As is so often true in the case of antimicrobial strategy, nature appears to know best.

3.2. The Cell Wall

In the dawn of antimicrobial research, the cell wall was recognized as an ideal target for chemotherapy *(13)*. The cell was has two significant advantages as a target for an antibacterial agent: its building blocks primarily consist of components external to the cell, thus allowing easy target access, and it has no eukaryotic counterpart, ensuring selectivity and a large therapeutic window *(14)*. Among the earliest antibiotics recognized for human use was penicillin, a member of the beta-lactam antibiotics, which targets components involved in the final step of cell wall assembly, called for obvious reasons, penicillin-binding proteins. These proteins have been exploited by multiple generations of beta-lactam antibiotics, both natural and derived from target-based design, and represent a cornerstone of antimicrobial chemotherapy. More recently, resistance mechanisms involving beta-lactamases, enzymes capable of degrading beta-lactam antibiotics, have presented a challenge for this class of agents *(15)*. This has led to combination therapy in which an agent targeting a penicillin-binding protein is combined with an agent that inhibits its corresponding beta-lacatamase, that is, a two-target attack in tandem.

The penicillin-binding proteins are by no means the only viable targets within the cell wall. One of the most essential antibiotics in the modern arsenal, vancomycin, is a glycopeptide natural product that binds to the D-Ala-D-Ala dipeptide terminus of peptidoglycan precursor lipid II *(16)*. Other glycopeptides or related antibiotics currently in use or under development similarly target the transglycosylation step that occurs external to the membrane *(17)*. Other accessible targets include MurG, which is membrane associated and inhibited by ramoplanin. Perhaps more difficult to target and thus less exploited are enzymes involved in the earlier steps of cell wall synthesis catalyzed by MurA-MurF. In large part, a reason these enzymes have been less exploited by "cell wall" drug discovery efforts is that they are internally located. Nonetheless, as described below, most commercial antimicrobials target enzymes/pathways within the cell, suggesting that MurA-MurF are excellent targets. Indeed, an inhibitor of this pathway, fosfomycin, has in fact been developed *(18)*.

3.3. Translation

The most fertile source of antibacterial targets to date is the ribosome *(14,19)*. From the early aminoglycosides to linezolid, a member of one of the newest classes of antibiotics, the bacterial protein synthetic machinery has been widely exploited. While the level of homology between bacteria and eukaryotic organisms is strong enough to support conserved function *(20)*, the divergence is enough to afford a tolerable window of safety and to make the ribosome a rich source for target-driven antimicrobial drug discovery.

Different classes of antibiotics have been defined that target various structures and functions of the protein synthetic apparatus. For example, the aminoglycosides target the 30S ribosomal subunit, whereas the 50S subunit is the target of many antibiotic classes, including the phenyl propanoids, streptogramins, the macrolides, and the ketolides to name a few *(19)*. An interesting aspect of ribosomal-based antimicrobial antibiotics is the critical role of agent-RNA interaction in this macromolecular structure. In addition, the fact that the ribosomal RNA is primarily the target and that it is so remarkably conserved among bacteria makes ribosome-based antimicrobial chemotherapy amenable to empirical use. The exquisitely detailed understanding of ribosomal three-dimensional structure makes this a particularly rich source for target-based drug design *(21)*. Additional components of the translational machinery have also recently been explored, and these will be described below.

3.4. Transcription and Replication

Two multipart molecular machines (holoenzymes) that continue to be exploited for potential antimicrobial inhibition are RNA polymerase, an enzyme comprised of multiple conserved subunits and required for the transcription of the bacterial genome *(22)*, and the replisome, required for its replication *(23)*. The 1950s witnessed the identification of the rifamycins, which target the RpoB subunit of RNA polymerase and of which rifampicin is one of the most potent antibiotics yet discovered *(14)*. Unfortunately, resistance to this class of antibiotics can occur readily, limiting their usefulness. Thus far, this enzyme complex has remained somewhat intractable to target-based drug discovery but may yet prove to have an Achilles' heel. Somewhat later than the rifamycins, members of the quinolone class were described, which target DNA gyrase and topoisomerase activities of the replisome. Quinolones have been exploited and expanded extensively by synthetic modification; however, it is likely that much could be done to identify novel inhibitors of the replicative machinery as better tools to probe the dynamic structure of the complexes come into use.

3.5. Other Validated Antibacterial Targets

Two other validated targets for antimicrobial chemotherapy will be mentioned here, both of which are defined by pathway inhibition. The earliest antibiotics, the sulfonamides, are inhibitors of the folate biosynthetic pathway *(24)*. Sulfamethoxazole and trimethoprim, targeting dihydropteroate synthase and dihydrofolate reductase, respectively, are each susceptible to rapid emergence of resistance but have been used successfully in combination. Recently, other enzymes in the pathway have begun to engender interest for target-based screening. Fatty acid biosynthesis is another pathway that has previously been validated for drug discovery and may yet succumb to targeted screening *(25)*. The fatty acid synthase complex is distinct between bacteria and eukaryotes, and the finding that triclosan, thiolactomycin, and cerulenin all target components of the pathway suggest that the pathway should be considered further.

3.6. A Selection of Novel Targets—Innovative Approaches

As described above, the genomic age has resulted in a wealth of gene products that have been validated as potential targets, at least in terms of their essential nature and conservation. In many cases, these are the members of the same well-explored groups highlighted above, but in many others, newly identified targets have been culled. In addition, as genomics and other techniques yield a deeper understanding of bacterial physiology and better ways to manipulate it, researchers are afforded new opportunities to screen against previously known physiological targets. Certainly not all of the targets currently under consideration can be vetted here (and not even that subset that has been divulged to the public), but a selected subset of interest will be touched upon.

Several specific enzymes in different biosynthetic pathways have been described in terms of potential drug inhibition in the past few years. UDP-GlcNac deacetylase (LpxC) is a hydrolase critical for the biosynthesis of lipopolysaccharide lipid A, required for outer membrane synthesis in Gram-negative bacteria *(16)*. Several series of compounds have been designed that relate to the substrate structure, some of which have demonstrated limited antibacterial activity. More recently, a whole-cell screen has identified sub-micromolar inhibitors of the enzyme but with limited spectrum. Nevertheless, it would appear that this target has been validated. Similarly, the enzyme phospho-pantetheine adenylyltransferase, which catalyzes the last step in the biosynthesis of the essential molecule CoA, has been investigated as a novel target, in one case by a high-throughput screen for pyrophosphate release inhibition *(26)*. This screen did in fact identify inhibitors of the enzyme; however, the inhibition did not translate into antibacterial activity (a common obstacle for molecules identified by in vitro high-throughput screening). An additional pathway being mined for druggable targets is that of lysine biosynthesis *(27)*. In bacteria, de novo lysine synthesis is of course required for protein synthesis, but lysine is also a critical component of peptidoglycan. The pathway leading to lysine production is the diaminopimelate pathway and contains steps common to lysine, threonine, isoleucine, and methionine production. Efforts have been made to synthesize small inhibitors of several enzymes in the pathway, primarily as mimics of small amino acid-derived substrates and products. These efforts have yet to yield agents with antibacterial activity but further an understanding of potential sites for disruption of the pathway.

3.7. Well-Defined Complex Macromolecules–New Looks at Old Friends

DNA replication and the complex structural and catalytic components involved remain clear targets for new antimicrobial efforts and numerous researchers are investigating them. Recently, a number of novel and selective inhibitors of DNA polymerase III in Gram-positive bacteria have been described in the literature. A recent report describes the use of hexapeptides that theoretically bind to the Holliday junctions formed during DNA repair *(28)*. Treatment with the hexapeptides appears to inhibit resolution of the junctions and results in accumulation of DNA breaks, leading to filamentation, erroneous DNA partitioning, and anucleate cells. Researchers demonstrated antimicrobial activities associated with these inhibitors in vitro but to date have not tested them in animal models. Nevertheless, they suggest that the inhibitors might be useful either as antimicrobial agents alone or in combination or at least as tools to aid in understanding DNA repair.

It is widely believed that as the structure of the translational machinery is elucidated even further, the ribosome will yield even more opportunities for antimicrobial intervention *(21)*. Moreover, other enzymes required for efficient and accurate protein synthesis are being investigated as targets. Aminoacyl tRNA synthetases are essential enzymes required for the charging of tRNAs with their respective amino acids. Because the catalytic mechanism is well understood, the design of substrate or reaction-intermediate mimics can be utilized *(29)*. Both natural product and synthetic libraries have been screened for inhibitors and have yielded multiple candidate inhibitors. While many have demonstrated significant activity in vitro, limited success has been demonstrated to date against whole cells, once again highlighting the challenge of translating enzymatic inhibition into actual antibacterial activity. However, limited reports of activity in infection models maintain interest in these enzymes for use as chemotherapeutic targets.

Another target of great interest in the protein synthetic pathway is peptide deformylase *(30,31)*. This metalloenzyme is required for the removal of the N-terminal formyl group of newly synthesized proteins. This process is absent in eukaryotes, and hence this enzyme is an attractive target. Several peptide deformylase inhibitory compound series have been identified, some exhibiting activity in animal models, suggesting that this target might prove fruitful in the search for novel antimicrobials.

3.8. Virulence as a Target – a Kinder and Gentler Form of Disarmament

In the search for new targets, recently thoughts have turned to a unique group of gene products that are not inherently essential in vitro but that are critical for pathogenic bacterial survival in the host, that is, gene products involved in virulence itself. Numerous targets have been identified that are required for the establishment and maintenance of infection, including those involved in the processes of adherence, invasion, and avoidance of the host defenses. While these may serve better as vaccine candidates, multiple companies are exploring targets such as type III secretion, quorum sensing, two-component regulatory systems, and sortase-mediated attachment of Gram-positive virulence factors to name only a few *(1,32–35)*. It has been postulated that inhibition of such targets would debilitate the invading organism and allow the host to clear the infection immunologically. It has further been suggested that this type of inhibition could in some cases spare the commensal flora and hence be a more gentle approach to antimicrobial chemotherapy. Targeting virulence has an added challenge in that inherently a number standard secondary assays for activity (e.g., antibacterial activity) are impossible. While proof of concept has been established using knock-outs of target genes and animal models of infection and while numerous inhibiting agents have been identified, to date this line of research has yet to deliver on its potential. Further, their usefulness in patients with debilitated immune systems has to be considered, which when coupled to the fact that there are a plethora of targets with essential function(s) has made virulence components a controversial area of drug discovery exploration.

4. Challenges for Target-Driven Antibacterial Drug Discovery – How to Hit the Bullseye

Once a target for antibacterial intervention has been selected and validated, a path to obtaining inhibitors must be chosen. Assays can be cell based or in vitro, actual or virtual, and screening methods can be high-throughput, focused, or anywhere in between. The size and type of the net to be cast will determine not only the quantity of "hits" available for characterization but often will determine the quality and type.

Current technologies allow thousands of inhibition assays to be run simultaneously and in vanishingly small volumes *(36)*. These high-throughput screening campaigns have become shorter and shorter with more sophisticated detection methods *(37)* and now can include secondary assays to improve reproducibility and determine selectivity. However, this bulk approach has its own limitations. The robotics inherent in high-throughput screening cannot make fine adjustments to create perfect assay conditions. Large amounts of materials are required because of large void volumes, and ideal temperatures may not be maintained. Assay components must be stable over long periods of time. Nevertheless, high-throughput screening undoubtedly offers a staggering improvement in assaying large collections of potential inhibitors.

In vitro assays in which the target has been purified or partially isolated are widely used in high-throughput screens. Advantages to having purified components are obvious. One can adjust parameters to optimize substrate concentrations and conditions *(38)*. Activity kinetics can be characterized, and often binding curves can be generated to thoroughly understand the mechanism of inhibition and optimize inhibitory measurements. However, assay development time can be longer than a cell-based screen in that purified materials are required and may be difficult to obtain. Ideal conditions must be determined, and those conditions may not reproduce the cellular milieu in a biologically meaningful way. In addition, in vitro assays for antimicrobial agents may not ultimately be active in vivo if the inhibitor cannot penetrate or are modified by the cell. Lipinski's rules are often utilized to predict whether an identified inhibitor will have a chance of actual "drug-like" activity *(39)*. It should be noted that in vitro schemes for identifying inhibitors require secondary assays with alternative detection methods to ensure that target modulation and biological activity correlate *(40,41)*.

In vitro assays are often designed to detect inhibitors of enzyme activity and are ideal if the purified enzyme is available. Depending on reaction rate, continuous detection or stopped reaction times may be used. Rate-based assays are highly sensitive, and the generation of multiple fluorescent probes has provided exceptional tools for their development *(37)*. Often, product detection is the measure used to determine inhibition. In vitro assays may not always only measure the activity of a single enzyme—it is quite possible to examine the activities of linked components of a pathway. A complicating component of fluorescent readout assays that must be considered is that compounds (putative inhibitors) may alter the fluorescent properties of the probe being measured necessitating the need for secondary assays (as described above).

In vitro inhibition assays need not be only for enzymatic targets. Purified structural components may also be investigated although assay development and interpretation of results may be more challenging. Assays to determine affinity of binding, both of protein-protein and nucleic acid interactions, have been developed and can be used in moderate throughput *(42)*. These can include nuclear magnetic resonance or mass spectrophotometric techniques, as well as fluorescent modulation or calorimetric methods. Obvious drawbacks to these methods are the need for pure and stable components and challenges in identifying inhibitors capable of disrupting high-affinity interactions that will actually be antibacterial and non-toxic at the end of the day. These methods are, however, highly useful in the characterization of the inhibitor–target interface.

In vivo or whole-cell assays have certain advantages over in vitro assays in that the complexities of biological systems can be addressed directly *(2)*. While targets with multiple components can be assayed simultaneously in vitro, assessments of antibacterial activity and target accessibility are intrinsic to a whole-cell method. However, issues plague this assay scheme as well. "Off target" activity is extremely possible (or even likely). Determining selectivity of inhibition is a challenge. Just as in vitro assays require secondary assays to ensure target modulation is the root cause of antibacterial activity, in vivo screens require multiple secondary assays to confirm mechanism of action *(43)*.

Readouts for in vivo drug discovery are multiple and diverse. Spectrophotometric methods to assay calorimetric reagents, changes in metabolites, or optical density have been used. Reporter gene fusion systems in which modulation of a target or target expression results in a detectable change in, for example, nutrient requirements or color, have been used repeatedly *(2)*. For pathway screens in which inhibition of any one of a number of gene products can result in antibacterial activity, induction of a reporter signal generated downstream in the pathway is ideal. One parameter that must be considered for these screens is the timing of signal detection versus cell death, as death will be the ultimate outcome of successful target inhibition and may inherently result in signal loss.

A final type of screening is of course virtual, in which the scientist utilizes known structural information to predict what type of molecule will bind to a given target and presumably alter its function *(44,45)*. Prediction programs and advances in crystallographic methods have made this a useful method, where reagents, time, or other resources may be limiting. Virtual screening is also often used to take advantage of a known inhibitor and model structure–activity relationships. As attractive as this method can be, it cannot replace the wet laboratory approach and any predicted structures need to be tested at the bench for at the least antimicrobial activity if not target interaction.

5. Concluding Thoughts

This discussion has attempted to provide a top-level view of target types currently and historically under investigation and some thoughts on various methods for targeting them. While the fact that the genomic revolution has provided a plethora of targets known to be essential for bacterial viability is undeniable, it is not clear to date how many of them will provide any use in drug discovery. Have all the low-hanging fruit been gathered such that the search for new targets must continue? As noted before, it is interesting that many of the targets identified and currently under investigation are the same as those inhibited by the venerable agents of old. Is it perhaps true that these really are the most conducive to antimicrobial drug discovery and that efforts should focus on novel methods rather than novel targets? An interesting endeavor recently reported utilizes defining essentiality by chemical inhibition, thus identifying target and inhibitor simultaneously *(46,47)*. Perhaps this approach will yield a next cohort of antimicrobials. However, as noted earlier, natural products have been the most fruitful source of inhibitors, and many of these were identified by good old bacterial death *(11,48)*. It remains to be seen how the fruits of genomics can be combined with lessons learned over the last 60 years to supplement our antimicrobial arsenal.

References

1. Lerner, C. G. and Beutel, B. A. (2002) Antibacterial drug discovery in the post-genomics era. *Curr. Drug Targets Infect. Disord.* **2**, 109–119.
2. Miesel, L., Greene, J., and Black, T. A. (2003) Genetic Strategies for antibacterial drug discovery. *Nat. Rev. Genet.* **4**, 442–456.
3. Monaghan, R. L. and Barrett, J. F. (2006) Antibacterial drug discovery–then, now and the genomics future. *Biochem. Pharmacol.* **71**, 901–909.
4. Vila, J., Sanchez-Cespedes, J., and Giralt, E. (2005) Old and new strategies for the discovery of antibacterial agents. *Curr. Med. Chem. Antiinfect. Agents* **4**, 337–353.
5. Schmid, M. B. (2006) Do targets limit antibiotic discovery? *Nat. Biotechnol.* **24**, 419–420.
6. Black, M. T. and Hodgson, J. (2005) Novel target sites in bacteria for overcoming antibiotic resistance. *Adv. Drug Deliv. Rev.* **57**, 1528–1538.
7. Pucci, M. (2006) Use of genomics to select antibacterial targets. *Biochem. Pharmacol.* **71**, 1066–1072.
8. Berg, C. M., Berg, D. E., and Groisman, E. A. (1989) Transposable elements and the genetic engineering of bacteria, in *Mobile DNA* (Berg, D. and Howe, M., eds.), American Society for Microbiology, Washington, DC., pp. 879–925.
9. Gerdes, S., Scholle, M., Campbell, J., et al. (2003) Experimental determination and system level analysis of essential genes in *Escherichia coli* MG1655. *J. Bacteriol.* **185**, 5673–5684.
10. Salama, N., Shepherd, B., and Falkow, S. (2004) Global transposon mutagenesis and essential gene analysis of Helicobacter pylori. *J. Bacteriol.* **186**, 7926–7935.
11. Butler, M. S. and Buss, A. D. (2006) Natural products – the future scaffolds for novel antibiotics? *Biochem. Pharmacol.* **71**, 919–929.
12. Becker, D., Selbach, M., Rollenhagen, C., Ballmaier, M., Meyer, T. F., Mann, M., and Bumann, D. (2006) Robust Salmonella metabolism limits possibilities for new antimicrobials. *Nature* **440**, 303–307.
13. Silver, L. (2006) Does the cell wall of bacteria remain a viable source of targets for novel antibiotics? *Biochem. Pharmacol.* **71**, 996–1005.
14. Projan, S. J. (2002) New (and not so new) antibacterial targets - from where and when will the novel drugs come? *Curr. Opin. Pharmacol.* **2**, 513–522.

15. Georgopapadakou, N. H. (2004) Beta-lactamase inhibitors: evolving compounds for evolving resistance targets. *Expert Opin. Investig. Drugs* **13**, 1307–1318.
16. Rogers, B. L. (2004) Bacterial targets to antimicrobial leads and development candidates. *Curr. Opin. Drug Discov. Dev.* **7**, 211–222.
17. Halliday, J., McKeveney, D., Muldoon, C., Rajaratnam, P. and Meutermans, W. (2006) Targeting the forgotten transglycosylases. *Biochem. Pharmacol.* **71**, 957–967.
18. Silver, L. L. (2003) Novel inhibitors of bacterial cell wall synthesis. *Curr. Opin. Microbiol.* **6**, 431–438.
19. Sutcliffe, J. A. (2005). Improving on nature: antibiotics that target the ribosome. *Curr. Opin. Microbiol.* **8**, 534–542.
20. Alksne, L. E., Anthony, R. A., Liebman, S. W., and Warner, J. R. (1993). An accuracy center in the ribosome conserved over 2 billion years. *Proc. Natl. Acad. Sci. U.S.A.* **90**, 9538–9541.
21. Franceschi, F. and Duffy, E. M. (2006) Structure-based drug design meets the ribosome. *Biochem. Pharmacol.* **71**, 1016–1025.
22. Ebright, R. H. (2000) RNA polymerase: structural similarities between bacterial RNA polymerase and eukaryotic RNA polymerase II. *J. Mol. Biol.* **304**, 687–98.
23. Bruck, I. and O'Donnell, M. (2000) The DNA replication machine of a gram-positive organism. *J. Biol. Chem.* **275**, 28971–28983.
24. Bermingham, A. and Derrick, J. P. (2002) The folic acid biosynthesis pathway in bacteria: evaluation of potential for antibacterial drug discovery. *Bioessays* **24**, 637–648.
25. Heath, R. J., White, S. W., and Rock, C. O. (2002) Inhibitors of fatty acid synthesis as antimicrobial chemotherapeutics. *Appl. Microbiol. Biotechnol.* **58**, 695–703.
26. Zhao, L., Allanson, N. M., Thomson, S. P., Maclean, J. K., Barker, J. J., Primrose, W. U., Tyler, P. D., and Lewendon, A. (2003) Inhibitors of phosphopantetheine adenylyltransferase. *Eur. J. Med. Chem.* **38**, 345–349.
27. Hutton, C. A., Southwood, T. J., and Turner, J. J. (2003) Inhibitors of lysine biosynthesis as antibacterial agents. *Mini Rev. Med. Chem.* **3**, 115–127.
28. Gunderson, C. W. and Segall, A. M. (2006) DNA repair, a novel antibacterial target: Holliday junction-trapping peptides induce DNA damage and chromosome segregation defects. *Mol. Microbiol.* **59**, 1129–1148.
29. Pohlmann, J. and Brotz-Oesterhelt, H. (2004) New aminoacyl-tRNA synthetase inhibitors as antibacterial agents. *Curr. Drug Targets Infect. Disord.* **4**, 261–272.
30. Clements, J. M., Ayscough, A. P., Keavey, K., and East, S. P. (2002) Peptide deformylase inhibitors, potential for a new class of broad spectrum antibacterials. *Curr. Med. Chem. Antiinfect. Agents* **1**, 239–249.
31. Johnson, K. W., Lofland, D., and Moser, H. E. (2005) PDF inhibitors: an emerging class of antibacterial drugs. *Curr. Drug Targets Infect. Disord.* **5**, 39–52.
32. Leeds, J. A., Schmitt, E. K., and Krastel, P. (2006) Recent developments in antibacterial drug discovery: microbe-derived natural products–from collection to the clinic. *Expert Opin. Investig. Drugs* **15**, 211–226.

33. Alksne, L. E. and Projan, S. J. (2000) Bacterial virulence as a target for antimicrobial chemotherapy.[erratum appears in *Curr. Opin. Biotechnol.* 2001 Feb;12(1):112]. *Curr. Opin. Biotechnol.* **11**, 625–636.

34. Otto, M. (2004) Quorum-sensing control in Staphylococci – a target for antimicrobial drug therapy? *FEMS Microbiol. Lett.* **241**, 135–141.

35. Nordfelth, R., Kauppi, A. M., Norberg, H. A., Wolf-Watz, H., and Elofsson, M. (2005) Small-molecule inhibitors specifically targeting type III secretion. *Infect. Immun.* **73**, 3104–3114.

36. Posner, B. A. (2005) High-throughput screening-driven lead discovery: meeting the challenges of finding new therapeutics. *Curr. Opin. Drug Discov. Dev.* **8**, 487–494.

37. Goddard, J. -P. and Reymond, J. -L. (2004) Enzyme assays for high-throughput screening. *Curr. Opin. Biotechnol.* **15**, 314–322.

38. Allison, R. D. (1997) Kinetic assay methods in *Current Protocols in Molecular Biology*, Supplement 40, Appendix 3H, John Wiley & Sons, Inc., pp. A.3H. 1–A.3H.10.

39. Lipinski, C. A. (2000) Drug-like properties and the causes of poor solubility and poor permeability. *J. Pharmacol. Toxicol. Methods* **44**, 235–249.

40. Donadio, S., Carrano, L., Brandi, L., Serina, S., Soffientini, A., Raimondi, E., Montanini, N., Sosio, M., and Gualerzi, C. O. (2002) Targets and assays for discovering novel antibacterial agents. *J. Biotechnol.* **99**, 175–185.

41. Thomson, C. J., Power, E., Ruebsamen-Waigmann, H., and Labischinski, H. (2004) Antibacterial research and development in the 21(st) Century–an industry perspective of the challenges. *Curr. Opin. Microbiol.* **7**, 445–450.

42. Lundqvist, T. (2005) The devil is still in the details–driving early drug discovery forward with biophysical experimental methods. *Curr. Opin. Drug Discov. Devel.* **8**, 513–519.

43. Mills, S. D. (2006) When will the genomics investment pay off for antibacterial discovery? *Biochem. Pharmacol.* **71**, 1096–1102.

44. Krumrine, J., Raubacher, F., Brooijmans, N., and Kuntz, I. (2003) Principles and methods of docking and ligand design, in *Structural Bioinformatics* (Weissig, H. and Bourne, P. E., eds.), Wiley, Indianapolis, IN, pp. 441–476.

45. Schmid, M. B. (2004) Seeing is believing: the impact of structural genomics on antimicrobial drug discovery. *Nat. Rev. Microbiol.* **2**, 739–746.

46. Poole, K. (2004) Uninhibited antibiotic target discovery via chemical genetics. *Nat. Biotechnol.* **22**, 1528–1529.

47. Li, X., Zolli-Juran, M., Cechetto, J. D., Daigle, D. M., Wright, G. D., and Brown, E. D. (2004) Multicopy suppressors for novel antibacterial compounds reveal targets and drug efflux susceptibility. *Chem. Biol.* **11**, 1423–1430.

48. Singh, S. B. and Barrett, J. F. (2006) Empirical antibacterial drug discovery–foundation in natural products. *Biochem. Pharmacol.* **71**, 1006–1015.

22

Sequence Analysis of the Variable Number Tandem Repeat in *Staphylococcus aureus* Protein A Gene

spa Typing

Barun Mathema, Jose Mediavilla, and Barry N. Kreiswirth

Summary

The analyses of numerous prokaryotic and eukaryotic genomes have revealed the presence of variable number tandem repeats (VNTRs). VNTR analysis is currently widely used to sub-speciate many bacterial, fungal, and viral pathogens and has facilitated a number of molecular epidemiology studies. In this chapter, we focus on *spa* typing which is based on sequence analysis of VNTRs in the polymorphic X region of the *Staphylococcus aureus* protein A gene *Staphylococcus aureus*. As the specific methods for *spa* typing, detailed in this chapter, are well-established and routine procedures (e.g., DNA isolation, PCR and DNA sequencing) for most molecular biology laboratories, we highlight the analytic methods used to interpret the genotyping data generated by sequence analysis and their potential applications in local and global epidemiologic investigations.

Key Words: *Staphylococcus aureus*; genotyping; variable number tandem repeats; *spa* typing; epidemiology; DNA sequencing.

1. Introduction

Staphylococcus aureus has long been recognized as a major human pathogen and remains a frequent cause of morbidity and mortality (*1*). According to the National Nosocomial Surveillance System (NNIS), *S. aureus* is the most common cause of nosocomial infections (*2*). These infections include pneumonia, surgical site, and bloodstream infections, which can be complicated by endocarditis, osteomyelitis, or septic shock (*1*). The versatile tissue tropism

From: *Methods in Molecular Biology, vol. 431: Bacterial Pathogenesis*
Edited by: F. DeLeo and M. Otto © Humana Press, Totowa, NJ

displayed by *S. aureus* is attributed to a remarkable array of cell-associated and secreted virulence factors involved in pathogenesis, many of which are acquired on mobile genetic elements. The control and prevention of *S. aureus* infections in recent years have been further complicated by the wide spread of multidrug-resistant *S. aureus* strains known as methicillin-resistant *S. aureus* (MRSA) in nosocomial settings. Since the introduction of methicillin into clinical use in 1961, the occurrence of MRSA has steadily increased in health care institutions worldwide *(1)*. Recently, the epidemiology of MRSA, which is typically associated with nosocomial environments, has changed dramatically, causing an inordinate number of skin and soft tissue infections among an otherwise healthy population in the community *(3,4)*. Referred to as community acquired (CA-) MRSA, it represents a contemporary epidemic of increasing global occurrence.

Control and prevention methods for methicillin-susceptible *S. aureus* (MSSA) and MRSA have relied largely on prompt and appropriate antibiotic treatment, use of universal precautions, and cohorting. The introduction of molecular typing methods to sub-speciate and track strains of *S. aureus* in institutional settings (e.g., hospital, nursing home) or in the community has had a dramatic impact on epidemiologic analysis and in elucidating transmission pathways, thereby aiding control efforts. Differentiation of *S. aureus* isolates not only clarifies the epidemiologic scenario but also sets the stage to pose biological questions about mechanisms of resistance, molecular evolution, and pathogenesis. As such, the last decade has seen a plethora of molecular tools become available for the sub-speciation of pathogens including *S. aureus*. Common techniques include pulsed-field gel electrophoresis (PFGE) of genomic macrorestriction fragments and multilocus sequence typing (MLST) methods *(5,6)*. The former is often used in short-term epidemiologic investigations (e.g., outbreaks) while the later is applied to long-term epidemiologic studies (e.g., phylogenetic studies). Recently, genotyping based on repetitive DNA, and specifically the analysis of variable number tandem repeats (VNTRs), has been used to provide resolution for both short- and long-term studies. In this chapter, we focus on the methods, analysis, and application of VNTR analysis in *S. aureus*.

1.1. Variable Number Tandem Repeats

The recent availability of numerous annotated bacterial genome sequences has generated a rich source of data for the recognition of new and robust genotyping targets. Many of these targets have given rise to methodological approaches that are amenable to rapid, affordable, and high-throughput systems that facilitate data storage and interlaboratory comparisons. However, a current challenge is to identify a single-locus sequencing target that is conserved in a given species, hypervariable, and provides epidemiologically and/or phylogenetically robust data. Repetitive DNA sequences in bacterial genomes offer such possibilities at both the species and subspecies level. In fact, repetitive DNA and VNTRs in particular have been used since the early 1990s in human genetics and forensic studies *(7,8)*. More recently, VNTR analysis has been employed in a number of studies of pathogenic organisms, including *S. aureus*, *Mycobacterium tuberculosis*, *Bacillus anthracis,* and *Yersinia pestis (9–12)*.

Numerous prokaryotic genomes contain monomeric sequences (repeat units) that repeat periodically and are arranged in a head-to-tail configuration. These DNA regions are catalogued on the basis of their repeat unit size, which ranges from a single nucleotide to sequences >100 bp in length. Prokaryotic genomes generally have microsatellite DNA that has repeat units ranging in size from 1–10 bp. These sequences are abundant throughout most bacterial genomes and have been shown to play a significant role in both transcriptional and translational control of gene expression. Many of the microsatellite sequences, as well as larger minisatellite DNA (which has repeat units ranging in size from 10 to 100), are commonly referred to as VNTRs. Such repeats have been found in intergenic regions, gene expression control regions, and within open reading frames *(13,14)*.

In *S. aureus*, minisatellite VNTRs have been identified in two well-conserved genes: protein A and coagulase. These genes are unique to *S. aureus* and contain an in-frame region composed of VNTRs. Protein A (*spa*) has a 24-bp repeating unit near the 3′ terminus (*see* **Fig. 1A**), and coagulase (*coa*) has an 81-bp repeating unit in the 5′ region *(10,15)*. Genetic alterations in these repetitive sequences include point mutations as well as intragenic recombination events that presumably arise by slipped-strand mispairing during chromosomal replication *(16)* and, in the case of protein A repeats, result in a high degree of genetic polymorphism. In contrast to most VNTR analysis where the total number of repeats is counted, in the case of coagulase and protein A, the repeat units differ in DNA sequence, and the overall genotypic diversity is reflected in repeat content, number of repeats, and organization *(10,15,17,18)*.

A comparative genotyping study conducted by the Centers for Disease Control and Prevention evaluated 8 methods using a blinded collection of *S. aureus* strains from three separate outbreaks. The blinded samples included 59 staphylococci, of which 58 were *S. aureus*, and 37 were methicillin resistant; 29 isolates had been previously grouped into four identifiable clusters based on sound epidemiological links *(10)*. When compared to PFGE, ribotyping, and other genotyping methods, *spa* typing produced results better than the mean score of 25 correct classifications and 5 misclassifications and showed that it could correctly group epidemiologically related strains *(10,19)*. Analysis using *spa* and *coa* as target sequences to sub-speciate *S. aureus* isolates showed

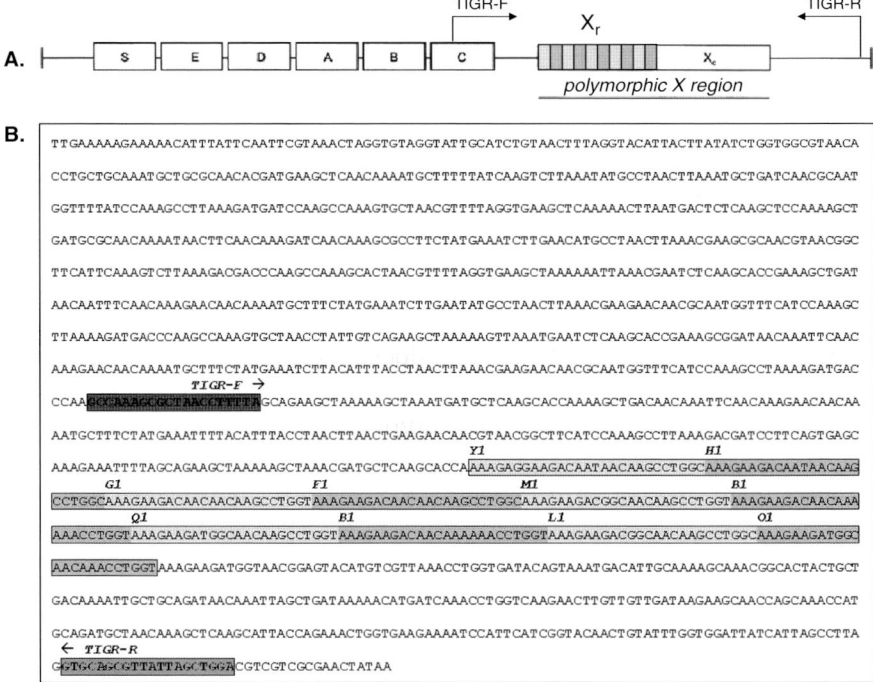

Fig. 1. **(A)** Protein A gene organization. Boxes indicate segments of the gene coding for the signal sequence (S); the immunoglobulin G-binding regions (A–D); a region homologous to A–D (E); and the polymorphic X region, which includes (Xr) the variable *spa* repeat region, shaded in alternating segments and (Xc) the cell wall attachment sequence. Locations of primers described in the "Methods" section are also shown *(50)*. **(B)** Nucleotide sequence of *Staphylococcus aureus* USA300-FPR3757 protein A gene (SAUSA300_0113) showing locations of *spa* type 1 repeats and sequencing primers referred to in "Methods" section. Alternating repeats, are highlighted as in the Xr region in **Fig. 1(A)**. Individual repeats are named according to eGenomics database nomenclature.

concordant results; however, *spa* typing was more discriminating, and the smaller amplicon sizes were more amenable to single sequencing runs *(10,20)*. The variance in discriminatory power between *spa* and *coa* is largely attributed to the evolutionary rate of the specific genetic locus, consistent with the observation that the level of discrimination afforded by VNTR analysis is proportional to the molecular "clock-speed" of the target *(21)*.

As epidemiologically related isolates are typically considered descendants of a common precursor cell, outbreak investigations of *S. aureus* and other pathogens are viewed as short-term events or cases of local epidemiology that are both temporally and geographically restricted *(22)*. In these settings, the primary objective is to identify common strains that are spreading within a larger population (i.e., clonal spread) of unrelated isolates; consequently, the degree of polymorphism among temporally restricted outbreak strains is expected to be small. A number of studies have confirmed the validity and utility of *spa* typing for local epidemiological investigations *(23–26)*. However, as *spa* typing is a single-locus genotyping method, it was not clear whether it would be amenable to long-term or global epidemiologic studies, such as understanding the relatedness of strains isolated from distal regions, the nature and extent of genetic variation within the species, and the phylogenetic relationships among all strains *(21)*. In other words, a single-locus target may not accurately reflect the relationships between distantly related isolates that have had substantial time to diversify due to convergence and/or recombination events.

Staphylococcus aureus is a heterogenous (polymorphic) species that was recently reported to possess a clonal population structure *(27,28)*. It is thought that this species diversifies largely by point mutations rather than undergoing extensive recombination or horizontal transfer and displays a high degree of linkage disequilibrium (non-random associations between genetic loci). Therefore, to study global epidemiology and population genetics, a highly discriminating marker that accumulates genetic variation relatively slowly is required. Typically, multilocus enzyme electrophoresis (MLEE) is used to discern amino acid changes in metabolic enzymes *(29)*; more recently, MLST, an analogous method that compares sequence variation in numerous house-keeping gene targets is used for such studies *(5)*. However, a single-locus DNA sequence-based method that can simultaneously index micro- and macro-variation by two independent mechanisms would offer a rapid and cost-effective alternative.

A recent study reexamined a collection of 36 strains previously characterized by DNA microarray analysis based on 90% of the annotated genome of the COL strain *(21,28)*. Isolates were selected to be representative of the most abundant lineages derived from over 2000 spatio-temporally diverse isolates shown by MLEE to provide a likely population structure of *S. aureus*. The strains were genotyped by sequencing VNTRs in *coa* and *spa*, as well as by PFGE analysis, and the extent of genetic clustering was compared to the phylogenetic structure generated by MLEE and DNA microarray data. The results revealed *spa* typing to be highly discriminating in grouping strains based on sequence changes in the repeat region, which appears to have a clock-speed in register with the overall evolutionary clock of this species. That is, point mutations occur at a lower rate (clock speed) than repeat number variation, and therefore, the dual dynamics of slow point mutations in combination with faster changes in repeat number enable *spa* typing to be used for both micro- (local) and macro- (global) epidemiologic questions *(21)*. Direct comparison against the current gold standard, MLST analysis, suggests that *spa* typing is predictive of the *S. aureus* genotype, and its resolution approaches the discrimination of PFGE *(21,30–33)*.

Although the exact biological function of protein A is not well understood, there are no examples in the literature of a naturally occurring protein A-negative *S. aureus* strain. While the inactivation of protein A does not appear to affect the strain's "fitness," it is noteworthy that as a result of their in-frame organization, the size and content of the repeat units appear to have biological restrictions. Among the non-synonymous changes, the overall amino acid composition of the repeat region appears to be maintained, suggesting that the substitutions are not under strong selective pressures (*see* **Fig. 2**). This observation is further supported by the dS/dN value, which is the ratio of the number of synonymous substitutions per potential synonymous site to the number of non-synonymous substitutions per potential non-synonymous site. Analysis of 38 *spa* repeats showed a dS/dN value of 6.4, where a ratio of <1 indicates positive selection pressure, a value of 1 indicates neutral evolution,

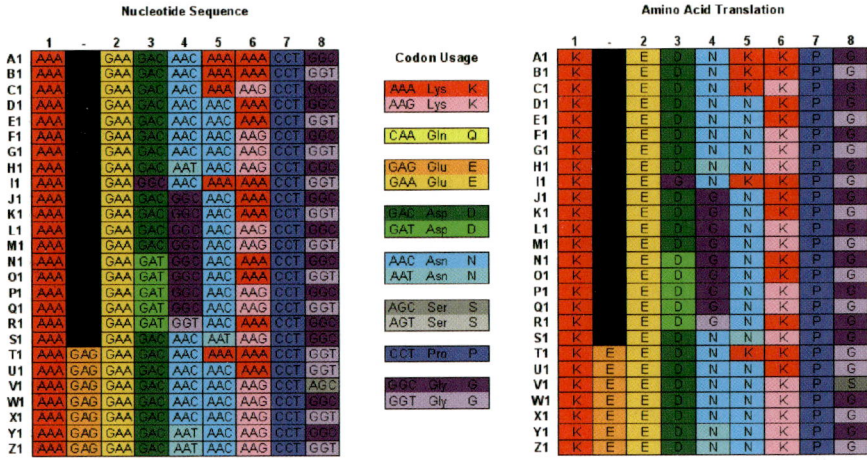

Fig. 2. Sequence content of *spa* repeats at nucleotide and amino acid levels. Repeats differing at the nucleotide level may be identical at the amino acid level. Identical codons are depicted using the same colors; equivalent codons corresponding to synonymous mutations are colored with similar hues. The translated sequence reveals a conserved octapeptide motif comprised almost exclusively of charged amino acids and is suggestive of a surface-exposed loop region with repeating turns. Repeats T1–Z1 bear a single codon insertion between positions 1 and 2; most common *spa* types begin with one of these repeats, such that the initial repeat is usually 27-bp in length (vs. 24-bp).

and a value of >1 is indicative of purifying or stabilizing selection *(21)*. This result strongly suggests that the variation seen in the VNTR region is not the result of outside influences and that the polymorphism is reflective of intrinsic changes in the evolution of the species. Therefore, *spa* typing while appropriate for short-term investigations can also be useful for global epidemiological studies.

2. Materials
2.1. DNA Isolation
1. Solid media for isolation of *S. aureus*.
2. Sterile pipet tips.
3. Sterile water.
4. 1.5-mL tube microfuge tubes (or 96-well microtiter plate).
5. Boiling water bath (or thermocycler).
6. Centrifuge.
7. Lysostaphin (optional)

2.2. PCR
1. PCR primers as follows (*see* **Fig. 1B**):
   a. (TIGR-F) 5´-GCCAAAGCGCTAACCTTTTA-3´
   b. (TIGR-R) 5´-TCCAGCTAATAACGCTGCAC-3´
2. PCR reaction components (dNTPs, buffer, etc.).
3. 0.2-mL PCR reaction tubes (or 96-well microtiter plates).
4. Thermocycler.
5. Materials and apparatus for agarose electrophoresis.
6. PCR purification method (user-specific).

2.3. DNA Sequencing
1. Either or both primers used for PCR (*see* **Subheading 2.2, step 1**).
2. User-defined DNA sequencing platform and materials.
3. Software for trace chromatogram viewing.

3. Methods
3.1. DNA Isolation
spa typing is a simple, fast, and accurate method that involves the isolation, amplification, and sequencing of the VNTR region near the 3´ end of the protein A gene. The efficient lysis of *S. aureus* cells requires incubation with lysostaphin, an endopeptidase that specifically cleaves the penta-glycine cross

bridges found in the staphylococcal cell wall. For *spa* typing, however, a simple boiling procedure in sterile water is sufficient to isolate enough target DNA to produce robust amplicons for direct sequencing; for large numbers of samples, however, greater lysis efficiency may be obtained by prior treatment with lyso-staphin, as described below.

1. Grow cells overnight on solid media. *S. aureus* grows well on tryptic soy agar, Columbia agar, GL agar, 5% sheep blood agar, and mannitol salt agar (a selective medium on which *S. aureus* colonies produce yellow halos).
2. Isolate a small amount of bacterial colony using a sterile pipet tip; resuspend the colony in 100 μL of sterile water in a 1.5-mL tube.
3. Boil contents of tube for 15 min in a water bath and microcentrifuge for 5 min at maximum speed. The supernatant may be used directly for PCR, or it may be diluted anywhere from 10- to 100-fold in sterile water. This minimizes the effects of potential inhibitors or excess template concentration.

3.1.1. High-Throughput DNA Isolation

For higher throughput, up to 12 strains may be streaked in thin lines on individual agar plates (*see* **Fig. 3A**), with each plate corresponding to one row of a 96-well microtiter plate.

1. Aliquot 100 μL of a 100 μg/mL lysostaphin-water mixture into each well; inoculate with 96 isolates using pipet tips. Mix well.
2. Seal plate and incubate at 37 °C for 15–30 mins. Boil for 15 mins in water bath, or place in thermal cycler for 15 mins at 100 °C.
3. Centrifuge plate at 2500 rpm for 5–10 mins; dilute 5 μL of supernatant (1:40) in sterile water in a new plate. Use a diluted template for PCR.

3.2. PCR

PCR reactions are typically set up in 30-μL volumes to ensure sufficient volume for gel analysis, amplicon purification, and DNA sequencing (*see* **Fig. 3B**). Any standard commercial DNA polymerase mixture containing dNTPs may be utilized. Reaction mixtures include 0.1 μ*M* of each of the following primers shown in **Fig. 1B**: (TIGR-F) 5´-GCCAAAGCGCTA ACCTTTTA-3´ and (TIGR-R) 5´-TCCAGCTAATAACGCTGCAC-3´.

1. Dispense reaction mixtures in 25 μL volumes into 0.2-mL PCR reaction tubes (or into a 96-well microtiter plate) to which 5 μL of the 1:40 diluted lysate is then added.
2. Set cycling parameters as follows: An initial 2-min denaturation step at 94 °C, followed by 35 cycles of denaturation at 94 °C for 30 s, annealing at 58 °C for 1 min, and extension at 72 °C for 1 min. Although the overall amplicon size ranges from 100–500 bp, the longer annealing time seems to improve product yield.

A. DNA Isolation

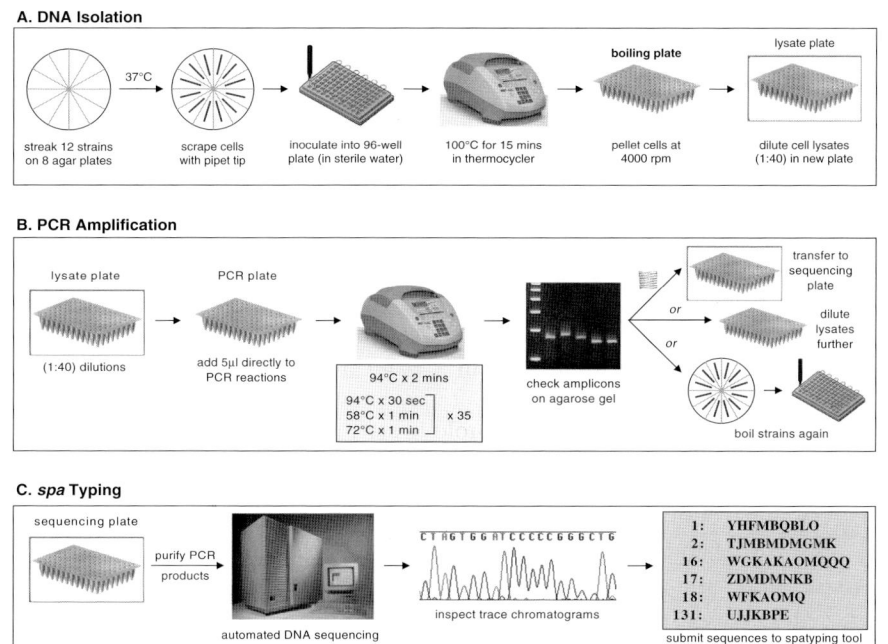

B. PCR Amplification

C. *spa* Typing

Fig. 3. High-throughput *spa*-typing protocol. (**A**) Isolation of DNA from boiled lysates of *Staphylococcus aurus* culture in 96-well format; (**B**) PCR amplification and verification by agarose gel electrophoresis; and (**C**) DNA sequencing and *spa*-type assignment.

The cycling procedure lasts for approximately 2 h, following which 5 μL of the reaction products is run on a 1% agarose gel, stained in ethidium bromide, and visualized by UV light. The presence of a single amplicon ranging in size from ~100 to 500 bp is evidence of a successful amplification reaction; faintly staining bands are sufficient to produce quality sequence data. Reactions that produce multiple bands are suggestive of sample contamination with more than one strain of *S. aureus*. Additionally, experience in our laboratory has suggested potential specificity issues with primer TIGR-F; an alternative forward primer may be useful for samples that fail to amplify as expected. Reactions that fail to produce amplicons may be reboiled in individual tubes utilizing the water bath method, diluted further (up to 1:100) in sterile water and reamplified. Samples that fail to amplify multiple times should be set aside for further analysis and should be confirmed to be *S. aureus* by other methods, such as isolation on mannitol salt agar followed by coagulase testing, 16S rRNA sequencing, or real-time PCR detection of the *spa* gene.

3.3. Sequencing

Sequencing is performed using either of the primers used for PCR. Prior to sequencing, PCR products must be purified to remove un-incorporated reaction components. This may be performed in-house using commercially available column-based methods or else by newer methods that utilize magnetic silica-based technology. As a result of cost, speed, and quality, the sequencing of PCR amplicons is frequently out-sourced to university-based core facilities or commercial laboratories where the cost ranges from 3–12 dollars per sequence. Some facilities currently include purification of PCR products as part of a discounted package for high-throughput sequencing; in this case, 20 µL of reaction product may be transferred to a new microtiter plate after verification of amplification and shipped according to the facility's specifications (*see* **Fig. 3C**).

Following retrieval of sequence data, trace chromatograms should be inspected for each sequence to ensure quality; individual samples may also be sequenced in both directions for purposes of confirmation. As individual *spa* repeats are defined by single nucleotide polymorphisms, sequences with high background or ambiguities should not be utilized for *spa*-typing analysis. Good-quality sequences may be submitted directly for *spa*-typing analysis using any of several available software platforms (e.g., eGenomics: http://www.egenomics.com, New York; Ridom: http://www.ridom.de, Würzburg, Germany); results are returned in text-based format that is highly amenable to personalized database management. New or atypical *spa* types should be confirmed by sequencing in the other direction, especially those that include novel repeats.

3.4. Analysis/Interpretation

As the main source of variation within the polymorphic X region of the *spa* gene seems to be driven by duplication or deletion of the repeat units, strain lineages cannot be constructed by direct sequence comparison using standard alignment algorithms *(10)*. As such, visual depictions of strain typing results using dendrograms are precluded as they rely primarily on sequence alignment. Therefore, to examine strain relatedness, all possible variations of the repeat units are first identified, following which the organization of the repeats in the polymorphic X region is compared between different isolates. In this manner, each *spa* type denotes a collection of specific repeat units organized in a particular pattern to the exclusion of other known types. Other algorithms to denote relatedness of *spa* types have been developed, such as based upon repeat pattern (BURP) *(34)*, a method that is similar to based upon related sequence types (BURST) that is used for MLST analysis *(35)*.

As is the case for MLST, *spa* typing provides objective information that is easily scored and stored in a relational database. In MLST analysis, sequencing results of seven gene fragments are analyzed, whereby each unique genetic alteration (nucleotide change, deletions, or insertions), regardless of whether it is a synonymous or non-synonymous change, is assigned an allelic number *(5,20)*. The catenation of the allelic number for the seven gene fragments defines the sequence type (ST); and the assignment of the allelic number and ST is performed using a free web-based site (http://www.mlst.net) maintained in London, England.

Analogous to the assignment of allelic numbers to genetic variants in MLST, each repeat unit in the polymorphic X region of *spa* is given a unique identifier (*see* **Fig. 2**). Based on two large international *spa* databases (eGenomics and Ridom), there have been over 150 unique repeats identified to date. The alterations in the repeats are primarily single nucleotide substitutions in the third position (synonymous), more rarely codon insertions/deletions, and in each case, the protein A sequence remains in-frame producing a functional protein. Currently, of over 2100 different *spa* types from more than 30,000 *S. aureus* isolates that have been identified in the eGenomics and Ridom databases, all isolates contain at least one repeat unit and some strains contain up to 16 repeat units. Conversely, there are no examples where the protein A gene is void of repeats, where individual repeats are truncated, or where repeat sequences are found out-of-frame. Although untested, this observation suggests a potential biological significance underlying the intact, in-frame repeat sequences in the polymorphic X region of the protein A gene. While protein A is a well-known virulence factor in murine models of *S. aureus* infections including pneumonia *(36,37)*, the biological function of the polymorphic X region is not well understood. It is thought that the protein A domain encoded by the X region may serve to extend the N-terminal immunoglobulin G binding portion of the protein through the cell wall *(38)*. However, the biological function and significance of the diverse repeat content and organization in isolates of *S. aureus* are not clear.

As mentioned earlier, *spa* type determination is based on differences between strains in the number and content of the VNTRs. Therefore, two strains with the identical repeat sequence are assigned the same *spa* type (i.e., same repeat content and organization) and considered genetically related; in short-term investigations, identical *spa* types may be indicative of transmission events. As *spa* typing does not have the resolving power of PFGE subtyping, identical *spa* types can in some instances exhibit similar yet non-identical PFGE profiles *(33)*. Nevertheless, identical *spa* types are strongly suggestive of clonality, and different *spa* types that share a similar repeat motif may likewise be related by descent. For example, *spa* types 1 and 7 share a near identical repeat motif expect for one unit, which differs by a single point mutation and changes the amino acid (*see* **Fig. 4**). Therefore, while *spa* types 1 and 7 are not identical, they are closely related and most likely arose directly from one another although the directionality of ancestry cannot be determined by *spa* type comparison only. Furthermore, a number of studies have grouped together strains with similar repeat motifs into the same *spa* lineages or *spa* clonal complexes, supporting the use of this method to examine macrovariation *(10,21,33,39)*. Although *S. aureus* as a whole does not undergo extensive recombination, the occurrence of such events within the protein A gene can misalign *spa* types relative to PFGE or MLST results; similarly, recombination events can also alter MLST analysis *(32,33)*.

There are currently three similar software programs (eGenomics, Ridom, and Bionumerics) that are able to identify and name repeats, recognize repeat motifs or order, and assign a *spa* type from a raw sequence file of the polymorphic X region. The two most widely used applications, Ridom and eGenomics, use

spa Type: 1
Motif: Y1-H1-G1-F1-M1-B1-Q1-B1-L1-O1

Name	Sequence	Annotation	Len
Y1	AAAGAGGAAGACAATAACAAGCCTGGC	AAAGAGGAAGAC---AATAAC---AAGCCTGGC	27
H1	AAAGAAGACAATAACAAGCCTGGT	AAA---GAAGAC---AATAAC---AAGCCTGGT	24
G1	AAAGAAGACAACAACAAGCCTGGT	AAA---GAAGAC---AACAAC---AAGCCTGGT	24
➜ F1	AAAGAAGACAACAACAAGCCTGGC	AAA---GAAGAC---AACAAC---AAGCCTGGC	24
M1	AAAGAAGACGGCAACAAGCCTGGT	AAA---GAAGACGGCAAC------AAGCCTGGT	24
B1	AAAGAAGACAACAAAAAACCTGGT	AAA---GAAGAC---AAC---AAAAAACCTGGT	24
Q1	AAAGAAGATGGCAACAAGCCTGGT	AAA---GAAGATGGCAAC------AAGCCTGGT	24
B1	AAAGAAGACAACAAAAAACCTGGT	AAA---GAAGAC---AAC---AAAAAACCTGGT	24
L1	AAAGAAGACGGCAACAAGCCTGGC	AAA---GAAGACGGCAAC------AAGCCTGGC	24
O1	AAAGAAGATGGCAACAAACCTGGT	AAA---GAAGATGGCAAC---AAA---CCTGGT	24

spa Type: 7
Motif: Y1-H1-G1-C1-M1-B1-Q1-B1-L1-O1

Name	Sequence	Annotation	Len
Y1	AAAGAGGAAGACAATAACAAGCCTGGC	AAAGAGGAAGAC---AATAAC---AAGCCTGGC	27
H1	AAAGAAGACAATAACAAGCCTGGT	AAA---GAAGAC---AATAAC---AAGCCTGGT	24
G1	AAAGAAGACAACAACAAGCCTGGT	AAA---GAAGAC---AACAAC---AAGCCTGGT	24
➜ C1	AAAGAAGACAACAAAAAGCCTGGT	AAA---GAAGAC---AAC---AAAAAGCCTGGT	24
M1	AAAGAAGACGGCAACAAGCCTGGT	AAA---GAAGACGGCAAC------AAGCCTGGT	24
B1	AAAGAAGACAACAAAAAACCTGGT	AAA---GAAGAC---AAC---AAAAAACCTGGT	24
Q1	AAAGAAGATGGCAACAAGCCTGGT	AAA---GAAGATGGCAAC------AAGCCTGGT	24
B1	AAAGAAGACAACAAAAAACCTGGT	AAA---GAAGAC---AAC---AAAAAACCTGGT	24
L1	AAAGAAGACGGCAACAAGCCTGGC	AAA---GAAGACGGCAAC------AAGCCTGGC	24
O1	AAAGAAGATGGCAACAAACCTGGT	AAA---GAAGATGGCAAC---AAA---CCTGGT	24

Fig. 4. eGenomics (http://epigene.egenomics.com) *spa*-typing tool screenshots showing aligned repeat sequences for the closely related *spa* types 1 and 7. The C → A point mutation is highlighted, corresponding to the non-synonymous amino acid replacement Asn (AAC) → Lys (AAA) and underscoring the need for quality assurance in DNA sequencing data.

similar algorithms to identify and name repeats and assign *spa* types; however, they use different nomenclature. It is important to note that the numerical notation used to express the various *spa* types is not an index of relatedness. That is, *spa* types 1, 2, 3, and so on are named in the order that they were analyzed. Different nomenclature systems can confuse the literature and make interlaboratory comparison more difficult, therefore a consensus nomenclature method is needed. Furthermore, a consensus nomenclature that indicates, along with the *spa* type, information on the phylogenetic position of a particular strain within the overall population structure of *S. aureus* would be helpful in furthering our understanding of the relationship between the micro- and macrovariation within this species.

3.5. Other Genotyping Techniques

The early 1980s witnessed the initial integration of molecular techniques to discriminate between different isolates of *S. aureus*. While previous methods that have historically relied on characterizing phenotypic properties such as colony morphology, serology, phage typing, toxin production, and antibiotic susceptibility profiles were useful, they did not provide sufficient discrimination, thereby limiting their application in epidemiological studies. The transition from phenotypic characterization to genotyping saw the introduction of numerous DNA-based methods, including plasmid profiling, electrophoretic profiling of restriction enzyme digests of the chromosome, and random PCR typing. Southern blot hybridization using a rRNA gene probe, or ribotyping, has become a commonly used genotyping method as it is applicable to all bacteria although the limited number of ribosomal gene copies in most species limits the number of restriction fragment length polymorphisms and the overall discriminatory capabilities. Given the array of molecular tools available, it is important to choose an appropriate method(s) to address a particular study question, for example, outbreak investigations, transmission dynamics, or phylogenetics. In general, key aspects of choosing a molecular technique to study *S. aureus* epidemiology include the observed rate of polymorphism(s) of the biomarker, the genetic diversity of the strains in the population and the spatial and temporal distance between the study isolates.

3.5.1. PFGE

The most accepted bacterial genotyping method is PFGE, as it is applicable to all bacteria and the DNA fingerprints are highly discriminating. In this approach, restriction enzymes that cut infrequently in a given species are used to generate a macrorestriction profile of the entire bacterial chromosome. The resulting DNA fingerprint patterns are excellent tools in outbreak investigations, discriminating epidemiologically related isolates from those that are unrelated isolates. Although the discriminatory power of PFGE is most likely higher than that of *spa* typing, a number of studies have shown that the Simpson's index of diversity of the two typing techniques approximate one another *(21)*. However, PFGE analysis requires unique and costly equipment, and the method is considerably more time and labor intensive compared to *spa* typing. Furthermore, as PFGE produces image-based results, the storage and analysis of fingerprints require sophisticated software and dedicated personnel to digitize and interpret the image as well as maintain the database *(6,40,41)*. In addition, the interpretation is somewhat subjective, and protocol standardization poses difficulty in establishing interlaboratory reproducibility.

Numerous networks have used PFGE analysis as the primary genotyping tool for surveillance and monitoring of outbreak strains, including food-borne pathogens and multidrug-resistant organisms such as MRSA *(42–44)*. Here, even with the assistance of sophisticated, commercially available, pattern-matching software packages *(45)*, such as BioImage and BioNumerics, there has been limited success in generating reproducible and comparable images between laboratories and in assessing genetic relatedness on the basis of subtle restriction fragment length polymorphisms. Therefore, while this method is useful for distinguishing recent or closely related *S. aureus* isolates from unrelated ones, image-based systems are limited in their ability to objectively discern the degree of relatedness between geographically and temporally distal isolates. In contrast, a recent international multicenter study demonstrated that *spa* typing showed 100% intra- and interlaboratory reproducibility without extensive synchronization of protocols between the laboratories *(39)*.

3.5.2. MLST

The limitations noted above have paved the way for genotyping methods that are based on comparative DNA sequence analysis. Currently, the "gold-standard" of DNA sequence-based genotyping methods is MLST, which compares the DNA sequence variation of a number of housekeeping genes dispersed around the bacterial genome and has been widely used to decipher the population structure of bacterial pathogens, such as *Streptococcus pneumoniae* and *Neisseria meningitidis* *(5,46)*. While the target genes used for MLST are similar to those in MLEE, the former generates objective and unambiguous data that is highly amenable to database storage and analysis *(47,48)*. However, although MLST analysis provides a sound approach to assess genetic relatedness, the limited sequence variation within each target requires that numerous genes be compared to increase discrimination. The number of MLST target gene fragments needed to genotype bacterial isolates is dependent on the extent of homogeneity of the given species and on the level of discrimination needed to answer a given epidemiological or evolutionary hypothesis. Among monomorphic species such as *M. tuberculosis* or *B. anthracis*, where the genetic clock-speed or rate of change in housekeeping genes is relatively slow, the discriminatory power of MLST is limited. In contrast, *S. aureus* is an ideal species for MLST analysis as it is relatively diverse, with limited evidence of recombination *(27)*. Consequently, MLST is able to address global epidemiological questions where the isolates have had substantial time to diversify but is less discriminating in detecting changes in closely related strains. It should also be stressed that due to the sequencing of seven housekeeping genes for *S. aureus* MLST analysis, this method, while objective and highly portable, is also labor-intensive, time-consuming, and costly to use routinely in a clinical or surveillance setting. In contrast, the results generated by *spa* typing have been shown to approximate the population structure of *S. aureus* afforded by MLST analysis while remaining objective, highly portable, and significantly less costly.

3.6. Discussion

As a single locus sequence-based typing method, *spa* typing has many advantages over other genotyping methods including cost, labor, ease of analysis, and quality assurance. Initial concerns that *spa* typing was too variable and not rooted in the phylogeny of *S. aureus* led to its limited acceptance as a robust genotyping method, especially in regard to population-based studies. However, numerous comparative genotyping studies have repeatedly showed that *spa* typing is concordant with classical MLEE typing, MLST analysis, and genome-wide microarray data. In fact, *spa* typing has proven to be superior to MLST in discriminating MRSA strains, as its faster molecular clock provides higher resolution that approaches the level provided by PFGE. For example, the two closely related CA-MRSA clones, termed USA300 and USA500 by PFGE, are distinguished by *spa* typing as types 1 and 7, respectively, but clustered together as sequence type 8 (ST8) by MLST *(3)* (*see* **Figs. 4** and **5**). Consistent with the fact that these two clones are members of the same clonal complex and descendants of a common ancestral strain, the sequence of the 10 repeats in *spa* types 1 and 7 are highly conserved and differ only by a single non-synonymous point mutation *(21,33)*. **Figure 5** (bubble) shows an example where *spa* typing is able to further resolve the MLST grouping among ST8 strains.

A current limitation of *spa* typing involves understanding the relatedness among types with similar repeat organization, along with the weighting of

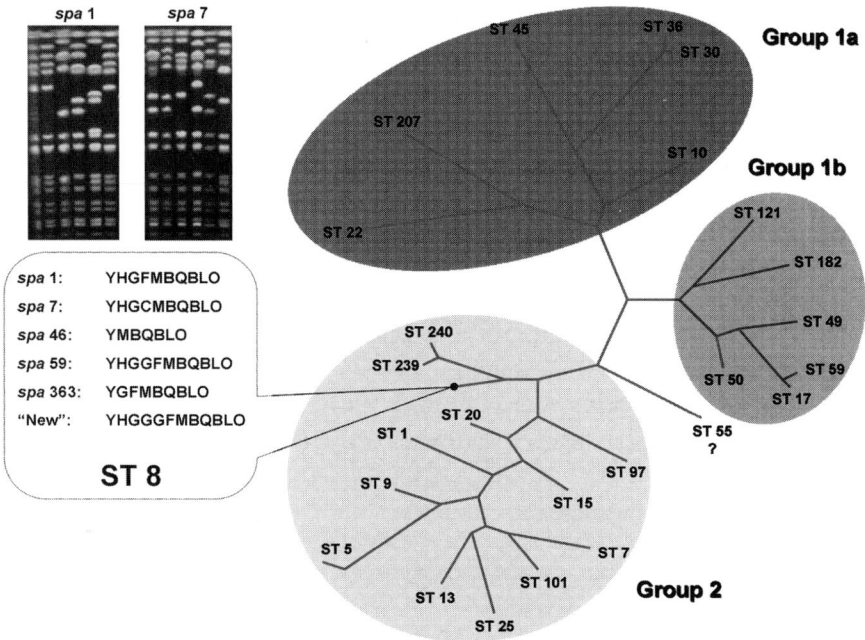

Fig. 5. Bayesian reconstruction of *Staphylococcus aureus* phylogeny based on concatenated sequences of MLST gene fragments, adapted from reference *(49)*. The three proposed groups (1a, 1b, and 2) are highlighted and the MLST sequence types corresponding to each branch labeled. Individual *spa* types associated with ST8 are indicated, corresponding to the "tips" of the ST8 branch; the phylogenetic relationships are undetermined, so no topology is depicted. Additional *spa* types can be assigned similarly to the other branches (not shown). Pulsed-field gel electrophoresis patterns for different *spa* type 1 and 7 strains are shown, depicting the potential variation underlying defined *spa* types, and corresponding to the "leaves" on the tips of the branches.

observed changes. That is, are two strains that differ by a single point mutation, such as *spa* types 1 and 7, more closely related than strains 1 and 363 (which differ by a single repeat) or strains 1 and 59 (which differ by two repeats) or are they equivalent under the hypothesis that each of these genetic alterations represent single events? A study by Robinson and Enright examined the relationship between seven housekeeping MLST gene targets and eight surface protein encoding genes (7 *sas* genes and *spa*) among 220 isolates representing major lineages of nosocomial MRSA, MSSA, and CA-MRSA *(32)*. Here, results of *spa* typing were concordant with the grouping of MLST and *sas* gene targets. This study also lends credence to the idea that one can infer MLST results on the basis of *spa*-typing data, an assumption that will be evaluated over time as both methods continue to be used by *S. aureus* researchers.

The recent study by Cooper and Feil revisited the population structure of *S. aureus*, as derived by MLST analysis, by additionally comparing 33 gene loci representing various functional categories in 30 genetically characterized strains *(49)*. Approximately 17.8 kb of DNA sequence data for each strain produced an unrooted Bayesian tree that resembles the population structure established by MLST but with higher resolution and without unresolved branches. Their data sub-divides the *S. aureus* species into three genetic groups (1a, 1b, and 2), and within each group, MLST branches can be mapped, thus providing a robust framework onto which *spa* types may be assigned (*see* **Fig. 5**). One could envision the future naming of *spa* types to include both the genetic group as well as the MLST branch designation, thereby providing phylogenetic structure to *spa*-typing results. Molecular approaches such as *spa* typing can therefore add further resolution to phylogenetically determined lineages or clusters to elucidate microevolutionary processes that may have biomedical significance; in turn, phylogenetic analysis can root molecular epidemiology-derived genotypes (e.g., USA300) into broad groups that share informative biologic traits and further our understanding of genotype–phenotype relationships.

References

1. Lowy, F. D. (1998) *Staphylococcus aureus* infections. *N. Engl. J. Med.* **339**, 520–532.
2. Centers for Disease Control and Prevention. (2004) National Nosocomial Infections Surveillance (NNIS) System Report, data summary from January 1992 through June 2004, issued October 2004. *Am. J. Infect. Control* **32**, 470–485.
3. Said-Salim, B., Mathema, B., and Kreiswirth, B. N. (2003) Community-acquired methicillin-resistant *Staphylococcus aureus*: an emerging pathogen. *Infect. Control Hosp. Epidemiol.* **24**, 451–455.
4. Grundmann, H., Aires-de-Sousa, M., Boyce, J., and Tiemersma, E. (2006) Emergence and resurgence of meticillin-resistant *Staphylococcus aureus* as a public-health threat. *Lancet* **368**, 874–885.
5. Maiden, M. C., Bygraves, J. A., Feil, E., Morelli, G., Russell, J. E., Urwin, R., Zhang, Q., Zhou, J., Zurth, K., Caugant, D. A., Feavers, I. M., Achtman, M. and Spratt, B. G. (1998) Multilocus sequence typing: a portable approach to the identification of clones within populations of pathogenic microorganisms. *Proc. Natl. Acad. Sci. U.S.A.* **95**, 3140–3145.
6. Tenover, F. C., Arbeit, R. D., Goering, R. V., Mickelsen, P. A., Murray, B. E., Persing, D. H., and Swaminathan, B. (1995) Interpreting chromosomal DNA restriction patterns produced by pulsed-field gel electrophoresis: criteria for bacterial strain typing. *J. Clin. Microbiol.* **33**, 2233–2239.

7. Nakamura, Y., Leppert, M., O'Connell, P., Wolff, R., Holm, T., Culver, M., Martin, C., Fujimoto, E., Hoff, M., Kumlin, E., et al. (1987) Variable number of tandem repeat (VNTR) markers for human gene mapping. *Science* **235**, 1616–1622.
8. Kasai, K., Nakamura, Y., and White, R. (1990) Amplification of a variable number of tandem repeats (VNTR) locus (pMCT118) by the polymerase chain reaction (PCR) and its application to forensic science. *J. Forensic. Sci.* **35**, 1196–1200.
9. Keim, P., Price, L. B., Klevytska, A. M., Smith, K. L., Schupp, J. M., Okinaka, R., Jackson, P. J., and Hugh-Jones, M. E. (2000) Multiple-locus variable-number tandem repeat analysis reveals genetic relationships within *Bacillus anthracis. J. Bacteriol.* **182**, 2928–2936.
10. Shopsin, B., Gomez, M., Montgomery, S. O., Smith, D. H., Waddington, M., Dodge, D. E., Bost, D. A., Riehman, M., Naidich, S., and Kreiswirth, B. N. (1999) Evaluation of protein A gene polymorphic region DNA sequencing for typing of *Staphylococcus aureus* strains. *J. Clin. Microbiol.* **37**, 3556–3563.
11. Supply, P., Mazars, E., Lesjean, S., Vincent, V., Gicquel, B., and Locht, C. (2000) Variable human minisatellite-like regions in the Mycobacterium tuberculosis genome. *Mol. Microbiol.* **36**, 762–771.
12. Klevytska, A. M., Price, L. B., Schupp, J. M., Worsham, P. L., Wong, J., and Keim, P. (2001) Identification and characterization of variable-number tandem repeats in the *Yersinia pestis* genome. *J. Clin. Microbiol.* **39**, 3179–3185.
13. van Belkum, A. (1999) Short sequence repeats in microbial pathogenesis and evolution. *Cell. Mol. Life Sci.* **56**, 729–734.
14. Yeramian, E. and Buc, H. (1999) Tandem repeats in complete bacterial genome sequences: sequence and structural analyses for comparative studies. *Res. Microbiol.* **150**, 745–754.
15. Shopsin, B., Gomez, M., Waddington, M., Riehman, M., and Kreiswirth, B. N. (2000) The use of coagulase gene (coa) repeat region nucleotide sequences for the typing of methicillin-resistant *Staphylococcus aureus. J. Clin. Microbiol.* **38**, 3453–3456.
16. Brigido, M. D. M., Barardi, C. R., Bonjardin, C. A., Santos, C. L., Junqueira, M. L., and Brentani, R. R. (1991) Nucleotide sequence of a variant protein A of Staphylococcus aureus suggests molecular heterogeneity among strains. *J. Basic Microbiol.* **31**, 337–345.
17. Mazars, E., Lesjean, S., Banuls, A. L., Gilbert, M., Vincent, V., Gicquel, B., Tibayrenc, M., Locht, C., and Supply, P. (2001) High-resolution minisatellite-based typing as a portable approach to global analysis of *Mycobacterium tuberculosis* molecular epidemiology. *Proc. Natl. Acad. Sci. U.S.A.* **98**, 1901–1906.
18. Farlow, J., Smith, K. L., Wong, J., Abrams, M., Lytle, M., and Keim, P. (2001) *Francisella tularensis* strain typing using multiple-locus, variable-number tandem repeat analysis. *J. Clin. Microbiol.* **39**, 3186–3192.
19. Tenover, F. C., Arbeit, R., Archer, G., Biddle, J., Byrne, S., Goering, R., Hancock, G., Hebert, G. A., Hill, B., Hollis, R., et al. (1994) Comparison of traditional and molecular methods of typing isolates of *Staphylococcus aureus. J. Clin. Microbiol.* **32**, 407–415.

20. Enright, M. C., Day, N. P., Davies, C. E., Peacock, S. J., and Spratt, B. G. (2000) Multilocus sequence typing for characterization of methicillin-resistant and methicillin-susceptible clones of *Staphylococcus aureus*. *J. Clin. Microbiol.* **38**, 1008–1015.
21. Koreen, L., Ramaswamy, S. V., Graviss, E. A., Naidich, S., Musser, J. M., and Kreiswirth, B. N. (2004) *spa* typing method for discriminating among Staphylococcus aureus isolates: implications for use of a single marker to detect genetic micro- and macrovariation. *J. Clin. Microbiol.* **42**, 792–799.
22. Levin, B. R., Lipsitch, M., and Bonhoeffer, S. (1999). Population biology, evolution, and infectious disease: convergence and synthesis. *Science* 283, 806–809.
23. Fey, P. D., Said-Salim, B., Rupp, M. E., Hinrichs, S. H., Boxrud, D. J., Davis, C. C., Kreiswirth, B. N., and Schlievert, P. M. (2003) Comparative molecular analysis of community- or hospital-acquired methicillin-resistant *Staphylococcus aureus*. *Antimicrob. Agents Chemother.* **47**, 196–203.
24. Saiman, L., Cronquist, A., Wu, F., Zhou, J., Rubenstein, D., Eisner, W., Kreiswirth, B. N., and Della-Latta, P. (2003) An outbreak of methicillin-resistant *Staphylococcus aureus* in a neonatal intensive care unit. *Infect. Control Hosp. Epidemiol.* **24**, 317–321.
25. Weese, J. S., Archambault, M., Willey, B. M., Hearn, P., Kreiswirth, B. N., Said-Salim, B., McGeer, A., Likhoshvay, Y., Prescott, J. F., and Low, D. E. (2005) Methicillin-resistant *Staphylococcus aureus* in horses and horse personnel, 2000–2002. *Emerg. Infect. Dis.* **11**, 430–435.
26. Layer, F., Ghebremedhin, B., Konig, W., and Konig, B. (2006) Heterogeneity of methicillin-susceptible *Staphylococcus aureus* strains at a German University Hospital implicates the circulating-strain pool as a potential source of emerging methicillin-resistant *S. aureus* clones. *J. Clin. Microbiol.* **44**, 2179–2185.
27. Feil, E. J., Cooper, J. E., Grundmann, H., Robinson, D. A., Enright, M. C., Berendt, T., Peacock, S. J., Smith, J. M., Murphy, M., Spratt, B. G., Moore, C. E., and Day, N. P. (2003) How clonal is *Staphylococcus aureus*? *J. Bacteriol.* **185**, 3307–3316.
28. Fitzgerald, J. R., Sturdevant, D. E., Mackie, S. M., Gill, S. R., and Musser, J. M. (2001) Evolutionary genomics of *Staphylococcus aureus*: insights into the origin of methicillin-resistant strains and the toxic shock syndrome epidemic. *Proc. Natl. Acad. Sci. U.S.A.* **98**, 8821–8826.
29. Selander, R. K., Caugant, D. A., Ochman, H., Musser, J. M., Gilmour, M. N., and Whittam, T. S. (1986) Methods of multilocus enzyme electrophoresis for bacterial population genetics and systematics. *Appl. Environ. Microbiol.* **51**, 873–884.
30. Crisostomo, M. I., Westh, H., Tomasz, A., Chung, M., Oliveira, D. C., and de Lencastre, H. (2001) The evolution of methicillin resistance in *Staphylococcus aureus*: similarity of genetic backgrounds in historically early methicillin-susceptible and -resistant isolates and contemporary epidemic clones. *Proc. Natl. Acad. Sci. U.S.A.* **98**, 9865–9870.

31. Oliveira, D. C., Tomasz, A., and de Lencastre, H. (2001) The evolution of pandemic clones of methicillin-resistant *Staphylococcus aureus*: identification of two ancestral genetic backgrounds and the associated *mec* elements. *Microb. Drug Resist.* **7**, 349–361.

32. Robinson, D. A. and Enright, M. C. (2004) Evolution of *Staphylococcus aureus* by large chromosomal replacements. *J. Bacteriol.* **186**, 1060–1064.

33. Strommenger, B., Kettlitz, C., Weniger, T., Harmsen, D., Friedrich, A. W., and Witte, W. (2006) Assignment of *Staphylococcus* isolates to groups by *spa* typing, SmaI macrorestriction analysis, and multilocus sequence typing. *J. Clin. Microbiol.* **44**, 2533–2540.

34. Harmsen, D., Claus, H., Witte, W., Rothganger, J., Claus, H., Turnwald, D., and Vogel, U. (2003) Typing of methicillin-resistant *Staphylococcus aureus* in a university hospital setting by using novel software for spa repeat determination and database management. *J. Clin. Microbiol.* **41**, 5442–5448.

35. Feil, E. J., Li, B. C., Aanensen, D. M., Hanage, W. P., and Spratt, B. G. (2004) eBURST: inferring patterns of evolutionary descent among clusters of related bacterial genotypes from multilocus sequence typing data. *J. Bacteriol.* **186**, 1518–1530.

36. Gomez, M. I., Lee, A., Reddy, B., Muir, A., Soong, G., Pitt, A., Cheung, A., and Prince, A. (2004) *Staphylococcus aureus* protein A induces airway epithelial inflammatory responses by activating TNFR1. *Nat. Med.* **10**, 842–848.

37. Labandeira-Rey, M., Couzon, F., Boisset, S., Brown, E. L., Bes, M., Benito, Y., Barbu, E. M., Vazquez, V., Hook, M., Etienne, J., Vandenesch, F., and Bowden, M. G. (2007) *Staphylococcus aureus* Panton-Valentine leukocidin causes necrotizing pneumonia. *Science* **315**, 1130–1133.

38. von Heijne, G. and Uhlen, M. (1987) Homology to region X from staphylococcal protein A is not unique to cell surface proteins. *J. Theor. Biol.* **127**, 373–376.

39. Aires de Sousa, M., Conceicao, T., Simas, C., and de Lencastre, H. (2005). Comparison of genetic backgrounds of methicillin-resistant and -susceptible *Staphylococcus aureus* isolates from Portuguese hospitals and the community. *J. Clin. Microbiol.* **43**, 5150–5157.

40. de Lencastre, H., Severina, E. P., Roberts, R. B., Kreiswirth, B. N., and Tomasz, A. (1996) Testing the efficacy of a molecular surveillance network: methicillin-resistant *Staphylococcus aureus* (MRSA) and vancomycin-resistant *Enterococcus faecium* (VREF) genotypes in six hospitals in the metropolitan New York City area. The BARG Initiative Pilot Study Group. Bacterial Antibiotic Resistance Group. *Microb. Drug Resist.* **2**, 343–351.

41. Maslow, J. N., Mulligan, M. E., and Arbeit, R. D. (1993) Molecular epidemiology: application of contemporary techniques to the typing of microorganisms. *Clin. Infect. Dis.* **17**, 153–162; quiz 163–164.

42. Chung, M., de Lencastre, H., Matthews, P., Tomasz, A., Adamsson, I., Aries de Sousa, M., Camou, T., Cocuzza, C., Corso, A., Couto, I., Dominguez, A., et al. (2000). Molecular typing of methicillin-resistant *Staphylococcus aureus* by pulsed-

field gel electrophoresis: comparison of results obtained in a multilaboratory effort using identical protocols and MRSA strains. *Microb. Drug Resist.* **6**, 189–198.

43. Mulvey, M. R., Chui, L., Ismail, J., Louie, L., Murphy, C., Chang, N., and Alfa, M. (2001) Development of a Canadian standardized protocol for subtyping methicillin-resistant *Staphylococcus aureus* using pulsed-field gel electrophoresis. *J. Clin. Microbiol.* **39**, 3481–3485.

44. Swaminathan, B., Barrett, T. J., Hunter, S. B., and Tauxe, R. V. (2001) PulseNet: the molecular subtyping network for foodborne bacterial disease surveillance, United States. *Emerg. Infect. Dis.* **7**, 382–389.

45. Gerner-Smidt, P., Graves, L. M., Hunter, S., and Swaminathan, B. (1998) Computerized analysis of restriction fragment length polymorphism patterns: comparative evaluation of two commercial software packages. *J. Clin. Microbiol.* **36**, 1318–1323.

46. Feil, E. J., Enright, M. C., and Spratt, B. G. (2000) Estimating the relative contributions of mutation and recombination to clonal diversification: a comparison between *Neisseria meningitidis* and *Streptococcus pneumoniae. Res. Microbiol.* **151**, 465–469.

47. Chan, M. S., Maiden, M. C., and Spratt, B. G. (2001) Database-driven multi locus sequence typing (MLST) of bacterial pathogens. *Bioinformatics* **17**, 1077–1083.

48. Enright, M. C. and Spratt, B. G. (1999) Multilocus sequence typing. *Trends Microbiol.* **7**, 482–487.

49. Cooper, J. E. and Feil, E. J. (2006) The phylogeny of *Staphylococcus aureus* – which genes make the best intra-species markers? *Microbiology* **152**, 1297–1305.

50. Diep, B. A., Gill, S. R., Chang, R. F., Phan, T. H., Chen, J. H., Davidson, M. G., Lin, J., Carleton, H. A., Mongodin, E. F., Sensabaugh, G. F., Perdreau-Remington, F. (2006) Complete geneome sequence of USA300, an epidemic clone of community-acquired meticillin-resistant *Staphylococcus aureus. Lancet* **367**, 731–9.

Index